FUNDAMENTALS OF MID-TERTIARY STRATIGRAPHICAL CORRELATION

FUNDAMENTALS OF MID-TERTIARY STRATIGRAPHICAL CORRELATION

BY

F. E. EAMES, A.R.C.S., D.Sc. (Lond.), F.G.S.

F. T. BANNER, Ph.D. (Lond.)   W. H. BLOW, Ph.D. (Lond.), F.G.S.

W. J. CLARKE, Ph.D. (Edin.)

WITH A CONTRIBUTION BY

L. R. COX, Sc.D. (Cantab.), F.R.S., F.G.S.

CAMBRIDGE
AT THE UNIVERSITY PRESS
1962

CAMBRIDGE UNIVERSITY PRESS
Cambridge, New York, Melbourne, Madrid, Cape Town, Singapore,
São Paulo, Delhi, Dubai, Tokyo, Mexico City

Cambridge University Press
The Edinburgh Building, Cambridge CB2 8RU, UK

Published in the United States of America by Cambridge University Press, New York

www.cambridge.org
Information on this title: www.cambridge.org/9780521172295

First published 1962
First paperback edition 2010

A catalogue record for this publication is available from the British Library

ISBN 978-0-521-04863-7 Hardback
ISBN 978-0-521-17229-5 Paperback

Additional resources for this publication at www.cambridge.org/9780521172295

CONTENTS

PART 1: REVIEW AND REVISION

BY F. E. EAMES, F. T. BANNER, W. H. BLOW & W. J. CLARKE

PART 2: THE MID-TERTIARY
(UPPER EOCENE TO AQUITANIAN)
GLOBIGERINACEAE

BY W. H. BLOW & F. T. BANNER

Cassigerinella
C. chipolensis
Globigerina
G. ampliapertura ampliapertura
G. ampliapertura euapertura
G. angulisuturalis
G. angustiumbilicata
G. linaperta linaperta
G. linaperta pseudoeocaena
G. officinalis
G. oligocaenica sp.nov.
G. ouachitaensis ouachitaensis
G. ouachitaensis ciperoensis
G. ouachitaensis gnaucki subsp.nov.
G. praebulloides praebulloides
G. praebulloides leroyi subsp.nov.
G. praebulloides occlusa subsp.nov.
G. pseudoampliapertura sp.nov.
G. senilis
G. tripartita tripartita emended
G. tripartita tapuriensis subsp.nov.

G. turritilina turritilina sp.nov. et subsp. nov.
G. turritilina praeturritilina subsp.nov.
G. yeguaensis yeguaensis
G. yeguaensis pseudovenezuelana subsp. nov.
G. aff. *yeguaensis*
Globigerinita emended
G. africana sp.nov.
G. dissimilis dissimilis
G. dissimilis ciperoensis subsp.nov.
G. globiformis sp.nov.
G. howei sp.nov.
G. martini martini sp.nov. et subsp.nov.
G. martini scandretti subsp.nov.
G. pera
G. unicava unicava
G. unicava primitiva subsp.nov.
Globigerinoides
G. quadrilobatus primordius subsp.nov.
Globoquadrina

CONTENTS

FIGURES

   * Figures 4 and 5 are available for download from
     www.cambridge.org/9780521172295

vii

CONTENTS

PART 1

REVIEW AND REVISION

PREFACE. In the course of extensive economic palaeontological researches based on world-wide investigations, it has been found necessary to review and re-assess the bases of mid-Tertiary correlation and dating; the principles involved are briefly reviewed, as are also the status and limits of some Tertiary stages.

The validity of the fivefold subdivision of the Oligocene and Lower Miocene as represented by the faunas of the type localities of the Lattorfian, Rupelian, Chattian, Aquitanian and Burdigalian is confirmed, and it is shown that these five faunal assemblages can readily be correlated through to southern Europe, the Mediterranean, the Middle East, East Africa, Pakistan and India, and the Far East as a whole. Dr L. R. Cox has independently and objectively reviewed the molluscan evidence from Europe, including the northern part of the Mediterranean region, in which latter area foraminifera first constitute a significant element of the faunas, and enable the correlation to be carried through to the Far East. During the course of the re-assessment, it has been necessary to propose a few relatively minor modifications to the dating of some beds in the Tethyan region; some of these adjustments are in accord with the views of earlier workers.

Throughout the above region the evidence of benthonic and planktonic foraminifera has been co-ordinated, and the biostratigraphical significance of important genera and species has been evaluated; in particular, the evidence of the Miogypsinidae and the Orbulininae is considered in detail. As a consequence, we show that the succession of faunas within the Oligocene and Lower Miocene of the whole vast region follows the same pattern which is not out of phase in any one place.

The consistent uniformity of these faunal successions in the whole of the tropical and subtropical Old World has led us to re-investigate the evidence and biostratigraphical conclusions concerning equivalent horizons in tropical and subtropical Central America, the only region in which the faunal successions were apparently out of phase with the sequences uniformly seen elsewhere.

Our review of the successions of marine faunas within the Central American region very largely confirms previous opinions concerning intercorrelation within the region. However, in comparing the mid-Tertiary marine faunas with those of the Old World, we not only find that few of them can be correlated with any Oligocene faunas in the Old World, but that almost all can be correlated with Old World Miocene faunas. No restricted Old World Oligocene fossils of any group have as yet been recorded from marine beds in Central America; on the other hand, typical Neogene fossils (e.g. *Anadara*, *Chione*) and Miocene genera (e.g. *Pliolepidina*—which we show to be the same as *Multilepidina*) do occur; the genus *Clypeaster* attains its acme simultaneously in the Miocene of the two regions. The diastrophic stratigraphical breaks, which have been so variously dated within the Central American region, are shown to belong to one major period of tectonic activity, of much greater stratigraphical duration and geographical extent than has hitherto been realized. This period of tectonic activity has often caused much reworking of older sediments and their fossils into beds which are subsequent to the Oligocene orogeny, and which we now consider to be of Miocene age. The disconformity of the American Gulf States, between the Vicksburgian and the Jacksonian, merges into the flagrant unconformity of the Caribbean, where the orogeny was most intense. Distinctive Oligocene planktonic foraminiferal faunas are now described from East Africa, and these are missing in the marine Tertiary succession in Central America precisely where we place this widespread hiatus.

It is shown that both the well-known names *Operculinella* Yabe, 1918, and *Operculinoides* Hanzawa, 1935, are synonyms of *Palaeonummulites* Schubert, 1908. New species of firmly dated Oligocene Globigerinaceae are described, and their evolutionary relationships are discussed in relation to other species from the Upper Eocene and Lower Miocene faunas of the Caribbean, Mediterranean and Indo-Pacific regions.

I. Introduction

Before proceeding with the detailed systematics of the genera and subgenera of planktonic foraminifera defined in *Palaeontology* (vol. II, no. 1), it is essential that consideration be given to the dating and correlation of certain Tertiary beds in various parts of the world. In order that our opinions on the stratigraphical distribution of the genera and species to be considered may be expressed without ambiguity, and that the reader may understand precisely what is meant by the ages and ranges given (thereby ensuring a proper appreciation of the evolutionary history of planktonic foraminifera) it is essential that a number of conceptions and interpretations, some of which have persisted for over half a century, be critically reviewed.

Two basic principles are stressed:

(1) The type region for a stage or series of stages must be regarded as constituting the fundamental basic standard.

(2) ALL the palaeontological evidence (not merely that of the foraminifera alone) must be considered and assessed when using the faunas for dating and for subsequent correlations.

The need for a proper application of these basic premises has led us to reconsider:

(1) The subdivision of the Miocene, including the placing of the Aquitanian.

(2) The age of the so-called 'Oligocene' of the Central American region (including the Caribbean, the southern part of the United States, and the northern part of South America).

Although the nomenclature of the Globigerinaceae has been revised in Part 2, current usage has been retained for stratigraphical convenience in Part 1.

II. Acknowledgments

We are indebted to the Chairman and Directors of The British Petroleum Co., Ltd, for permission to publish this work. We also wish to acknowledge advice and assistance during discussions with Dr H. Dighton Thomas of the British Museum (Natural History) (coral, foraminiferal and bryozoan faunas), with Dr L. R. Cox of the British Museum (Natural History) (molluscan faunas), with Dr A. H. Smout of the Iraq Petroleum Co., Ltd (foraminiferal faunas), with Mr W. E. Crews of the Shell Petroleum Co., Ltd (distribution of foraminiferal faunas in general), with Dr T. Barnard of University College, London (foraminiferal faunas), and with Mr T. F. Grimsdale (foraminiferal faunas); also stimulating discussions with our colleagues, in many parts of the world, over a period of many years. Finally, we are greatly indebted to Dr H. G. Kugler for reading the original manuscript, for useful and helpful comments and suggestions, and for a chart showing the latest opinions on the post-Eocene successions in Trinidad, which latter has been incorporated in our text-figure; and for helpful comment and suggestion from various sources in America, especially from Dr F. S. MacNeil, who has contributed Fig. 3.

Dr L. R. Cox has also contributed a section dealing with the Mollusca of the type Lattorfian, Rupelian, Chattian and Aquitanian and their relationships.

We wish to acknowledge the assistance of Dr C. G. Adams, British Museum (Natural History), in facilitating the study of the type material of *Palaeonummulites pristinus* (Brady) and in lending us the negative of Plate I F; E. M. Finch, British Petroleum Co. Ltd, has given valuable help in the preparation of the remaining photographs.

III. Type Regions for Stages of the Tertiary

All stages of the Tertiary have their type localities within the region comprised by west and central Europe and the Mediterranean area, and these stages constitute the acknowledged, basic, fundamental standard of classification. These must remain the standard; all age-determinations from other areas must be considered in direct relation to them, and are always open to reconsideration. While some terms (e.g. 'Oligocene') have been used in the Western Hemisphere for more than half a century, it has always to be remembered that, although the *local* classification, terminology and successions may be reasonably well evaluated, the *correlation* with the standard European succession is open to modification as knowledge accumulates. We consider that the time has now come for the review of opinions which have, in some cases, persisted for many decades, and have, we are convinced, caused a perpetuation of confusion concerning the evolutional history of faunas and floras, with a correlative misinterpretation of palaeogeography. Furthermore, there has been a tendency, during the last decade, to regard particular groups of fossils occurring in certain stages or series in extra-European areas as being 'typical' of certain European stages or series; here again, this is purely dependent upon correlation, assessment of faunas, and a wide knowledge of palaeontological evolution, and the interpretation is open to modification.

IV. Assessment of Fossil Faunas for Dating and Correlation

A review of world-wide stratigraphical palaeontology reveals many cases where beds have been dated on the current opinion concerning one group of fossils only and not on the total palaeontological evidence available. Conclusions reached have, in the absence of a critical approach, been perpetuated and have completely obscured the real nature of the evolutionary succession of fossil faunas. It is considered opportune to reconsider such cases in the light of modern knowledge. It is emphasized that in all cases of dating and correlation the entire faunas of the units concerned must be taken into account. Only in this way is it possible to reduce to a minimum incorrect conclusions arising from such considerations as:

(*a*) The influence of facies on the local ranges of a species in two or more different localities; even if facies control is acknowledged, correlations adduced may still be incorrect.

(*b*) The obsolescence of some identifications meaning that a modern interpretation of the morphological characters of certain taxa will involve nomenclatorial changes resulting in either: (i) the increased correlative value of some taxa owing to their considerably reduced range in time; or (ii) the reduced correlative value of some taxa owing to their considerably increased range in time.

(*c*) The assessment of the values previously assigned to certain taxa, forming a portion only of the fauna under consideration, has changed, so that their dating and ranges require revision. Here again, proper recognition may result in either reduced or increased correlative value. Non-recognition of these values in a series of correctly correlated occurrences of a single fossil group, the first record being incorrectly dated, will result in an *apparent*, but actually misleading, massive accumulation of incorrect records of a time unit; following upon this, it is evident that: (i) some fossils may have a longer time-range than has been assigned to them; and (ii) some fossils, even some in association with those in such cases as in (i), may actually have a shorter time-range than was believed.

(*d*) Even over very extensive areas, the succession need not necessarily be complete, and that important disconformities or unconformities (not always easily discernible locally in the field) may be present; the appreciation that such important hiatuses really did exist would materially affect concepts of the evolutionary succession of faunas. In some cases, preconceived ideas of the palaeogeography have prevented a true age assessment being reached.

It has become evident that some investigators have, in describing faunas, merely accepted current opinions as to the ages of the beds concerned, without any attempt to assess their own palaeontological evidence. This applies, for example, to the works of Dall (1890–1903) on Mollusca; the changes of age given (e.g. Upper Eocene to Oligocene, Miocene to Oligocene) in different parts of his descriptive works are unaccompanied by any indication of a reason or assessment of evidence, except for some discussion at the end of Part 6 in which, owing to misconceptions concerning the stratigraphy, the problems involved are not appreciated; in spite of subsequent contributions to the knowledge of the Vicksburgian Molluscan faunas (e.g. by Mansfield), we consider that their correct age in terms of European type stages has not yet been generally appreciated.

Finally, it has been acknowledged by competent authorities that post-Eocene sediments, in the Gulf Coast States (including Florida) and in Trinidad, contain reworked Eocene fossils (e.g. *Hantkenina*) or boulders of rock material up to 24 feet in one dimension; furthermore, beautifully preserved, delicate Eocene planktonic microforaminifera are acknowledged to occur in post-Eocene sediments. Our writing in subsequent pages will, we hope, make it clear that the extent of the earth-movements involved and the amount of reworking and redeposition of both rock material and fossils has not yet been fully recognized. In many cases the excellent preservation of the fossils (e.g. *Hantkenina* in the Red Bluff clay and Cretaceous *Globotruncana* and Palaeocene *Globorotalia* in the San Fernando formation) is no guide to the age of the enclosing sediments. We consider that it is only a logical extension of these considerations, combined with a knowledge of world-wide Tertiary faunal evolution, that leads to the conclusions we have drawn concerning such areas as Panama, Trinidad, and Cuba. We suggest that the extensive redeposition of Upper Eocene material into Vicksburgian sediments was the probable cause of the anomalous dating of these beds by the potassium/argon method (Evernden, 1959) as equivalent to the Upper Eocene of California.

V. Dating of Beds within the Limits Oligocene and Aquitanian in the Old World

A. GENERAL REMARKS

We are in complete agreement with the established usage of the terms Lattorfian, Rupelian and Chattian as constituting the tripartite subdivision of the Oligocene in Europe. There are many Mollusca and other fossils, the species of which are confined to the Oligocene, and, where larger foraminifera are found, many of the species do not occur in the underlying Eocene; of these latter, the reticulate *Nummulites intermedius-fichteli* is well known. Furthermore, the top of the underlying Eocene is marked by the extinction of many well-known forms (e.g. *Velates, Discocyclina,* pillared *Nummulites, Pellatispira*). The overlying Miocene is marked by the sudden appearance of numerous genera and species of modern (Neogene) type; such forms as *Miogypsina* (*s.s.*) (see Drooger, 1954*a*), *Globoquadrina, Cyllene, Nassarius* (*s.s.*), *Chione, Timoclea* (see A. M. Davies, 1934, p. 163)

belong to this evolutional surge. A careful study of the faunas listed, for example, by Haug (1908–11) from Lattorfian, Rupelian, and Chattian horizons in Europe will show how distinctive they are from Aquitanian faunas.

Much confusion has recently been caused by unnecessary amendments to earlier well established dating and correlation, and by the further suggestion that much of the basic earlier work requires revision. In the case of Szöts (1956), for example, the suggested amendments, mainly concerned with brackish and freshwater faunas, are, in our opinion, erroneous. The study and assessment of the faunas recorded from the beds he tabulated as 'Aquitanian (= Chattian)' clearly shows that two ages, not one, are involved. The Upper Oligocene (Chattian) faunas of the Mainz Basin, northern Germany (including the Cassel sands, which are the type Chattian), Slovenia, Budapest, and Transylvania are distinct from those of the Aquitanian of Aquitaine and the Vienna Basin. Haug pointed out that of 300 species of Mollusca in the type Aquitanian, only 10 per cent are common to the Oligocene, whereas 53 per cent are common to the Burdigalian. We have no doubt that the Chattian is to be regarded as the highest stage of the Oligocene and the Aquitanian as the lowest stage of the Miocene and that they are *not* equivalents.

Beds, the faunas of which are reliably correlated with the Chattian, while containing no true *Nummulites*, which apparently became extinct at the end of the Rupelian (*incrassatus*-like forms belong to the genus *Palaeonummulites*—appendix 1), do contain some forms persisting from the Rupelian, and underlie (where both are present) beds which contain faunas including a new Neogene element (e.g. *Miogypsina* (*s.s.*)) and which can be reliably correlated with the Aquitanian.[1]

[1] We have noticed that Bolli, Loeblich & Tappan (1957, p. 22) use the term Vindobonian as being absolutely synonymous with Miocene, and include in it the stages Burdigalian, Helvetian, Tortonian and Sahelian. Quite apart from the fact that the Aquitanian should be included in the Neogene, the study of rich European Miocene faunas has indicated a subdivision into Lower Miocene (Aquitanian plus Burdigalian), Middle Miocene or Vindobonian (Helvetian plus Tortonian) and Upper Miocene (Sarmatian plus Pontian), which we see no reason to modify. All these subdivisions are, for example, present in ascending order in the Vienna Basin which is the type area for the Vindobonian part only of the succession.

The evidence of the Miogypsinids and *Orbulina* is dealt with in §IXF in full, to which reference should now be made.

Since the dating and correlation of the Lattorfian, Rupelian, Chattian and Aquitanian stages in Europe depends so much upon the evidence of Mollusca, we have asked Dr L. R. Cox to review the evidence; his contribution follows.

B. THE OLIGOCENE AND AQUITANIAN MARINE MOLLUSCAN FAUNAS OF PARTS OF EUROPE AND THEIR BEARING ON CORRELATION (by Dr L. R. COX)

A survey of the relevant literature has been carried out for the purpose of re-examining the arguments for the recognition of the Oligocene as a distinct geological system and of investigating (in the light of our present knowledge of molluscan faunas) the soundness of the correlation of beds regarded as Oligocene in southern Europe with those of the type area; and, further, of examining the molluscan faunistic grounds for regarding the type Aquitanian fauna as Miocene rather than Oligocene, and of ascertaining to what extent molluscan evidence supports the correlation of supposedly Aquitanian deposits in the Mediterranean area with those of Aquitaine.

Beyrich (1854) considered that it was preferable to refer a series of beds in Germany and Belgium, intermediate in age between the typical Eocene and the typical Miocene, to a distinct geological period, which he proposed to designate as 'Oligocän', than to regard them either as Upper Eocene or as Lower Miocene. Starting with northern Germany, the Oligocene and basal Miocene molluscan faunas of various areas will be reviewed in turn, and the facts briefly presented, by which the soundness of correlations between those areas may be judged.

Northern Germany

The stage name 'Lattorfian', for the Lower Oligocene, is derived from Lattorf, in this area. The Lattorfian beds, laid down in a marine transgression, overlie lignite deposits considered to be Upper Eocene or basal Oligocene in age. Their fauna has been exhaustively monographed by von

Koenen (1893–94). Its 721 molluscan species include 42 which range up from the Eocene; 72 species persist to the Middle Oligocene, 37 to the Upper Oligocene, and only three to the Miocene.

The most important fossiliferous formations of northern Germany included in the Middle Oligocene are the Magdeburg sands, the Septarian clay, and the Stettin sands. The molluscan fauna of these deposits was described by von Koenen (1867–68), whose distribution tables show that, of a total molluscan fauna of 190 species, 13 range up from the Eocene and 77 from the Lower Oligocene,[1] while as many as 101 persist into the Upper Oligocene and 17 into the Miocene.

The marine sands of Kassel constitute the typical Upper Oligocene, the Chattian stage of Fuchs. In the latest revision of the molluscan fauna of this formation Görges (1952) recognizes 240 species. Ninety-seven range up from the Middle Oligocene, 46 from the Lower Oligocene, and only four from the Eocene; 47 are recorded from the Miocene.

At various localities in northern Germany and Denmark, below lignite-bearing sands attributed to the Miocene, are marine beds referred to the Lower Miocene. These beds have been termed the Vierländer stage and have been thought by some authors to represent the Aquitanian. They have yielded a molluscan fauna of 223 species listed by Gripp (1916, pp. 18–29). Out of this total, only 19 have been recorded from the Oligocene, whereas most of the others persist to higher horizons of the Miocene. Ten species are restricted to this stage in the region under consideration, but not one of these is found in the Aquitanian of the type area. Hence the molluscan evidence does not suffice to date the Vierländer stage as Aquitanian, and, indeed, it may be younger as suggested by Kautsky (1925).

Belgium

This is the type area of the Tongrian and Rupelian stages (Lower and Middle Oligocene

---

[1] This figure of 77 differs slightly from that of 72 given above. The discrepancy may be due to changes in ideas regarding the synonymy of a few species during the twenty-six years that elapsed between the appearance of von Koenen's two monographs, or to differences in the areas of distribution taken into account.

respectively), while the term Chattian is used by Belgian geologists for the Upper Oligocene. Two recent monographs by Glibert (1954, in Glibert & Heinzelin de Braucourt; 1957) describe respectively the Mollusca of the Tongrian and Lower Rupelian, and of the Upper Rupelian and Chattian. Of the 202 species recognized in the Tongrian and Lower Rupelian, eight range up from the Eocene, 50 persist into the Upper Rupelian, and 35 into the Chattian. The Upper Rupelian fauna consists of 68 species of which 30 are found only in this stage; while the Chattian fauna consists of 104 species, 63 of which are confined to this stage. Of 13 species ranging up into the Miocene, two start in the Tongrian, two in the Rupelian, and nine in the Chattian. It is quite clear from the relatively small numbers of species that come up from the Eocene or continue into the Miocene that the fauna of the Belgian Oligocene is a very distinct one stratigraphically. There are no marine deposits in Belgium which have been attributed to the Aquitanian. In all three stages of the Oligocene species too numerous to mention are common to Belgium and northern Germany.

Paris Basin

In this region and in southern England the stratigraphical relationship of the Oligocene beds to the Eocene is seen more clearly than in northern Germany and Belgium. In the Paris Basin the uppermost Eocene (Ludian stage) is represented by variable beds which are partly of marine and partly of lagoonal and freshwater origin. The Lower Oligocene beds (here termed the Sannoisian stage) are largely non-marine and it was not until the Middle Oligocene (here termed the Stampian stage) that marine conditions became more general. These marine Stampian beds have yielded about 300 molluscan species, 64 of which are known from one or more stages of the German Oligocene (25 are found in the Lower Oligocene, 33 in the Middle Oligocene, and 40 in the Upper Oligocene). Affinities with the Belgian Oligocene are equally close. Only two species range up from the Eocene, and not one is known to occur in the Miocene. Among species establishing correlation with the North German Oligocene are the following:

Glycymeris obliterata (Deshayes), *Crassostrea callifera* (Lamarck), *Chlamys picta* (Goldfuss), *Spondylus tenuispina* Sandberger, *Tellina nysti* Deshayes, *Laevicardium tenuisulcatum* (Nyst), *Drepanochilus speciosus* (Schlotheim) [*pescarbonis* (Brongniart)], *Euspira achatensis* (Récluz, in de Koninck), *Charonia flandrica* (de Koninck), *Dentalium kickxii* Nyst. The presence of the gastropod *Globularia* (*Ampullinopsis*) *crassatina* (Lamarck) is significant for correlation with more southerly areas.

Brittany

A marine Oligocene deposit near Rennes has yielded a molluscan fauna of 56 species described by Cossmann (1919). Thirty-two of the species are known only from this locality, 16 are found in the Middle Oligocene of the Paris Basin, and five are found in the German Oligocene. *Globularia* (*Ampullinopsis*) *crassatina* and *Crommium angustatum* (Grateloup) are common to this fauna and that of the Oligocene of south-western France. Roger (1944) has pointed out that, whereas the Oligocene Pectinidae of Germany, Holland and Belgium show little relationship to those of the Mediterranean area, the beds at Rennes have yielded two species, *Pecten sylvestreisacyi* Cossmann and *Chlamys* (*Aequipecten*) *gregoriensis* Cossmann, closely related to species of the more southerly province.

South-western France

Beds referable to the Aquitanian in this, the type area of that stage, rest at most localities on lacustrine limestones considered to be Upper Oligocene in age, and these succeed marine beds which have yielded a rich fauna at some localities (notably around Gaas) and are referred to the Middle Oligocene. The lamellibranch fauna of the Oligocene beds has been revised by Cossmann (1921–22), but for an account of the gastropods it is necessary to consult the old monograph of Grateloup (1847–48) and one or two subsequent faunal lists, particularly that of Raulin (1896).

Of 131 lamellibranch species recorded by Cossmann from the Middle Oligocene of this area, 126 are confined to this stage locally, according to the data given by Cossmann. The fauna as a whole differs considerably from that of the Middle Oligocene of the Paris Basin, but nine lamellibranchs and five gastropods are common to the two regions, among them *Crassostrea cyathula* (Lamarck), *C. longirostris* (Lamarck), *Spondylus tenuispina* Sandberger, *Corbulomya nysti* (Deshayes), *Corbula subpisum* d'Orbigny, and *Globularia (Ampullinopsis) crassatina* (Lamarck). Faunal affinities with the Oligocene of the Mediterranean region are more pronounced, 39 species being common to the Gaas fauna and that of Liguria, according to data given by Rovereto (1900) and others. Among these species are *Pecten arcuatus* Brocchi and *Antigona aglaurae* (Brongniart). At Biarritz the Middle Oligocene beds with *Pecten arcuatus* and *Crassostrea cyathula* contain *Nummulites intermedius* and *N. vascus*. The succession upwards from beds with Upper Eocene nummulites can be well seen at this locality.

There has been some controversy about the age of the marine 'faluns bleus' of Saint-Géours-de-Maremne, Saint-Étienne-d'Orthe, Peyrère, and other localities west of Dax, first brought to notice by Raulin (1891). About 130 species from Saint-Étienne are described in the monograph of Cossmann & Peyrot (1909–35), who give their age as Helvetian in the earlier parts of the work and as Aquitanian in the later parts. The assemblage, which includes such forms as *Chama gryphoides mioasperella* Sacco, *Myrtea spinifera tenuicardinata* Cossmann & Peyrot, *Phacoides orbicularis* (Deshayes), *Lucina columbella basteroti* Agassiz, *Cardita auingeri* Hoernes, *Anadara diluvii* (Lamarck), *Pecten subarcuatus* Tournouer, *Xenophora crispa* (Koenig), *Terebralia bidentata* (Defrance), *Canarium grateloupi* (d'Orbigny), *Roxania burdigalensis* (d'Orbigny), *Siphonaria vasconiensis* Michelin, etc., and lacks any characteristic Oligocene species, is unquestionably a Miocene one. Peyrot (1933, p. 40) records that the number of species from the 'faluns bleus' described in the monograph just cited amounts to 201; 147 of these occur only in this formation, 54 occur elsewhere in various stages of the Miocene (12 in the Aquitanian only), and scarcely a single one in the Oligocene. Entirely contradictory evidence, however, was afforded by a series of 11 molluscan species which were said to come from the 'faluns bleus' of localities which had yielded these Miocene fossils and which were described by Dollfus (1918). Among these species were such characteristic Oligocene forms as *Pecten arcuatus* Brocchi and *Chlamys (Aequipecten) deleta* (Michelotti), and in consequence Dollfus concluded that their age was Upper Oligocene—a conclusion accepted by Peyrot (1933, p. 48) and applied to the whole fauna of the 'faluns bleus'.

The explanation of the apparently conflicting evidence as to the age of the 'faluns bleus' is, no doubt, that the small series of fossils described by Dollfus came from a lower horizon than that of the large molluscan assemblage described by Cossmann & Peyrot. This supposition seems to be confirmed by Daguin (1948, pp. 133, 134), who states that in the marl-pit at Escornebéou (2 km east of Saint-Géours) two distinct horizons are present, the lower a grey marl, yielding Middle Oligocene fossils, and the upper, a more sandy formation, belonging to the Lower Aquitanian. The marl-pits at Saint-Étienne-d'Orthe and other localities from which Raulin made his collections are now filled in, but it is reasonable to assume that the rich series of fossils described by Cossmann & Peyrot came from the higher of two horizons and that the smaller series described by Dollfus, which had probably been kept apart, came from a lower horizon, belonging to the Middle or Upper Oligocene. Descriptions of the mollusca found in the typical Aquitanian beds are included in the monograph by Cossmann & Peyrot (1909–35). The total fauna consists of 721 species, a few of which are non-marine. Two hundred and forty-seven are peculiar to the Aquitanian stage, 470 persist to later stages of the Miocene, and only four come up from Oligocene beds older than the 'faluns bleus'. These figures give overwhelming support to the recognition of the Aquitanian as a distinct stage, and its inclusion in the Miocene rather than in the Oligocene. The following is a translation of remarks by Peyrot (1933, p. 49): 'Thus there took place in the Aquitanian an almost complete renovation of the molluscan fauna, due no doubt to a migration (from where it is difficult to determine) coinciding with the start of the marine transgression which took place in all the geosynclinal regions. The passage of the Aquitanian to the Burdigalian fauna, on the other hand, took place in a much less abrupt manner.... The Aquitanian is incontestably at the base of the Neogene.'

South-eastern France

Mollusca-bearing deposits referred to the Lower Oligocene are exposed at several inland localities in Provence, notably Castellane and Barrême. Boussac (1911; 1912, pp. 169, 180, etc.) has described and listed the species occurring here. This fauna has more affinities with that of the Oligocene of northern Europe than has the Oligocene fauna of south-western France. This is shown by the presence of the following species: *Glycymeris obliterata* (Deshayes) [Paris Basin, northern Germany, Belgium], *Chlamys bellicostata* (Wood) [England, Belgium, northern Germany], *Spondylus bifrons* Münster [Osnabrück area of Germany], *Laevicardium tenuicostatum* (Nyst) [Paris Basin, Belgium], *Panopea heberti* Bosquet [Paris Basin, Belgium], *Drepanochilus speciosus* (Schlotheim) [Paris Basin, northern Germany], *Athleta rathieri* (Hébert) [Paris Basin, England, Belgium], *Hemipleurotoma odontella* (Edwards) [England, northern Germany], *Borsonia costulata* (von Koenen) [northern Germany], *Borsonia sulcata* (Edwards) [England]. Associated with these forms are such species as *Pecten arcuatus* Brocchi, *Spondylus cisalpinus* Brongniart, *Globularia (Ampullinopsis) crassatina* (Lamarck), *Proadusta splendens* (Grateloup), *Pugilina laxicarinata* (Michelotti), *Eburna caronis* (Brongniart), and *Conus (Leptoconus) grateloupi* d'Orbigny, all abundant in the Oligocene of northern Italy. Several species, including some of those just mentioned, are common to this area and Gaas and other localities in south-western France.

The beds in the Castellane-Barrême area referred to the Aquitanian are of non-marine origin, but marine Neogene beds from the Aquitanian upwards are well displayed along the coast west of Marseilles, where they rest on Cretaceous rocks. The fauna of the Aquitanian beds at this locality has been described by Depéret (in Fontannes &

Depéret, 1889, pp. 50–90), who cites a number of species which are characteristic also of the Aquitanian of the type area. Unfortunately, the assemblage of species found here provides little evidence for correlation with supposedly Aquitanian beds in northern Italy, where, as seen below, the molluscan fauna consists largely of Pectinidae. However, *Amusium subpleuronectes* (d'Orbigny) not only occurs in great abundance in the deposits near Marseilles, but it has recently (Colom, 1958) been found in Lower Miocene beds on Majorca in close association with planktonic foraminifera which we believe to be of Aquitanian age.

Mainz Basin and Alsatian Plain

In these areas of south-western Germany marine Oligocene beds belong mostly to the Rupelian stage. Their molluscan fauna is closely related to that of contemporaneous beds in the Paris Basin and includes *Globularia (Ampullinopsis) crassatina*, which is not found in northern Germany. It seems unnecessary to give further details.

Southern Bavaria and Austria

About half-way between Innsbruck and Salzburg lie the localities Reit-im-Winkel and Häring, which have yielded an invertebrate fauna considered to be Lower Oligocene in age. The mollusca, described by Dreger (1892, 1903) and Deninger (1901), have been discussed by Boussac (1912, pp. 597, 604) and revised by Schlosser (1922). They belong to about 120 species, a few not distinguishable from Eocene forms known to be long-ranging. Both northern European and Mediterranean elements are present, the latter being slightly predominant. The following species found in the Oligocene of northern Germany or the Paris Basin are present: *Xenophora (Trochotugurium) subextensa* (d'Orbigny) [northern Germany, Belgium], *Drepanochilus speciosus* (Schlotheim) [northern Germany, Paris Basin], *Globularia (Ampullinopsis) crassatina* (Lamarck) [Paris Basin], *Galeodea buchi* (Boll) [northern Germany, Paris Basin].

Beds considered to represent the Upper Oligocene and the Aquitanian occur in southern Bavaria in the Flysch zone of the Alpine foreland.

The lower marine molasse (Untere Meeresmolasse) has yielded over 100 molluscan species, described by Wolff (1897). The fauna, considered to be Chattian in age, includes many species found in the Oligocene of northern Germany, among them *Crassostrea callifera* (Lamarck), *Laevicardium tenuisulcatum* (Nyst) [as *cingulatum* (Goldfuss)], *Drepanochilus speciosus* (Schlotheim), *Galeodea buchi* (Boll), and *Dentalium kickxii* Nyst. *Globularia* (*Ampullinopsis*) *crassatina* occurs here, but species particularly characteristic of the Mediterranean Oligocene are absent. The overlying 'Cyrena beds', of brackish and freshwater facies, are referred to the Aquitanian, but fossil evidence for their direct correlation with the type Aquitanian is wanting.

Piedmont and Liguria

Rovereto (1900) has described a large series of molluscan fossils from beds in this area which he calls Lower and Upper Tongrian, but which, according to Haug, are of Middle and Upper Oligocene age. A species distribution list, compiled in collaboration with Sacco and Bellardi, is appended to his paper. This list includes 443 species, 73 of which are found in the Oligocene of the Vicentino, farther east, 39 in the Middle Oligocene of Gaas (south-western France), and 26, 22, and 21 in the Oligocene of the Paris Basin, Belgium, and northern Germany, respectively. Fifty-one of the species range up from the Eocene and 97 persist into the Miocene. Among the widely distributed species found in the Ligurian Oligocene are *Globularia* (*Ampullinopsis*) *crassatina* (Lamarck), *Euspira achatensis* (Récluz, in de Koninck), *Dentalium kickxii* Nyst, *Crassostrea longirostris* (Lamarck), *Pecten arcuatus* Brocchi, *Panopea heberti* Bosquet, *Tellina nysti* Deshayes, and *Chionella splendida* (Merian).

Sacco (1906, p. 900) has published a list of mollusca from beds in the hills west of Turin assigned to the Aquitanian, where they overlie Upper Oligocene beds. The fauna seems to have a dominance of Miocene forms. A single species (*Glycymeris bormidiana* (Mayer)) has otherwise been found only in the Oligocene, and this belongs to a genus in which specific distinction is difficult;

the ranges of the four species *Turritella strangulata* Grateloup, *Chlamys tauroperstriata* Sacco, *Chlamys* (*Aequipecten*) *northamptoni* (Michelotti) and *Acesta miocenica* (Sismonda) start in the Oligocene and continue into later beds than Aquitanian; 12 species occur in the Miocene of south-western France, but not one of these is confined to the Aquitanian in that area.

Veneto

The Oligocene molluscan faunas of the Veneto are well known from the monographs of several authors, notably Brongniart (1823), Fuchs (1870), Oppenheim (1900), Canestrelli (1908), Kranz (1910, 1914) and Venzo (1937). Fabiani (1915, pp. 267–273) has listed the species recorded from the Lower, Middle and Upper Oligocene respectively, and the fauna of the last stage has since been dealt with more fully by Venzo. The Lower Oligocene fauna, for which the chief locality is Sangonini, consists of 128 species, 68 of which are found also in the Middle Oligocene, from which 187 species have been obtained. Of the Middle Oligocene fauna (Castelgomberto, etc.) about 25 per cent of the species occur also in the contemporaneous beds at Gaas, in south-western France. The number common to the Middle Oligocene of the Vicentino and the Paris Basin is about eight, among which may be noted *Globularia* (*Ampullinopsis*) *crassatina* (Lamarck), *Calyptraea striatella* Nyst, *Glycymeris obliterata* (Deshayes), *Crassostrea cyathula* (Lamarck), and *Laevicardium tenuisulcatum* (Nyst). The Upper Oligocene 'Glauconie Bellunesi' of this area has yielded 216 molluscan species or subspecies according to the monograph of Venzo (1937). Many of these range up from the Lower or Middle Oligocene. Eight occur in the Oligocene of the Paris Basin and twelve in the Oligocene of northern Germany. According to Venzo's distribution tables, only five of the species range up into the Aquitanian in any area.

These Upper Oligocene beds were included by earlier authors in the Schio beds, considered by Oppenheim (1903) to be entirely Aquitanian in age. The molluscan fauna of the part of the Schio beds which is still considered to be Aquitanian in age consists mainly of Pectinidae and is poor in

9

species. Unfortunately it does not suffice to date the beds accurately. Only one of the species, *Amussiopecten burdigalensis* (Lamarck), occurs in the Aquitanian of the type area, and there, as elsewhere, it ranges up into the Burdigalian. The other four recorded pectinid species also range up at least to the Burdigalian, but two of them, *Chlamys* (*Aequipecten*) *northamptoni* (Michelotti) and *C.* (*A.*) *praescabriuscula* (Fontannes), appear first in the Oligocene.

Accordi (1956) has recorded a small assemblage from the Oligocene of the eastern Trevigiano, and twenty-two molluscan species from beds regarded as Aquitanian. The Oligocene assemblage calls for no special comment. As at Schio, the Aquitanian fauna consists largely of Pectinidae. It is surprising to find in beds supposed to belong to this stage the species *Pecten arcuatus* Brocchi. It is, however, by no means certain that Accordi's identification is correct. The specimen figured by him (1956, pl. 4, fig. 14) as *P. arcuatus* much exceeds the size normally attained by that species and does not appear to have the conspicuous evenly spaced transverse threads in the intervals between its costae which characterize it. It may be suggested that the specimen belongs to *P. fuchsi* Fontannes subsp. *praecedens* Venzo (1934, p. 73, pl. 7, fig. 12), a Lower Aquitanian form, and in this connexion it may be pointed out that Depéret & Roman (1902, p. 14) recorded *P. fuchsi*, a Miocene species, from the same locality (Serravalle) as that of the specimen figured by Accordi. Accordi's list of species from these Aquitanian beds includes several other forms known to occur in the Oligocene, but with one exception these were already known to range up into the Aquitanian, Burdigalian, or later. The one exception is a form identified as *Peplum* (?) *oligopercostatum* Sacco, but as the specimens figured by Accordi are not well preserved and have fewer ribs than the syntypes figured by Sacco, their identification is to be queried.

Conclusions

(1) Two faunal provinces (northern European and Mediterranean) existed in Europe in Oligocene and Lower Miocene times. South-western France

may be considered to have belonged to an Atlantic sub-province of the Mediterranean province.

(2) In each province the Oligocene fauna is sufficiently distinctive to justify the separation of an Oligocene system from both the Miocene and the Eocene, although it includes a few species the ranges of which straddle either the upper or lower boundaries of the system.

(3) There are a few widely distributed molluscan species which enable the Oligocene deposits of the provinces mentioned to be correlated.

(4) In south-western France the molluscan fauna of the Aquitanian beds is a very distinctive one, but its affinities are much more with Miocene faunas than with those of the Oligocene.

(5) Precise correlation of Aquitanian deposits within the Lower Miocene does not at present seem possible on the sole basis of molluscan faunas, but in all faunas which have been regarded as Aquitanian, Miocene affinities predominate.

C. CORRELATION OF THE TYPE STAGES WITH THEIR EQUIVALENTS IN THE MEDITERRANEAN REGION

In the preceding pages Dr L. R. Cox has confirmed the long-standing succession and dating of the classical Oligocene and Lower Miocene stages and their correlation, step by step, with their equivalents in the northern part of the Mediterranean Region where larger foraminifera join the Mollusca in constituting the most important elements of the faunas.

In south-west France in the Biarritz area (e.g. Daguin, 1948), Lattorfian beds (underlying the Rupelian molluscan faunas of Gaas) contain the Lower–Middle Oligocene *Nummulites intermedius* and *N. vascus* but no *Eulepidina*.

The Rupelian faunas of Gaas and Liguria (northern Italy) possess Mollusca in common (e.g. *Pecten arcuatus*) associated with *Nummulites intermedius* and *N. vascus*; the Ligurian faunas contain the first *Eulepidina* as well. All the above-mentioned species and *Eulepidina* occur together in the Rupelian of North Africa, and the same *Nummulites* species occur in the Rupelian of Saint-Géours in Aquitaine. Oligocene planktonic foraminifera are referred to in Part 2.

In the Mediterranean Region, Chattian beds are either missing or poorly developed, often containing poor faunas which are frequently of brackish-water type; this is well known to be associated with the regression preceding the Neogene transgression, and well developed marine Chattian foraminiferal faunas, clearly younger than Rupelian and older than Aquitanian, are only to be found in the Middle East. We would mention here (see also p. 29) that there is no definite evidence that *Miogypsinella complanata* occurs in the Chattian of Aquitaine; although some recorded occurrences may be in beds of Chattian age, they may equally well be in beds of Aquitanian age which there also contain *Miolepidocyclina burdigalensis* as do beds of Aquitanian age in Sicily (cf. pp. 28, 76 and Drooger, 1955, pp. 14 and 45).

In Aquitaine, beds regarded as belonging to the Aquitanian contain *Timoclea*, *Eulepidina*, *Lepidocyclina (Nephrolepidina) morgani*, *Miogypsina globulina*, *M. gunteri*, *M. tani*, *Miogypsinella complanata* (according to many authors) and *Miolepidocyclina burdigalensis*. Planktonic foraminifera we have seen from low Aquitanian beds at Moulin de l'Église suggest that they belong to the Zone of *Globigerina ciperoensis ciperoensis* or slightly lower, but are younger than Oligocene planktonic faunas known from East Africa, northern Europe and Russia (see Part 2). One of the most characteristic fossils of the Aquitanian of the Rhone valley is *Amusium subpleuronectes*, and this species also occurs on Majorca (Colom, 1958) below the first appearance of *Orbulina* which is in assemblages which include faunas belonging to the Zone of *Globigerinatella insueta*. This same Zone is represented in Sicily (see p. 28), where its members are associated with *Austrotrillina howchini*, *Miogypsina globulina*, *Miogypsinella complanata*, *Praerhapydionina delicata* and *Spiroclypeus blanckenhorni ornata*. Immediately overlying this fauna the first *Orbulina* appear, in the highest part of the Zone. Equivalent beds also contain *Eulepidina* and *Miolepidocyclina burdigalensis*. On Malta, the Lower Coralline limestone, with *Austrotrillina howchini*, *Eulepidina*, *Miogypsina*,

Miogypsinella complanata, *Praerhapydionina delicata* and *Spiroclypeus blanckenhorni ornata*, obviously constitutes part of the Aquitanian. In North Africa, beds with *Miolepidocyclina burdigalensis* and *Spiroclypeus blanckenhorni ornata*, the first *Austrotrillina howchini*, *Lepidocyclina (Nephrolepidina) morgani*, *Meandropsina anahensis*, *Miogypsina* and *Orbulina* (high level), and the last *Eulepidina*, *Miogypsinella complanata* and *Praerhapydionina delicata* also belong to the Aquitanian. Equivalent beds contain *Globigerinatella insueta*. Drooger & Magné (1959, pp. 277–278) have recorded *Miogypsina cushmani* and *Miogypsinella complanata* in natural association in Algeria immediately below an *insueta* Zone planktonic fauna, and it is evident that they also are of Aquitanian age there.

In the type area for the Burdigalian in southwestern France there is a rich Neogene molluscan fauna closely related to that of the underlying Aquitanian; associated with these Burdigalian molluscan faunas are *Miogypsina globulina* and *Lepidocyclina (Nephrolepidina) morgani* (Douvillé, 1917), and Drooger (1955) has recorded *Miogypsina intermedia* from beds in the same area which he places in the Upper Burdigalian. In northern Italy *Pliolepidina subdilatata* has been recorded from beds at Rossignano which were dated as Upper Burdigalian. On Majorca, Colom (1958) has recorded molluscan-foraminiferal faunas of both Aquitanian and Burdigalian age, the lowest Burdigalian Zone of *Globorotalia fohsi fohsi* being probably present as we know it to be in Sicily. In Sicily, as in south-western France, neither *Eulepidina* nor *Miogypsinella complanata* survived the Aquitanian; they are succeeded in Sicily by planktonic foraminifera belonging to the *Globorotalia fohsi fohsi* Zone together with the first *Borelis melo* and *B. melo curdica*, which must be regarded as Burdigalian. The same sequence of faunas is found in North Africa, where the extinction of *Eulepidina*, *Miogypsinella complanata* and *Spiroclypeus* is immediately followed by the same Burdigalian faunas with the addition of *Taberina malabarica*.

11

	SOUTHERN EUROPE	NORTH AFRICA AND S.MEDITERRANEAN	MIDDLE EAST	EAST AFRICA	PAKISTAN AND INDIA	NETHERLANDS EAST INDIES, NEW GUINEA, AND FAR EAST
BURDIGALIAN	*An., T.;* *L.m., M.g., Pl.*	*An.;* *A.h., B.m., B.m.c., Me.a., M., Or., T.m.*	<u>L.Fars</u> (l. part):- *O.l.* <u>U.Asmari Lst.</u>:- *An., O.l.;* *A.h., B.m., B.m.c., Me.a., T.m.*	*An., O.l., T.;* *A.h., B.m., F.b., L.m., Me.a., M., Md.d., Or., Pl., T.m.*	<u>Gaj series</u>:- *An., O.l., T.;* *A.h., M., T.m.*	"f₁ – f₂":- *An., T.;* *A.h., B.m., F.b., M., Md.d., Or., Pl., T.m.*
AQUITANIAN	*An., A.s., T.;* *E., L.m., M.g., Ml.c., Mp.b.*	*An., A.s.;* *A.h., E., L.m., Me.a., M., Ml.c., Mp.b., Or.(h.l.), Pr.d., S.b.*	<u>M. Asmari Lst.</u>:- *Ar.h., A.h., Me.a., M., Ml.c., Pe.t., Pr.d., S.b.*	*A.h., E., F.r.(h.l.), Me.a., M., Ml.c., Md.d., Or.(h.l.), Pe.t., Pl., S.m., S.r.*	<u>U.Nari series</u>:- *An., T.;* *A.h., E., S.r.*	"e":- *A.h., E., F.r.(h.l.), M., Md.d., Ml., Mp., Or.(h.l.), Pl., Pr.d., S.m.*
CHATTIAN	*P.a.;* *E., Ml.c.*	(NOT PROVED)	<u>L. Asmari Lst.</u> (u. part):- *Ar.o., A.p., E., Pa.i., Pe.t., Pr.d.* *P.a.;* *Ml.c.*	(NOT KNOWN)	(NO SATISFACTORY PUBLISHED EVIDENCE)	(NOT PROVED)
RUPELIAN	*P.a.;* *E., N.i., N.v.*	*P.a.;* *E., N.i., N.v.*	<u>L.Asmari Lst.</u>(l.part):- *Ar.o., A.p., B.p.(1), E., N.i., N.v., Pa.i., Pe.t., Pr.d.*	*B.p.(1), E., N.i., N.v., Pa.i.*	<u>L.Nari series</u>:- Some Italian Olig: mollusca; *E., N.i., N.v.*	"d":- *B.p.(1), E., N.i., N.p., N.v.*
LATTORFIAN	*N.i., N.v., Pa.i.* (No *E.*)	(NOT PROVED)	<u>Brissopsis beds</u>:- *N.i., N.v., Pa.i.* (No *E.*)	(Possibly present :- *N.i., N.v.*. —— No *E.*)	(NO SATISFACTORY PUBLISHED EVIDENCE)	"c":- ?*B.p.(1).* *N.i., N.p., N.v.* (No *E.*)

Fig. 1. Correlation of Oligocene and Lower Miocene stages of Europe and the Mediterranean region with successions in the Middle East, East Africa, and the Indo-Pacific region.

An.	Anadara	*F.b.*	Flosculinella bontangensis	*N.p.*	N. pengaronensis
A.s.	Amusium subpleuronectes	*F.r.* (h.l.)	F. reicheli (high level only)	*N.v.*	N. vascus
O.l.	Ostrea latimarginata	*L.m.*	Lepidocyclina (Nephrolepidina) morgani	*Or.*	Orbulina
P.a.	Pecten arcuatus	*Me.a.*	Meandropsina anahensis	*Or.* (h.l.)	Orbulina (high level)
T.	Timoclea	*M.*	Miogypsina	*Pa.i.*	Palaeonummulites incrassatus
Ar.h.	Archaias hensoni	*M.g.*	M. globulina	*Pe.t.*	Peneroplis thomasi
Ar.o.	A. operculiniformis	*Ml.*	Miogypsinella	*Pl.*	Pliolepidina
A.h.	Austrotrillina howchini	*Ml.c.*	M. complanata	*Pr.d.*	Praerhapydionina delicata
A.p.	A. paucialveolata	*Md.*	Miogypsinoides	*S.b.*	Spiroclypeus blanckenhorni
B.m.	Borelis melo	*Md.d.*	M. dehaarti	*S.m.*	S. margaritatus
B.m.c.	B. melo curdica	*Mp.*	Miolepidocyclina	*S.r.*	S. ranjanae
B.p.(1)	B. pygmaea (first appearance)	*Mp.b.*	M. burdigalensis	*T.m.*	Taberina malabarica
E.	Eulepidina	*N.i.*	Nummulites intermedius-fichteli		

12

VI. Correlation of Oligocene and Lower Miocene Stages of Europe and the Mediterranean Region with Successions in the Middle East, East Africa and the Indo-Pacific Region

As will have been seen from §V, and will be realized after reading §VII, there exist five distinct faunal assemblages, in Europe and the Mediterranean region, which correspond to the type stages Lattorfian, Rupelian, Chattian, Aquitanian and Burdigalian (see Fig. 1). These type stages have been based mainly on the evidence of species of Mollusca and Echinoidea, which are largely restricted (especially in the Miocene) to Europe and the Mediterranean region. However, the foraminifera are of much wider geographical distribution, and enable the classical subdivisions easily to be followed through from the Mediterranean region to the Middle East, East Africa, India and Pakistan, and the Far East.

LATTORFIAN faunas are represented in the Middle East by the *Brissopsis* beds of Persia which contain the Lower-Middle Oligocene *Nummulites intermedius* and *N. vascus*, and *Palaeonummulites incrassatus* which ranges from Upper Eocene to Chattian; there are, however, no *Eulepidina*. In East Africa, the same two *Nummulites*, unaccompanied by *Eulepidina*, occur in beds immediately below others containing characteristic Rupelian faunas, and immediately above the highest Eocene faunas known (upper Bartonian), so that the Lattorfian may be present here also. In Pakistan and India, it has been suggested that part of the Lower Nari series is of Lattorfian age, but, since *Eulepidina* is known to occur well down in the series, there is no firm evidence for the presence of the Lattorfian. In the Far East, the *c* stage has been defined on the basis of the presence of the Lower-Middle Oligocene *Nummulites intermedius* and the Upper Eocene-Oligocene *N. pengaronensis*, unaccompanied by *Eulepidina*. From equivalent beds we have also identified the Lower-Middle Oligocene *N. vascus*. It therefore seems likely that the *c* stage is of Lattorfian age as has been suggested, though indications provided by the record of *Borelis* (which very probably belongs to the species *pygmaea*) from the

c stage prevent one from fully accepting this conclusion without further evidence.

RUPELIAN faunas of the lower part of the Lower Asmari limestone of Persia are characterized by the continuance of *Nummulites intermedius*, *N. vascus* and *Palaeonummulites incrassatus* and by the appearance of *Archaias operculiniformis*, *Austrotrillina paucialveolata*, *Borelis pygmaea*, *Eulepidina*, *Peneroplis thomasi* and *Praerhapydionina delicata*. The distribution of *Nummulites intermedius*, *N. vascus* and *Eulepidina* in the Lower and Middle Oligocene of Persia is therefore the same as in Europe. In East Africa beds equivalent to the Rupelian contain the last *Nummulites intermedius* and *N. vascus*, and the first *Borelis pygmaea* and *Eulepidina*, as well as *Palaeonummulites incrassatus*. In Pakistan and India the Rupelian Lower Nari series contains *Nummulites intermedius* and *Eulepidina* as well as several molluscan species (e.g. *Globularia* (*Ampullinopsis*) *crassatina* and *Ptychocerithium ighinai*) which in Italy occur only in the Oligocene. In the Far East the *d* stage is evidently of Rupelian age since it contains the last *Nummulites intermedius*, *N. vascus* and *N. pengaronensis* and the first definite *Borelis pygmaea* and *Eulepidina*.

Beds of CHATTIAN age are well represented in the Middle East where they contain faunas in which *Archaias operculiniformis*, *Austrotrillina paucialveolata* and *Palaeonummulites incrassatus* persist from the Rupelian and through which *Eulepidina*, *Peneroplis thomasi* and *Praerhapydionina delicata* range up into the Aquitanian; these faunas, occurring in the upper part of the Lower Asmari limestone, are comparable with those of the Chattian of Europe in that true *Nummulites* had become extinct, older Oligocene species persist up into it but not above, and characteristic Neogene elements had not yet appeared. *Pecten arcuatus* occurs in the Chattian of Palestine, and *Miogypsinella complanata* in the Chattian of Israel.

AQUITANIAN faunas are represented in the Middle East in the Middle Asmari limestone; they contain *Archaias hensoni* and *Spiroclypeus blanckenhorni*, and are characterized by the last *Miogypsinella complanata*, *Peneroplis thomasi* and *Prae-*

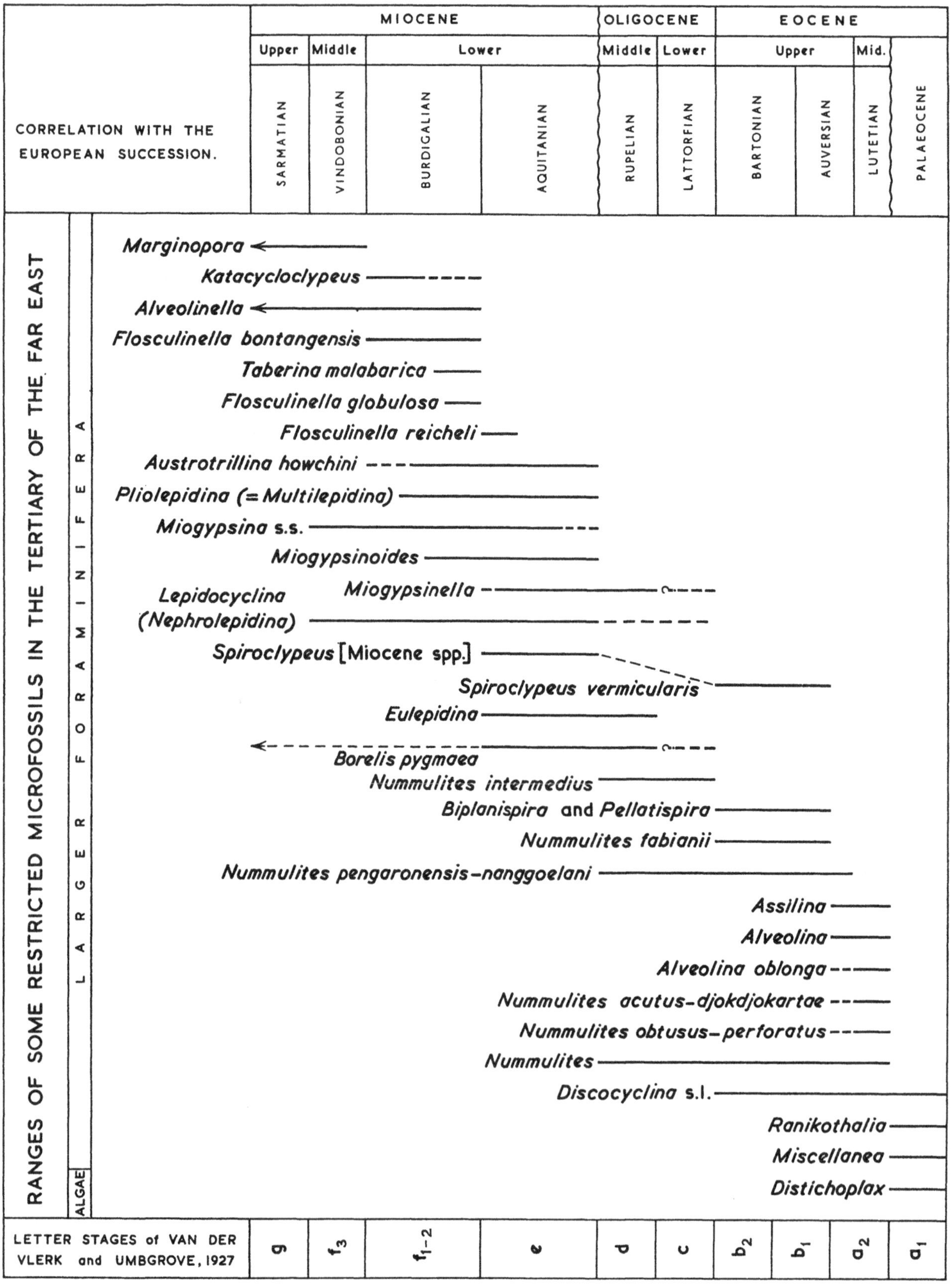

Fig. 2. The ranges of some stratigraphically restricted microfossils in the Tertiary of the Far East, which enable correlation between the letter-stages of van der Vlerk and the standard European succession.

rhapydionina delicata, and the first *Austrotrillina howchini* and *Meandropsina anahensis*; *Miogypsina* also occurs. The fundamental lagoonal nature of the beds is the reason for the absence of *Eulepidina*. Equivalent faunas in East Africa contain *Flosculinella reicheli* (high level only), *Miogypsina*, *Miogypsinoides dehaarti*, *Orbulina* (high level only), *Pliolepidina*, *Spiroclypeus margaritatus* and *S. ranjanae* together with the first *Austrotrillina howchini* and *Meandropsina anahensis* and the last *Eulepidina*, *Miogypsinella complanata* and *Peneroplis thomasi*. In India and Pakistan the Upper Nari series contains *Spiroclypeus ranjanae*, the last *Eulepidina* and the first *Austrotrillina howchini*. In the Far East the *e* stage contains *Flosculinella reicheli* (high level only), *Miolepidocyclina*, *Orbulina* (high level only) and *Spiroclypeus margaritatus* together with the last *Eulepidina*, *Miogypsinella* and *Praerhapydionina delicata* and the first *Austrotrillina howchini*, *Miogypsina*, *Miogypsinoides* and *Pliolepidina*; the Neogene elements are found at quite low levels in the *e* stage.

BURDIGALIAN faunas are represented in the Middle East in the Upper Asmari limestone and in the lower part of the Lower Fars; the Upper Asmari limestone contains *Ostrea latimarginata*, *Borelis melo curdica* and *Taberina malabarica*, the last *Austrotrillina howchini* and *Meandropsina anahensis*, and the first *Borelis melo* (which ranges up to Recent); the lower part of the Lower Fars also contains *Ostrea latimarginata*. In East Africa the equivalents of the Burdigalian contain *Ostrea latimarginata*, *Flosculinella bontangensis* and *Taberina malabarica*, the last *Austrotrillina howchini*, *Lepidocyclina* (*Nephrolepidina*) *morgani*, *Meandropsina anahensis*, *Miogypsinoides* and *Pliolepidina*, and the first *Borelis melo*, as well as *Miogypsina* and *Orbulina*. In Pakistan and India the Burdigalian Gaj series contains *Ostrea latimarginata* and *Taberina malabarica*, the last *Austrotrillina howchini*, and also *Timoclea* and *Miogypsina*. In the Far East the Burdigalian f_{1-2} substages contain *Flosculinella bontangensis* and *Taberina malabarica*, the last *Austrotrillina howchini*, *Miogypsinoides* and *Pliolepidina*, the first *Borelis melo*, and also *Timoclea*, *Miogypsina* and *Orbulina*.

VII. Consequent Commentary on and Minor Amendments to the Age of some Tethyan Beds and Remarks on *Lepidocyclina*.

In view of the conclusions of the previous two sections, it is necessary to refer here to a relatively small number of cases in the Tethyan region where the age-determination of beds requires minor revision; such revision is necessary in order to bring their ages into line with the vast bulk of the evidence from the Tethyan region as a whole. These cases constitute a very small proportion of the evidence available in the Tethyan region, concerning which there is a vast literature. Once these corrections are made, the orderly evolutionary succession of faunas in the Tethys becomes quite clear.

For the past thirty-three years the work of one of us (F.E.E.) with the Burmah Oil Co., Ltd, and with the British Petroleum Co., Ltd, has entailed intensive studies of Cretaceous and Tertiary faunas, first in the Burma–India–Pakistan region, secondly in the Middle East, and thirdly on a world-wide basis. During this period of time, the information that accumulated, and the interpretations arising therefrom, led to conclusions that were in some respects different from those that have been publicized.

There have been two classifications for the Tertiaries of Europe and of the Far East; the first consists of the standard stage names now in use, the second is the 'letter classification' first introduced by van der Vlerk & Umbgrove (1927) and subsequently revised by van der Vlerk (1955). There has always been doubt expressed about the precise correlation of the two. This doubt has persisted through an insufficient study of the faunas, and an inadequate appreciation of the evidence of the faunas of the Burma–India–Pakistan region. Work with the Burmah Oil Co., Ltd, in Burma, India and Pakistan, and with the British Petroleum Co., Ltd, in the Middle East, has enabled the correlation of the type European stages to be carried through from Europe and the Mediterranean region to the Middle East, to India and Pakistan, to Burma, and to the Netherlands East Indies. In Burma, the conclusions of Vredenburg, who established that there was an important unconformity, accompanied by a palaeontological break, between the Miocene and Oligocene portions of the Pegu system, have been fully confirmed (Eames, 1950). Since the latter can readily be correlated with the succession in India and Pakistan (Eames, 1950, p. 379), they, together with the succession in the Middle East, form an invaluable link between the European and Netherlands East Indies successions.

In ALGERIA, Flandrin (1938) could not tell to which stage of the Oligocene the Oligocene *Nummulites* and Mollusca he recorded belonged, their occurrences being recorded merely as 'Oligocene'. The author showed (pp. 106–109) how the distribution of the species of *Lepidocyclina* differed in Morocco and Algeria according to the ages assigned to the beds, and that *L. raulini* was the only species confined to a single stage, and this in Morocco only. Apart from this single species, all other nine species mentioned occurred in both the so-called 'Lattorfian' and 'Rupelian' of both Morocco and Algeria. Furthermore, four of these nine species (*dilatata*, *levis*, *roberti* and *formosoides*)—and *raulini* also—belong to the subgenus *Eulepidina*, which world-wide knowledge of the distribution of lepidocyclinas in the Tethys now indicates does not occur below the Rupelian. Flandrin also recorded three species belonging to the subgenus *Nephrolepidina* (*tournoueri*, *aquitaniae*—referred to as *simplex*—and *marginata*, which is really the microspheric form of *tournoueri*) and two species belonging to the subgenus *Eulepidina* (*dilatata* and *levis*) from the so-called 'Chattian' of Algeria, and all the above except *levis* from the so-called 'Chattian' of Morocco. Every one of the above species is known to occur in both the Rupelian and the Aquitanian. Consequently, it is not surprising that Dr E. Schijfsma and colleagues, who carried out considerable research on mid-Tertiary foraminiferal faunas of Tunisia and Algeria in the post-war years, failed to find any basis for subdivision of the Oligocene there. It is our opinion that the evidence indicates that all these Algerian and Moroccan beds with *Nummulites* are of Middle Oligocene (Rupelian) age.

In ALGERIA and MOROCCO, Drooger & Magné

(1959) have dated certain beds containing *Miogypsina mediterranea* as Helvetian in opposition to much other evidence indicating a Lower Miocene age, on the principle of nepionic acceleration in miogypsinids; it has even been suggested that as much as 90% of the faunas may be reworked, a suggestion which, in view of other evidence, is quite unnecessary (see our subsequent writing concerning the evidence of the miogypsinids and *Orbulina*). On MAJORCA, Colom (1958) has shown that *Miogypsina mediterranea* occurs in beds which can only be dated as Lower Miocene; with this we are in full agreement, and it is evident that in Algeria *M. mediterranea* and *Miogypsinoides* (Drooger & Magné place *complanata* under *Miogypsinoides*) occur together in normal association at locality ten immediately below an *insueta* Zone planktonic fauna, and must there be, in part, of Aquitanian age. We indicate below that the miogypsinids are to be regarded as an evolving plexus and that, on a regional synthesis, it is quite evident that the ranges of species do overlap. Consequently, we agree with Colom and many French palaeontologists that some, at any rate, of the above-mentioned beds in Morocco, Algeria, and Majorca are of Lower Miocene (Aquitanian and Burdigalian) age, not Helvetian.

On MALTA, Gregory (1891) dated the *Scutella* bed, near the base of the exposed Lower Coralline limestone, as Tongrian, on the basis of his identification of *Scutella striatula*, and he also made the unusual suggestion that the Tongrian was homotaxial with the Aquitanian. However, our conviction that the species of *Scutella* was incorrectly identified has been confirmed by Durham (1955), and Bather suggested (MS. writing on the label attached to the specimen in the British Museum (Natural History)) that it was *S. melitensis*, a Miocene Mediterranean species. Apart from two new species, the only other form identified from this horizon was a Miocene species of *Cidaris*. Haug also placed all the lower exposed beds on Malta as Lower Miocene on stratigraphical grounds. Furthermore, in wells recently drilled on the island, it has been established that the Aquitanian extends for at least 2000 feet below the *Globigerina* limestone (Burdigalian), which overlies the Lower

Coralline limestone, and therefore well below the *Scutella* bed.

With regard to the higher Miocene horizons exposed on Malta, evidence has recently been forthcoming to show that some, at least, of these are of Burdigalian (Lower Miocene) and not Middle Miocene age. The evidence of Foraminifera (e.g. *Spiroclypeus blanckenhorni ornata*, *Austrotrillina howchini*, etc.) confirms that the Lower Coralline limestone is of Aquitanian age, and the *Globigerina* limestone was consequently placed in the Burdigalian. However, on the basis of rather poor faunas, the overlying Blue clay was previously dated as lower Helvetian, the Greensand as upper Helvetian, and the overlying Upper Coralline limestone as Tortonian. Blow (1957) has recorded that the Blue clay contains *Orbulina*, *Biorbulina*, *Globigerina praebulloides*, and *Globorotalia fohsi barisanensis*, together with forms transitional to *G. fohsi fohsi*, thus indicating that the Blue clay is of very low Burdigalian age (not Helvetian). It therefore seems likely that the Greensand and Upper Coralline limestone are also probably of Burdigalian age.

In SYRIA, although some doubt seems to have been expressed later, it is worth recalling that David-Sylvain (Dubertret, Vautrin, Keller & David, 1933; David-Sylvain, 1937) dated certain beds as lower Burdigalian, evidently primarily on the evidence of Echinoidea recorded from the Burdigalian of Malta (e.g. *Lovenia gauthieri*, *Metalia melitensis*, *Hemiaster coranguinum*, *Opissaster scillae* and *Schizaster parkinsoni*). Quite apart from the fact that the preservation of these Syrian echinoids was not always good (e.g. in some cases it was mentioned that only traces of fascioles could be seen, and in others it was mentioned that the position of the apical disk was somewhat different from that of the typical form), it has to be remembered that the facies of the Aquitanian Lower Coralline limestone of Malta is not one in which one is at all likely to find determinable Echinoidea, and the Burdigalian species may, therefore, actually range up from the Aquitanian. The most important point, however, is that these Syrian Echinoidea are associated with the foraminifera *Lepidocyclina* (*Eulepidina*) (two species) and

Spiroclypeus margaritatus, forms which elsewhere are never found above the Aquitanian. We have no doubt that these Syrian beds are of Aquitanian, not Burdigalian, age.

In the MIDDLE EAST (PERSIA and IRAQ), the latest published evidence (Kent, Slinger & Thomas, 1951; van Bellen, 1956, 1959) has dated much of the Middle Asmari limestone and its equivalents in Iraq as Upper Oligocene (Chattian), but a more recent (unpublished) interpretation of the evidence shows that the Middle Asmari limestone and its equivalents in Iraq are really of Lower Miocene (Aquitanian) age. These beds contain *Miogypsina* (*s.s.*) and other forms such as *Archaias hensoni*, true *Austrotrillina howchini*, and *Meandropsina anahensis*, which have not been found elsewhere (e.g. East Africa, North Africa) below beds of Lower Miocene (Aquitanian) age (*A. howchini* and *M. anahensis* occur in the Burdigalian also); these beds constitute the lower portion of the First Mediterranean Miocene Transgressive series. Furthermore, in several places in Persia and Iraq, it has been established without doubt that the basal portion of the Middle Asmari limestone and its equivalents contains fossils reworked from the Oligocene. In addition, there is an earlier phase of deposition, that of the upper part of the Lower Asmari limestone, which can firmly be dated as Upper Oligocene. Whereas true *Nummulites* (the species being *N. intermedius-fichteli* and *N. vascus*) become extinct at the top of the lower part of the Lower Asmari limestone (i.e. the Rupelian), species of *Heterostegina*, *Carpenteria* and *Mesophyllum* associated with them persist to the top of the upper part of the Lower Asmari limestone, but are never associated with species known, on a regional basis, to be of Aquitanian age. In some parts of the Persia–Iraq region sedimentation seems to have been fairly continuous, although a palaeontological and lithological break is always to be found at the Miocene/Oligocene boundary, in spite of the fact that both Upper Oligocene and Aquitanian may be present. In the type section for the Asmari limestone, however, at Asmari Mountain, we consider that the Asmari limestone succession is far from complete, and that the Aquitanian Middle Asmari limestone rests non-sequentially (with a pink-coloured bleach-bed forming the boundary) upon the Lower Oligocene *Brissopsis* beds. Consequently, it would be desirable to redefine the Asmari limestone so as to include a supplementary standard section, such as that encountered in wells in the Gach Saran field, in which Middle Oligocene (nummulitic beds) and the Upper Oligocene are both also well developed.

In ITALIAN SOMALILAND, Azzaroli (1958) has followed van der Vlerk's recent suggestions when dating the Tavo beds as lower *e* stage and Chattian. These beds, however, in addition to numerous specimens of *Strombus* (a gastropod genus abundant in the Lower Miocene only of areas farther south—e.g. Tanganyika), contain the Mediterranean Aquitanian species *Spiroclypeus blanckenhorni* and a new species of *Multilepidina*, a Foraminifer which we show to be a synonym of *Pliolepidina* and to be of Miocene age only. These beds also constitute the basal portion of the great Miocene transgression, and we regard them as being unquestionably of Lower Miocene (Aquitanian) age.

In the INDIA–PAKISTAN region, the Aquitanian Upper Nari beds rest with a palaeontological break upon Oligocene beds which, since it is well known that they often contain *Nummulites intermedius-fichteli* and *Lepidocyclina* (*Eulepidina*), are mainly of Middle Oligocene (Rupelian) age.

In the NETHERLANDS EAST INDIES, the lower part of the *e* stage (i.e. e_1 and e_2) has recently been suggested (see, for example, van der Vlerk, 1955) to be of Chattian age, although we suspect that this may have been to fill in the succession. These beds were originally placed in the Lower Miocene, and a consideration of the succession of faunas in the Netherlands East Indies and adjacent regions does not indicate to us that there is any evidence that they should not be so dated. In these regions, the work of the geologists of the British Petroleum Co., Ltd, and its associates, has shown that beds undoubtedly belonging to the lower part of the *e* stage contain such characteristic Miocene forms as true *Austrotrillina howchini*, *Miogypsina* (*s.s.*), and *Spiroclypeus margaritatus*, but no forms believed to be characteristic of

the Oligocene. It has often been recognized that the five subdivisions of the *e* stage are difficult to identify, and that several of them are of local significance only; since all are evidently of Aquitanian age this is quite understandable. In addition, it is highly significant that these basal *e* stage horizons often contain much reworked Eocene and Oligocene rock fragments and fossils, and are in many places in the nature of basal conglomerates or basal breccias constituting the initial phases of an important Miocene transgression. The full appreciation of the importance of the mid-Tertiary orogeny in the Tethyan region leaves us in no doubt as to the dating of these beds; the unconformity between the *e* and *d* stages is marked by an important lithological and palaeontological break, and is the same one that can be recognized all round the world in the Tethys (the Mediterranean region, the Middle East, East Africa, the Indo-Pacific as a whole, and in Central America), that which is always marked by the same fundamental palaeontological break with Neogene faunas above and Palaeogene faunas below. We give (Fig. 2) the dating of some of the stages of the letter classification, showing the distribution of important foraminifera in them. We also show (Fig. 4) a revised dating of some beds in the Indo-Pacific region, modified from Glaessner (1959), to whom we are greatly indebted for permission to adapt his chart into our synthesis.

It is pertinent to refer here to the morphogenetic analysis of the genus *Lepidocyclina*, as outlined by Grimsdale (1959) and by van der Vlerk (1959). As van der Vlerk rightly stressed, these investigations are still in their experimental stage, the statistics given in individual cases being taken from a relatively small number of examples or even only one. Quite apart from the experimental nature of the evidence presented, although van der Vlerk pointed out that the specimens on which his statistical information was based were chosen from deposits whose age was sufficiently known, the evidence presented by us casts grave doubts upon the ages assigned to several of the occurrences quoted. The ages of the beds in question, from which these species of *Lepidocyclina* were obtained, actually depends, not upon the evolution gradus percentage given for the *Lepidocyclina* species in question, but upon the ages of the beds in which they were found as deduced by an analysis of the *total* faunas of the beds concerned in relation to the palaeontological and palaeogeographical history of the Tethys as a whole. Again, considering the great difference in size and nature of the embryonic apparatuses of the subgenera *Nephrolepidina* and *Eulepidina*, it may well be that the evolution gradus figure for the two is not at all comparable. In view of the complete absence of any Oligocene representatives of the genus in America (resulting in a considerable gap in the record—see §IX), and of the redating here of the lower *e* stage as Lower Miocene, the evidence that has been presented by van der Vlerk actually deals with only three examples from the Oligocene. Consequently, lines of evolution and lineages, on this evidence, must be regarded as highly speculative, as would be evident on a redrawing of van der Vlerk's figs. 2–4 on the basis of the dating given here. Again, examples of *Lepidocyclina rutteni* are known from high levels in the *e* stage, horizons which are below that of *L. besaiensis* from which it was suggested to originate. Further, in connexion with the study of lepidocyclinas, the use of the term 'stolon-system' does not seem to be applicable; in the *New Oxford Dictionary* the word 'stolon' is defined as being 'each of the connecting processes of the coenosarc of a compound organism', and as such it is not applicable to Protozoa, to which phylum *Lepidocyclina* belongs. We suggest that these structures in future be referred to as 'intercameral foramina', which is what they are. Finally, we suggest that all recent biological and palaeontological studies in evolution (see, for example, Piggott, 1959) indicate that simple 'straight-line' evolution, with direct and complete replacement in time of morphological forms, does not take place, but rather that populations display evolution by a 'shifting norm' (as outlined for the Orbitoididae by Renz & Küpper, 1947) accompanied by an overlap in time of individuals and populations of varied evolutionary morphological stages.

On SAIPAN, Todd (in Todd *et al.*, 1954, p. 674; 1957, p. 274) recorded the *insueta–bisphericus* sub-

zone fauna from the Fina–Sisu formation, almost immediately underlying the Tagpochau limestones. These limestones contained assemblages of larger foraminifera which were identified by Cole (1957c) as *L.* (*Nephrolepidina*) *sumatrensis, L.* (*N.*) *verbeeki, Cycloclypeus eidae*, '*Eorupertia*' (*vel Hofkerina*) *semiornata* (Howchin), *Miogypsina thecidaeformis*, etc., which clearly belong to the *f* stage. The higher horizons of the Tagpochau limestone are as young as f_3, which is shown by Cole's record of *Marginopora*. The Donni Sandstone facies of the Tagpochau limestones (see Todd, 1947, pp. 277–279) contains *Globorotalia menardii, Globigerina nepenthes* and *Sphaeroidinellopsis* spp., which together indicate a lower Vindobonian (probably Helvetian) age (see Part 2).

In the CAROLINE ISLANDS, Cole, Todd & Johnson (1960), writing on an apparent conflict of age indications, have described the association of the *insueta–bisphericus* subzone planktonic foraminifera with *Lepidocyclina* (*Nephrolepidina*) *martini* (although their figured specimen is probably more accurately referable to the stratigraphically older species *L.* (*N.*) *radiata* (Martin)), *L.* (*N.*) *sumatrensis*, and *Miogypsina polymorpha*. These larger foraminifera form an association which occurs typically in the f_{1-2} stage of Papua and the Netherlands East Indies, which is dated as Burdigalian; their occurrence with a planktonic foraminiferal assemblage which we believe to be characteristic of the uppermost Aquitanian (*Globigerinoides bisphericus, Globigerinatella insueta, Orbulina suturalis*) therefore indicates that the Burdigalian/ Aquitanian boundary is to be found very near to, or even within, the strata from which the sample (Cole *et al.* 1960, pp. 95, 103, sample YM-306) was taken. Thus, the age-determinations suggested by the above foraminifera do not, in our interpretation, conflict.

In the MARSHALL ISLANDS, Cole (1954, 1957d) and Todd & Post (1954) have recorded assemblages of larger and smaller foraminifera from drill-holes in Eniwetok Atoll and Bikini Island. At the highest levels in the drill-holes, assemblages with *Alveolinella quoii* and *Marginopora vertebralis* indicate that these beds cannot be older than the f_3 stage (Vindobonian). The immediately underlying beds in Eniwetok, which contained *Miogypsinoides* and *Miogypsina*, and, at a lower level, *Flosculinella globulosa*, correlate with the f_{1-2} stage of the Netherlands East Indies, and are dated by us as Burdigalian. Beneath these beds in Eniwetok, *Spiroclypeus margaritatus* and *Lepidocyclina* (*Eulepidina*) *ephippioides* were associated, indicating the presence of the *e* stage (Aquitanian); *Austrotrillina* and *Miogypsinella ubaghsi* also occurred within this interval. The lowest *Austrotrillina* were associated with *Borelis primitivus*, a species which is, as Glaessner (1959) has suggested, very closely related to the typically Neogene species *Borelis haueri* (d'Orbigny); we agree with Cole (1957*d*) that this assemblage, which is accompanied by the *e* stage *Lepidocyclina* (*Nephrolepidina*) *abdopustula* and *L.* (*N.*) *augusticamera*, is correlatable with the East Indies *e* stage and is of lowest Miocene age. Below this interval the faunas are unknown until the Eocene *Heterostegina saipanensis* was recognized. The presence of Oligocene beds on Eniwetok is, therefore, not yet proved.

In NEW GUINEA and MELANESIA, the very limited published accounts (e.g. Schubert, 1911; Chapman, 1914; Whipple, 1934; Crespin, Kicinski, Patterson & Belford, 1956) confirm the extensive and detailed observations (which are being published separately) made by geologists and palaeontologists of The British Petroleum Co., Ltd, and associated companies that the assemblages of larger foraminifera used in the Netherlands East Indies to define the letter-stages of the Tertiary succession (see also Glaessner, 1953, and Fig. 2) occur in the same sequence as elsewhere. Middle Oligocene *d* stage beds with *Nummulites fichteli* and *Eulepidina* are commonly followed by the *Spiroclypeus*-rich beds of the *e* stage, which also contain *Borelis pygmaea*, *Miogypsina*, *Miogypsinella*, *Miogypsinoides*, *Austrotrillina howchini* and *Eulepidina*, but no *Nummulites*. The *e* stage deposits are frequently clearly transgressive, resting upon Middle Oligocene, or Eocene or older beds, and they often commence with a basal conglomerate or microbreccia containing Palaeogene and Mesozoic pebbles and fossils. The higher parts of the *e* stage contain the first *Flosculinella* (*F. reicheli*), and the succeeding f_{1-2} stage (Burdigalian) beds are

characterized by *F. bontangensis*, *F. globulosa*, *Alveolinella fennemai* (which grades at the top of the stage into *A. quoii*), and *Katacycloclypeus*, associated with *Miogypsina*, *Miogypsinoides* (in the lower part), *Austrotrillina* and *Nephrolepidina*, but with no *Eulepidina* or *Spiroclypeus*. The essential unity of the f_{1-2} stage and *e* stage assemblages of species of *Lepidocyclina* (*Nephrolepidina*), for example, together with the transgressive nature of the base of the *e* stage and its lack of characteristic Oligocene *Nummulites*, indicate that the base of the *e* stage should be regarded as the base of the Miocene. *Orbulina* has been recorded in beds believed to be equivalent to the higher parts of the *e* stage.

In PORTUGUESE TIMOR, Glaessner (1959) has recorded *e* stage limestones with *Spiroclypeus*; shales interbedded between these limestones contain planktonic foraminifera of upper *Globorotalia kugleri* Zone or *Catapsydrax dissimilis* Zone age, confirming that they are Aquitanian.

In SOUTHERN AUSTRALIA, Carter (1958 *a*, *b*, *c*, 1959) has divided the Victorian Tertiary into a series of 'Faunal Units', correlatable with the standard Australian stages; these 'faunal units' were taken by Carter as the basis of a corresponding series of 'zones' based upon the local stratigraphical distribution of benthonic and planktonic foraminifera. His Faunal Unit 3 is probably of the same age as the *Globigerina turritilina turritilina* Zone of the Uppermost Eocene (see Part 2, p. 68), since this Unit is dominated by *Globigerina linaperta linaperta* whilst *Globigerapsis* and *Hantkenina* have become extinct. The overlying Faunal Unit 4, which is apparently present only in the St Vincent Basin (Chinaman's Gully) (see Fig. 4), but which may be present also in the lower Janjukian of Victoria, is characterized by an abundance of forms belonging to the *Globigerina ouachitaensis* group. Carter (1958*a*) has pointed out that *G. ouachitaensis* may have been responsible for many of the incorrect published records of *G. bulloides* in the lower Tertiary; this group is discussed in detail in Part 2, and it will be seen that the records may have referred to subspecies of *G. praebulloides* and/or *G. ouachitaensis*. These forms are characteristic of the smaller Globigerinaceae present in the Oligocene of East Africa and Europe, and also in the lower Aquitanian (*G. ampliapertura* Zone) of Europe and the West Indies. Consequently, it is not possible on the evidence available to say whether Faunal Unit 4 belongs to the Oligocene or to the basal Miocene. Faunal Unit 4 also contains the last *Chiloguembelina rugosa* (Parr), which would suggest an affinity with the Palaeogene, but it also contains the first *Sherbornina atkinsoni* (see also Wade & Carter, 1957) which might suggest a Neogene affinity.

Carter's Faunal Unit 5 contains *Globorotalia opima opima* (Carter, 1958; Glaessner, 1959, p. 62) and *Cassigerinella boudecensis* (see also Part 2, p. 69), and, according to Glaessner, 'rests disconformably on the Palaeocene to Lower (or Middle) Eocene Wangerrip Group' in the coastal section of Western Victoria. Faunal Unit 5 is considered here as being approximately equivalent to the *Globorotalia opima opima* Zone, and is thus dated by us as lower Aquitanian. The appearance of *Globoquadrina dehiscens* in Faunal Unit 6 (Lower Longfordian) is recorded by Carter (1958), but this may refer to *G. dehiscens praedehiscens* (see Part 2). *Austrotrillina howchini* is known to occur in Faunal Unit 6, and it persists as high as Faunal Unit 11 in southern Australia (Glaessner, 1959). The high Aquitanian age of the Longfordian is confirmed by the presence of the *Globigerinoides bisphericus* Subzone fauna in Faunal Unit 7: '*G. bisphericus* continues in Faunal Unit 8 and is abundant and associated with *Lepidocyclina* and *Cycloclypeus* in Faunal Unit 9' (Glaessner, 1959, p. 62), indicating that Faunal Units 7–9 may be correlated broadly with the *Globigerinatella insueta* Zone of the West Indies. The evolutionary forms leading to *Orbulina* and *Biorbulina* are recorded by Carter (1958) from the Balcombian (Faunal Unit 10), and *Orbulina suturalis* appears within this stage. It is probable that the base of the succeeding Faunal Unit 11 (Bairnsdalian) approximates to the base of the Burdigalian, and the *Globorotalia fohsi fohsi* Zone fauna may yet be found within it.

The '*Lepidocyclina–Cycloclypeus* fauna', which has been used (e.g. Glaessner, 1959, p. 63) to correlate the successions of southern and north-western Australia, may, in our opinion, be a diachronous assemblage which can occur at various horizons within the Aquitanian and Burdigalian, and which is broadly analogous to the '*Heterostegina* Zone' of the United States Gulf Coast. However, the Nullarbor limestone of the Eucla Basin contains *Flosculinella bontangensis* and *Austrotrillina howchini*, but no *Lepidocyclina* or *Cycloclypeus*, and belongs to the lower *f* stage (Burdigalian). The Nullarbor limestone may be broadly correlated with the Tulki limestone of north-western Australia, which contains *Katacycloclypeus* (characteristic of the lower *f* stage), associated with *Lepidocyclina*, *Cycloclypeus* and *Miogypsina polymorpha*, and which immediately succeeds the *e* stage Mandu limestones with *Eulepidina*.

In NEW ZEALAND, Hornibrook (1958, following the work of Finlay) has expressed doubt as to the precise placing of the Miocene/Oligocene boundary. However, he believed the whole of the Landon series to be Lower and Middle Oligocene (considered to be equivalent to the *c* and *d* stages of Indonesia) and equated the succeeding Pareora series, and the lower part of the Southland series (Altonian stage), with the *e* stage. The position is obscured by the rarity of larger foraminifera in New Zealand; however, *Lepidocyclina (Nephrolepidina)* first appears at the base of the Pareora, and *Miogypsina* (*s.s.*) is known from the middle Pareora to the Altonian stage; *Orbulina* makes its first clear appearance, following its evolution through *Porticulasphaera* (private letter from Hornibrook, 24 May 1960), within the Clifdenian stage, and within this stage the horizon may be broadly correlated with the '*Orbulina* surface' of LeRoy; the Clifdenian being referred to the upper part of the *insueta* Zone and to the *Globorotalia (Turborotalia) fohsi barisanensis* Zone. The relationship between the first occurrence of *Orbulina* and the distribution of *Globigerinoides bisphericus* (Altonian and Clifdenian stages) indicates that the Altonian is broadly correlatable with the *insueta* Zone of the Caribbean, and, consequently, with high Aquitanian in the Mediterranean. This is confirmed by the presence of *Eulepidina* and *Miogypsina intermedia* at other localities of the Altonian. The planktonic and larger foraminiferal faunas of lower (still post-Eocene) horizons are obviously influenced in their distribution by local and regional facies; the published first occurrence of *Globigerinoides trilobus* in the upper Pareora would suggest a correlation with its first appearance in the *kugleri* Zone of the Caribbean, but we know from our own observations that *G. trilobus* occurs at the base of the Pareora and ranges up. This would indicate, when considered with the occurrences of *Globoquadrina dehiscens* and *Catapsydrax dissimilis* in the uppermost Landon and Pareora, a *stainforthi* Zone age for these beds, and this, in turn, would suggest a considerable disconformity between the Oligocene lower and middle Landon and the Middle and Upper Aquitanian upper Landon–lower Southland interval. The faunas of the interval concerned are not yet sufficiently known to give more than a closely approximate correlation, but our studies show that when further work has been added to published knowledge the picture which will emerge will not conflict with that known from the Tethys and the Caribbean. Dr Fleming, of the New Zealand Geological Survey, has drawn our attention to the most significant fact that the Whaingaroan (lower Landon), where no Duntroonian (middle Landon) is present, has in some places quite large pipes, filled with sand, in its topmost layers, beneath the succeeding Waitakian (upper Landon).

VIII. The Dating of Marine So-called 'Oligocene' Beds in the Central American Region (see Fig. 5)

A. INTRODUCTION

To some of us for nearly a decade, and to one of us (F.E.E.) for nearly thirty years, it has been apparent that there were discrepancies in the interpretation of the evidence upon which the 'Oligocene' age of certain beds in the Central American region has been based. For over half a century the presence of the Oligocene in this region has been repeatedly accepted and has been extensively referred to in numerous well-known palaeontological and stratigraphical works. During the last decade, publications dealing with various fossil groups have brought to light a number of anomalous features, which have never been satisfactorily

explained because the presence of the Oligocene has been accepted and there has been no attempt to assess the faunas *ab initio*. In order to give correct stratigraphical distribution for the planktonic foraminifera under consideration elsewhere, we have found it essential to undertake this reassessment; as a result we have been led to the conclusion that the evolutionary sequence of faunas in this region agrees with that in all other parts of the world but that radical revision of the age-determination of many deposits in Central America is essential. Our conclusions are based not only upon the study of the literature but also upon the examination of samples from practically all the countries under consideration, including Burma, India and the Middle East, which form the link between the Australasian and Indo-Pacific areas and the Mediterranean and European areas.

B. HISTORICAL

Although Conrad, in the later part of the nineteenth century, was apparently the first to apply the term Oligocene to beds in America, the first application to be based upon any comprehensive evidence was that of Maury (1902) who based her opinion of the presence of the Oligocene in the southern United States on the identification of nummulites and on the apparent similarity of a few molluscan species to forms from the Oligocene of Europe. As far as we can ascertain, the only other major contribution was that of Vaughan (1919a) supplemented by other authors in the same volume; Vaughan correlated the coral faunas of the Antigua limestone and similar horizons with part of the Vicksburg formation and with the Middle Oligocene of Italy. Since 1919 practically all work on these and similar faunas in the region, including that on the Foraminifera, has tacitly accepted Maury's and Vaughan's Oligocene dating as being correct. Even MacNeil (1944), in his excellent review of the 'Oligocene' stratigraphy of the southern United States, gave no reason for the Oligocene age of the beds, but accepted current opinion. Furthermore, in the latest work on planktonic foraminifera (Bolli, 1957a), although some uncertainty has been expressed, the old school of thought is essentially maintained. However, a most important contribution by Kugler (1954) has outlined what reappraisal is necessary, and it is on these lines that our investigations have been carried out.

C. THE TYPE VICKSBURGIAN

As far as Central American stratigraphy is concerned, the Vicksburgian has come to be regarded as typical of the American Oligocene, and we have accordingly studied the published faunas in their relation to the known evolutionary succession of Tertiary faunas in other parts of the world. With the correlation of the various members (Red Bluff clays, Forest Hill sand, Mint Spring calcareous marls, Marianna limestone, Glendon limestone and Byram calcareous marls) of the Vicksburgian amongst themselves in Florida, Alabama, and Mississippi we are in complete agreement. The faunas have been well documented, are well known, and (apart from facies differences) are obviously closely related. This is well illustrated by the Foraminifera (which have been tabulated, for example, by Cooke, 1923) and by the Mollusca; quite a number of these (e.g. species of *Lepidocyclina*, and *Chione*) occur also in overlying beds. There are also corals which are of much the same age as those from the Antigua limestone of Antigua, and these will be considered later.

We have been led to the firm opinion that the over-all aspect of these faunas definitely places them as forming part of the evolutionary surge at the beginning of the Neogene, and hence as constituting part of the Lower Miocene. Arguments and points which might be raised against this contention have been considered and found to be capable of quite different and more reasonable interpretations.

First, one of the principal original reasons for dating the Vicksburgian as Oligocene was the reported presence of nummulites in the Vicksburgian beds which overlie strata thought to be of Upper Eocene age, the genus *Nummulites* ranging from Eocene to Oligocene. Later, in 1935, Hanzawa proposed the new generic name *Operculinoides* with *Nummulites willcoxi* Heilprin of the Eocene as type species, and subsequent workers (e.g. Cole, 1953a) have almost universally applied

Hanzawa's name to nearly all the American small 'nummulites'. Cole (1953a, p. 10) stated that 'most of the American species of camerinids belong to this genus', and, apart from a few Eocene forms which are typical *Nummulites*, this appears to be true. However, in dealing with these forms, Cole (1953a, p. 9) stated that '*Operculinella* is not a valid genus as its broadly flaring, complanate border is a gerontic development'. This conclusion cannot be accepted, and the presence of a gerontic complanate border (developed only on a few exceptionally large specimens) is no reason why the species *Amphistegina cumingi* Carpenter should not be accepted as the type species of the genus *Operculinella* Yabe, 1918, validly proposed according to the Rules of Nomenclature. One of us (Eames, 1953) has already drawn attention to what is, in our opinion, the correct use of the term *Operculinella*:

'**Miniature** *Nummulites*-like forms, involute, of small size, with a very small megalospheric nucleoconch, with very little difference in size between the two generations, with or without a tendency to flare in old age'... and it was recommended that it was 'very necessary to retain the name *Nummulites* only for those forms, such as *N. atacicus*, *N. obtusus*, *N. laevigatus*, *N. millecaput*, *N. intermedius* and *N. vascus*, which characterize the Palaeogene'.

However, it is evident to us that both the names *Operculinella* and *Operculinoides* must be regarded as synonyms of *Palaeonummulites* Schubert, 1908, the genus ranging from Eocene to Recent (see appendix 1). Cole (1959) has suggested that *Operculinella* and *Operculinoides* are synonyms of *Operculina*; this contention is based upon variation in what he considered to be one species (*Operculina ammonoides*), but we are of the opinion that the illustrations and text show that seven species and two genera are involved, and that there is no gradation; it is to be noted that although the typical *Operculina ammonoides* occurs at localities three, four, and seven, *Palaeonummulites*-like forms do not occur with it at locality seven (see Smout & Eames, in the press).

A consideration of American nummuloid Foraminifera further indicates that the only true *Nummulites* occur in the Eocene (Cushman's single specimen (1922, pl. 24, fig. 4) of *Nummulites* sp. from the Byram calcareous marls appears to us to be an *Amphistegina*), all post-Eocene, as well as some of the Eocene forms, belonging to the genus *Palaeonummulites*. There are, therefore, no true members of the genus *Nummulites* in the so-called 'Oligocene' beds of the Central American region, and one of the prime original arguments in support of such a dating becomes valueless. Finally, this leads to the most important consideration that, although beds of a 'fore-reef' facies (containing *Lepidocyclina* and *Heterostegina*) are of widespread occurrence in the region, not a single true *Nummulites* (apart, perhaps, from an occasional specimen reworked from the Eocene) occurs in them. True *Nummulites* (e.g. *N. macgillivrayi*) do occur in the American Eocene, and we are of the opinion that, if Oligocene beds in a 'fore-reef' facies did exist in the region, then other true *Nummulites* would be found in them; that they are not is, in combination with other evidence, a supporting factor in our contention that the so-called 'Oligocene' beds are actually of Lower Miocene age and that all search for true *Nummulites* in them will be fruitless because by then the genus had become extinct.

It has been pointed out to us that negative evidence is not proof, and that, for example, no *Heterostegina* has been found between the Ocala limestone of the Jacksonian and the *Heterostegina* Zone. We were quite aware of this, and have given very careful consideration to the distribution of different facies types in the whole Gulf Coast States–Caribbean–northern South American region before mentioning the absence of *Nummulites* from the Vicksburgian and its equivalents, which is just one point amongst many in leading us to the conclusion that the Vicksburgian faunas are of Lower Miocene age. In considering the whole of this region we believe that it must be agreed that, in beds previously assigned to the 'Oligocene', there are localities and horizons which are of the right 'fore-reef' facies type to have contained *Nummulites* should that genus have survived until those times. There are many records of *Lepidocyclina*-bearing samples in the so-called 'Oligocene', and, although in the Tethys *Lepidocyclina* and *Num-*

mulites do not occur together in many Lower and Middle Oligocene beds, it is well known that there are very numerous cases in which they have been found together in natural association in profusion. The absence of *Heterostegina* from Vicksburgian rocks seems to us to be due to the fact that there was no niche for such forms in the local ecological framework; *Cycloclypeus* is another related form with a similar sporadic distribution in the Tethys. Consequently, we feel that some reference to the absence of *Nummulites* from Vicksburgian beds is justified.

Secondly, it was considered by Maury (1902) that the molluscan faunas of the so-called 'Oligocene' of America showed some resemblance to those of the Oligocene of Europe. However, according to modern interpretation, it is doubtful if even one species is common to the two regions. As far as the general constitution of the fauna is concerned, we feel that it is of decided Neogene aspect. The lamellibranch venerid genus *Chione*, for example, is of common occurrence in the Neogene in many parts of the world, the only unquestionably correctly identified specimens from below the Miocene being from the so-called 'Oligocene' of America. Also, the common genus *Anadara*, apart from a few possible rare Oligocene occurrences known to us, appears to be confined to, and abundant in, the Neogene elsewhere. Again, the supposed 'Oligocene' index-fossil *Orthaulax pugnax* occurs in horizons above the Vicksburgian (e.g. the Tampa formation and the Anguilla formation) which are now acknowledged to be of Miocene age (see Schuchert, 1935).

There is, apparently, a distinct palaeontological break at the top of the Jackson formation, it having been reported that not more than 5 per cent of Eocene Mollusca range up into the Vicksburgian. Although it is acknowledged that Dall's work (1890–1903), as far as stratigraphy is concerned, is long outdated, the faunas he lists give the maximum opportunity for indicating the degree of relationship of the Vicksburgian molluscan faunas with those of the underlying Eocene. He lists twenty-seven forms as ranging up from the Eocene into the Vicksburgian or higher levels; there are considerable doubts about many of the above Vicksburgian records, both from the point of view of strati-graphical horizon and specific determination, and we anticipate that the revision of these forms (which we understand is now being undertaken) will confirm other opinions that very few, if any, Vicksburgian forms range up from the Eocene. On the other hand, there are seventeen firmly identified species which are cited by Dall as occurring in the Vicksburgian and higher beds, affording good evidence for the close relationship of the Vicks-burgian molluscan fauna with that of the overlying Miocene beds.

The genus *Clypeaster* (see Durham, 1955) first appeared in the upper Eocene, its representatives being usually small. Even in the Oligocene it was of rare occurrence, there being, for example, very few species recorded from the Lower Nari (Middle Oligocene) of West Pakistan and western India. It is only in the Miocene that the genus attained its acme and flourished, deposits in many parts of the world containing numerous large representatives of the genus. In view of the fact that it is frequently recorded from beds of Vicksburgian and equivalent age (which we regard as Miocene) in the Central American region, we feel that its vernacular name 'the Mexican hat' is appropriate.

Reverting now to the foraminiferal faunas, the distribution and frequency of such species of *Lepidocyclina* as *L. mantelli*, *L. supera* and *L. (Eulepidina) undosa* (often accompanied by the echinoid *Clypeaster rogersi* and such Lamelli-branchia as *Pecten poulsoni*) are well known, and the microforaminiferal faunas of the various mem-bers of the Vicksburgian, allowing for slight facies changes, constitute a fairly homogeneous unit the dominant benthonic specific component of which gives no reliable evidence for absolute dating. We have noted, however, that *Planorbulina mediter-ranensis* d'Orbigny, which is Neogene in Europe, occurs in all except the lowest of the so-called 'Oligocene' members which we would attribute to the Lower Miocene (Todd, 1952, p. 46). Incident-ally, *P. mediterranensis* has been reputed to occur in the Stampian or Rupelton in Germany, but the specimens we have from these beds are to be identified as *P. mangyschlakensis* Vassilenko, a Russian Oligocene species. Bolli (1957a) ampli-fying Akers (1955) correlated the Vicksburg 'stage' with the lowest Zone (*Globigerina ampliapertura*) of the Cipero formation of Trinidad, shown below to be of Aquitanian age from an assessment of its fauna (e.g. the occurrence together at this horizon of *Miogypsina* (*s.s.*) and *Lepidocyclina* (*Eulepidina*)). We also have examined Vicksburgian samples, and believe this correlation to be correct. From our own knowledge, we also agree with Bolli that the Chickasawhay marl is broadly equivalent to the *Globigerina ciperoensis* (*s.s.*) Zone of Trinidad. In spite of the records (Mornhinveg, 1941; Cushman & Todd, 1948) of the planktonic Eocene genus *Hantkenina* from the underlying Red Bluff clay (confirmed by our own observations), we consider that the complete faunas of the Red Bluff clay and of the Mint Spring calcareous marl are so similar that they must belong to the same major time unit. In any case, current American opinion is that the Red Bluff clay is of post-Eocene age (e.g. MacNeil, 1944; Todd, 1952), and, from world-wide know-ledge of the range of the genus *Hantkenina*, there can be no question but that the specimens of *Hantkenina* in the Red Bluff clay are reworked (an opinion held by Thalmann).

Mornhinveg (1941, p. 45) recalled that for some considerable time it has been realized that the Red Bluff clay contains 'both Vicksburg and Jackson macrofossils'. It will have been gathered by now that it is our contention that a major unconformity, accompanied by uplift and erosion of diverse types, followed upon the end of Eocene deposition in the Central American region, a view also held currently in America, but the implications of which have apparently not been fully appreciated; conse-quently, the presence of reworked Eocene fossils in subsequent deposits is only to be expected. Under this category we include not only the above-men-tioned occurrences of *Hantkenina*, but also the occurrences of the small *Turbinolia insignifica* Vaughan from the Mint Spring calcareous marl and of *?Turbinolia insignifica* from the Byram cal-careous marl, the genus *Turbinolia* being known on world-wide evidence to be restricted to the Eocene and the Oligocene. We shall draw attention else-where in this study to the acknowledged existence (e.g. in Trinidad) of reworked Eocene fossils in beds overlying this hiatus. Indeed, we suggest that much more thought be given to the varied and differing conditions of deposition following upon a major unconformity, not only from the point of view of the development of conglomerates and the presence of reworked macrofossils, but, in the case of a transgressive sea moving forward over a land mass of low relief and composed of richly fossiliferous soft rocks, from the point of view of the presence of reworked microforaminifera in an excellent state of preservation. Bolli (1957b), amongst others, has drawn attention to the presence, in Trinidad, of reworked blocks of Eocene material in the Miocene, these blocks having yielded an 'exceptionally well preserved fauna' on washing in the laboratory; the processes of nature could do this even better.

D. THE FAUNA OF THE ANTIGUA
LIMESTONE

In 1919, Vaughan described coral faunas from the Antigua limestone of Antigua and from beds of closely similar age elsewhere in Central America; he dated them as Oligocene, and remarked that the evidence from other groups of fossils agreed with such dating. We have been led, therefore, to re-investigate the evidence upon which this opinion was based, and in doing so we acknowledge the help and advice of Dr H. Dighton Thomas of the British Museum (Natural History).

First, of all the species recorded from these and equivalent beds in the Central American region, not a single species has been recorded as occurring elsewhere (e.g. Europe), so that there is no specific evidence for dating. Secondly, the (Middle) Oligocene dating is, according to Vaughan, dependent solely upon the fact that the following sixteen genera are common to the Antigua limestone and to the Middle Oligocene of Italy:

Actinacis	*Hydnophora*
Alveopora	*Leptomussa*
Antiguastrea	*Mesomorpha* (*vel Thamnasteria*)
Astreopora	*Orbicella* (*vel Montastrea*)
Astrocoenia	*Porites*
Euphyllia	*Stylocoenia*
Goniastrea	*Stylophora*
Goniopora	*Trochoseris*

According to the most recent information (*Treat. Invert. Pal.*, ed. Moore, (F)), the ranges of these genera are:

Thamnasteria	Middle Trias to Middle Cretaceous
Montastrea	Upper Jurassic to Recent
Hydnophora	Cretaceous to Recent
Actinacis	Middle Cretaceous to Oligocene
Stylophora	Eocene to Recent
Goniopora	Middle Cretaceous to Recent
Antiguastrea	Upper Cretaceous to Oligocene
Astreopora	Upper Cretaceous to Recent
Astrocoenia	Eocene to Miocene
Stylocoenia	Eocene to Miocene
Alveopora	Eocene to Recent
Euphyllia	Eocene to Recent
Goniastrea	Eocene to Recent
Porites	Eocene to Recent
Leptomussa	Oligocene

Concerning these occurrences, the following points may be noted.

(i) Eleven of the genera (*Alveopora*, *Astreopora*, *Astrocoenia*, *Euphyllia*, *Goniastrea*, *Goniopora*, *Hydnophora*, *Montastrea*, *Porites*, *Stylocoenia* and *Stylophora*) have long ranges, and have no bearing upon the question as to whether the Antigua limestone is Oligocene or Miocene.

(ii) Two genera (*Antiguastrea* and *Trochoseris*) have Oligocene as the upper limit of their recorded ranges, but since their ranges are long, and since some of this 'Oligocene' consists of the beds the age of which is now questioned, their evidence can hardly be accepted. Furthermore, Dr Dighton Thomas has kindly identified for us, from 'e' Stage Aquitanian (Lower Miocene) beds in Australasia, a new coral belonging to the genus *Antiguastrea*. In any case, the extension of the ranges of these two genera does not invalidate the Oligocene age of the Italian beds, which is based on evidence from fossils other than corals as well.

(iii) The genus *Mesomorpha* (*vel Thamnasteria*) has been misidentified; it is, incidentally, merely mentioned in passing as being present in the approximately equivalent Bainbridge limestone of Georgia, no species having been either described or illustrated.

(iv) The genus *Leptomussa* is merely mentioned in passing as being present in the Antigua limestone fauna, no species having been described or illustrated; the Treatise gives the distribution of the genus as Oligocene of Italy only.

(v) The genus *Actinacis* almost certainly includes the species *Dendrophyllia macroriana* Pascoe & Cotter (1908), which has been recorded from Burma from beds of both Upper Oligocene and lower Aquitanian age.

As a consequence of the above considerations, Dr Dighton Thomas and ourselves have been led to the firm conclusion that there is actually no definite evidence of precise dating from the coral faunas of the beds concerned; furthermore, it is hardly correct to say that the other groups of fossils associated with the corals yielded evidence in conformity with an Oligocene age, because most of the species are indigenous to the region, and some of the evidence has not yet been correctly interpreted. These latter points are considered later on.

Turning now to the foraminiferal faunas of the Antigua limestone, the close similarity of the 'fore-reef' *Lepidocyclina* fauna to those of the Culebra formation, of the Vicksburgian, and of similar horizons in the Central American region is well known. As in the case of these latter horizons, although the beds are in a 'fore-reef' facies, at least in part, not a single true *Nummulites* has been found; the species *antiguensis* Vaughan & Cole, regarded by them as an *Operculinoides*, but placed in *Palaeonummulites* by us, is indigenous to the region and does not necessarily indicate an Oligocene age. Amongst samples we have examined, however, some contain a species of *Sporadotrema* which we have found (abundantly) only in beds of Aquitanian to Recent age elsewhere (Indian Ocean region, Australasia).

Cooke (1919) considered the Antiguan molluscs to have their relationships with the Lower Miocene Tampa and Anguilla formations. Indeed, the coral fauna of the Antigua limestone also shows relationships with that of the Anguilla formation since six species (*Antiguastrea cellulosa, Goniopora clevei, G. cascadensis, Montastrea costata, Pironastrea anguillensis* and *Siderastrea conferta*) are common to the two. Hence, to us, all the indications are that the Antigua limestone is of Lower Miocene (Aquitanian) age—the fauna includes *Lepidocyclina* belonging to the subgenus *Eulepidina*, so it cannot be younger than Aquitanian.

E. OTHER CLOSELY RELATED CORAL FAUNAS IN THE CENTRAL AMERICAN REGION

In addition to the corals of the Antigua limestone, Vaughan (1919a) also recorded coral faunas from a number of beds of closely similar horizon in the Central American region; with the correlation of these amongst themselves we are in agreement, and we acknowledge the great amount of useful evidence he has placed on record. The horizons concerned (Vaughan, 1919a, pp. 203–207) are the Pepino formation of Porto Rico (eight out of twelve species also in the Antigua formation), the Guantanamo coral limestone of Cuba (seven out of twelve species also in the Antigua formation), the Bainbridge limestone which forms the basal part of the Chattahoochee formation in Georgia (nine out of thirteen forms also in the Antigua formation), the coral limestone forming the basal part of the Chattahoochee formation of Salt Mountain, Alabama (both species also in the Antigua formation),[1] the San Rafael formation of eastern Mexico (three out of five forms also in the Antigua formation), the Tonosi limestone of Panama (four out of five species also in the Antigua limestone) and the Arube coral fauna (both named species also in the Antigua formation). In conformity with our previous deductions, we date all these faunas as Lower Miocene; none of the fossils associated with these coral faunas yields, in our opinion, any evidence of an Oligocene or a pre-Miocene age. Vaughan (1919a, pp. 208, 209) has also pointed out the close relationship (based upon species in common) of the Antigua limestone coral fauna to those of the slightly younger Culebra formation and Emperador limestone. Finally, it will be recalled that Duncan in 1863 originally dated all of these West Indian coral faunas that he dealt with as Miocene.

F. THE EVIDENCE OF THE MIOGYPSINIDS AND '*ORBULINA*'

One of us (Eames, 1953) has drawn attention to the importance of *Miogypsina* and *Orbulina* in regional correlation and dating; in so far as the Central American Tertiary successions are concerned, the present work continues these earlier considerations to their logical conclusions in the light of worldwide evidence. Moreover, it seems that Drooger may have changed his opinion, since, although in 1954b he considered that a sequence of *Miogypsina* species, similar to that of the Aquitanian and

---

[1] Cooke (1935) pointed out that the Salt Mountain limestone (Coral limestone *auctt.*) contains abundant *Discocyclina* and is of Eocene age, being upthrust by the Jackson Fault into a position adjacent to Vicksburgian beds; these latter beds, or some other equivalent of the basal Chattahoochee formation, therefore seem to have been the source of the corals.

Burdigalian of Europe, was much older in the Western Hemisphere, and was there partly placed in the Oligocene, in 1958 (p. 116) he wrote 'from the occurrence of *Miogypsina* (*s.s.*) the only deduction that can be made is that the sediment in question is post-Oligocene in age', from which we infer that, in the Mediterranean area at any rate, he regarded *Miogypsina* (*s.s.*) as being confined to the Miocene; and Akers & Drooger (1957) date all *Miogypsina* (*s.s.*) of the Gulf Coast as Miocene.

We would make it clear here that we regard *Miogypsinella complanata* as being generically distinct from *Miogypsina* and from *dehaarti* (the type species of *Miogypsinoides*); the bases are that *Miogypsina* alone of the three has lateral chamber layers, and that in *Miogypsinoides* the solid outer layer is excessively thicker and the initial coil shorter than in *Miogypsinella*. These criteria are very easy to observe, and form a very easy basis for a distinction which is of stratigraphical value.

LeRoy (1948) drew attention to the fact that, whereas *Orbulina* was supposed to start in the upper Oligocene in the Caribbean region, in the Netherlands East Indies it first appears near the top of the 'e' stage which is known to be Aquitanian on account of the co-occurrence of such forms as *Austrotrillina howchini*, *Eulepidina*, *Flosculinella*, *Miogypsina*, *Spiroclypeus* and Miocene species of *Nephrolepidina* such as *Lepidocyclina* (*N.*) *sumatrensis*. LeRoy (1952, p. 581) further emphasized this point on a chart showing the anomalous Oligocene occurrences of *Orbulina universa* in the Central American region only. We can confirm that, in the Mediterranean and East African areas also, *Orbulina* first appears near the top of the Aquitanian (Eames & Clarke, 1957, p. 80).

Blow (1956) discussed the origin and evolution of *Orbulina* and pointed out that *Orbulina* and *Biorbulina* developed (in a very short interval of time) from *Globigerinoides trilobus* (*s.s.*) via *G. bisphericus* and *Porticulasphaera glomerosa* in the upper part of the *Globigerinatella insueta* Zone of Venezuela, Trinidad and Barbados. The same evolutionary sequence was also seen in the Mediterranean area (Blow, 1957). Geologists of The British Petroleum Co., Ltd, carefully sampled measured sections of Lower Miocene limestones in the Ragusa platform area in Sicily. Beds called the Upper Ragusa limestone in the Monte Casisia section include marly bands which contain *Porticulasphaera glomerosa*, *P. transitoria*, *Globigerinoides bisphericus*, *Globigerinatella insueta*, *Globorotalia* (*Turborotalia*) *fohsi barisanensis*, etc.; this assemblage clearly matches the *G. insueta* Zone of the Caribbean. The Upper Ragusa limestone may be dated as Aquitanian by its larger foraminiferal fauna of *Austrotrillina howchini*, *Miogypsina globulina*, *Miogypsinella complanata*, *Praerhapydionina delicata* and *Spiroclypeus blanckenhorni ornata*, all occurring in normal association. The Upper Ragusa limestone is succeeded in the Molino Gaetani section by beds called the Tellaro marls which contain *Globorotalia* (*Turborotalia*) *fohsi barisanensis* with forms transitional to *G.* (*T.*) *fohsi fohsi* associated with *Orbulina universa*, *O. suturalis* and *Biorbulina bilobata*, etc. This fauna has also been observed in the Blue clay of Malta. In the Melilli section of the Ragusa platform area the Tellaro marls pass laterally into the basal portion of beds there called the Palazzolo limestone which contains *Borelis melo*, a form unknown anywhere from below the Burdigalian; in this section, the underlying limestones equivalent to the Upper Ragusa limestone contain indigenous *Spiroclypeus blanckenhorni ornata* up to the top, and also *Miogypsinella complanata* and *Miogypsina* (*s.s.*) nearly to the top. We conclude, therefore, that the Tellaro marls and their equivalents are of basal Burdigalian and of *Globorotalia* (*T.*) *fohsi fohsi* Zone age, and that the underlying Caribbean *G.* (*T.*) *fohsi barisanensis* Zone is represented by the highest part of the Aquitanian Upper Ragusa limestone. These conclusions are in accord with the published findings of AGIP Mineraria (1957) in Italy. Colom (1958) has recognized on Majorca the evolutionary sequence from *Porticulasphaera* into *Orbulina* and *Biorbulina* in beds which he dated, on the basis of macrofossils in equivalent beds on the island, as Lower Miocene (Burdigalian and/or Aquitanian); his conclusion, with which we agree, was that the beds are not of Helvetian age (see also Stainforth, 1960).

Drooger & Socin (1959), in their excellent account of *Miogypsina globulina* and *Lepidocyclina*

(*Nephrolepidina*) *tournoueri* from the basal conglomeratic portion of the transgressive Miocene series at Rosignano, northern Italy, again, however, have based their dating on principles of nepionic acceleration. These beds have evidently been variously dated previously as Aquitanian, Burdigalian and Helvetian. However, we agree that the planktonic foraminifera indicate a correlation with the *Catapsydrax dissimilis* Zone of the Caribbean, and, since the Italian material also evidently contains *Lepidocyclina (Eulepidina)*, we have no doubt that it is of Aquitanian age. Since the *Catapsydrax dissimilis* Zone is below the highest Aquitanian at which level *Orbulina* first appears, the absence of *Orbulina* is quite understandable.

The presence of *Orbulina suturalis* (recorded as '*Candorbulina universa*') and *O. universa* in Lower Miocene sediments of southern Spain has been demonstrated by Colom (1952, p. 879); these species are accompanied by *Globorotalia menardii praemenardii*, indicating that these beds are either of highest Aquitanian or (as we believe) lowest Burdigalian age. These sediments are followed (p. 879) by acknowledged Burdigalian strata which contain many species of benthonic and planktonic foraminifera common to the Cipero and Vicksburg formations of America (pp. 880, 881).

The evolutionary series leading to *Orbulina* occupies a very short time-interval and has been recognized in many different parts of the world (e.g. in Australia by Jenkins, 1958; in Saipan by Todd, in Blow, 1956, p. 62); it corresponds in principle to the '*Orbulina* surface' of LeRoy. However, *Orbulina*, like all other animals, is influenced by ecological conditions, and there are places (e.g. the Vienna Basin) where its later appearance is not accompanied by its evolutionary precursors. As pointed out by Drooger (1956), *Orbulina* has never been recorded from the Lower Vindobonian (Helvetian) of the Vienna Basin, but first appears in the Tortonian of that area; our own Tortonian collections from Rohrbach, near Nussdorf, in the Vienna Basin contain *Orbulina* associated with *Globigerina bulloides* (*s.s.*), *G. bulloides concinna*, *G. juvenilis*, '*G*' *nepenthes* (here figured), *Globigerinoides quadrilobatus*, *Globorotalia (Turborotalia) minutissima*, *G. (T.) obesa*, *G. (T.) opima*

continuosa and *Hastigerina aequilateralis*; in addition, Marks (1951) has recorded the occurrence of *Globorotalia menardii*. This assemblage clearly correlates with that of the Lengua formation of Trinidad (see Bolli, 1957*a*). This latter formation is well above the first appearance of *Orbulina* in Trinidad. Consequently, we do not believe that the first appearance of *Orbulina* in the Tortonian of the Vienna Basin should be used as a criterion for world-wide dating (see also Stainforth, 1960, p. 226, and Socin, 1959).

Drooger (1956), etc., has stated that *Miogypsina* becomes extinct before the appearance of *Orbulina*. This misconception has arisen on account of the rarity of the warm-water *Orbulina* in the European and Mediterranean Lower Miocene. We have indigenous *Miogypsina* from horizons above those in which we have found *Orbulina* both in a well at Vittoria in Sicily and in a well at Zabbar on Malta, and we shall show below that east of the Mediterranean *Miogypsina* becomes abundant above the '*Orbulina* surface' of LeRoy. *Miogypsinella* (but not *Miogypsinoides*) appears to be confined to beds older than the '*Orbulina* surface' of LeRoy; although it has been stated that *Miogypsinella complanata* always preceded *Miogypsina* (*s.s.*), we have seen them in normal association, e.g. in the same rock slide of a sample from Sicily (see Plate III). Although *Miogypsinella complanata* undoubtedly does occur in beds of Upper Oligocene age, its association with *Miogypsina* (*s.s.*) can only indicate an Aquitanian age, and we agree with both Akers and Drooger that *Miogypsina* (*s.s.*) does not occur below the Miocene and with van der Vlerk that it does not occur below the *e* stage. The oldest authentic occurrence of *Miogypsina* in Europe is that of the morphologically extremely primitive *M. septentrionalis* in beds admitted to be younger than the type Chattian of Cassel (Drooger, 1960, p. 47). Of the Mediterranean Helvetian records of *Miogypsina* (*s.s.*) some may be correct, and we believe that the Helvetian may be the upper limit of its range.

The type locality for *Miogypsinella complanata* was originally given as Aquitanian of Saint-Étienne-d'Orthe in Aquitaine, although doubts have since been expressed about the age. Dr Cox has already shown (p. 7) that at Saint-Étienne-d'Orthe beds of both Aquitanian (Lower Miocene) and Oligocene (probably Chattian according to Dollfus—there are no *Nummulites*) age are present. Although Drooger (1955, p. 18) considered the type locality to be of Oligocene age (based upon his theory of nepionic acceleration in Miogypsinidae), there is actually no evidence to show whether the type horizon for the species was in the probable Chattian or the Aquitanian beds. It may be significant, however, that the Miocene horizons at Saint-Étienne-d'Orthe are the only ones from which larger foraminifera have been recorded (see Douvillé, 1917). Drooger (1955, p. 18) has also recorded *Miogypsinella complanata* from Saint-Géours, Christus and Abesse (near Dax) from horizons which he would date as '(Rupelian-) Chattian', but which other authors (e.g. Daguin, 1948) would date, in part at least, as Lower Miocene (Aquitanian). At Saint-Géours, again, it appears that beds of both Oligocene (probably Chattian) and Aquitanian age are present and are underlain by beds of Rupelian age with *Nummulites intermedius* and *N. vascus* (Daguin, 1948, p. 144), so that it cannot be ascertained, in the lack of precise information, whether the single museum specimen of *Miogypsinella complanata* recorded from this locality came from the Upper Oligocene or from the Aquitanian. However, Dr F. W. Anderson (H. M. Geological Survey of Great Britain) has allowed us to study samples collected by him, under the guidance of Professor J. Cuvillier and Dr Neumann, from beds at Escornebéou which were dated by these authorities as Lower Aquitanian. These samples contain *M. complanata* whose 'Mx coefficient' is of the order of 22 (Plate II) and much more 'primitive' than other specimens known to us from the true Oligocene of other areas (see also Part 2, p. 75). We consider that age-determination based solely on Drooger's theory of nepionic acceleration is not sufficient in itself to outweigh all the other palaeontological and geological evidence which indicates that the species ranges from at least the Chattian to the Aquitanian, as he himself admits (Drooger, 1955, p. 45).

On Malta, in a measured section near Ghar Lapsi, typical Miocene *Austrotrillina howchini* occurs below *Miogypsinella complanata*, and *M. complanata* occurs in natural association with *Spiroclypeus blanckenhorni ornata* (see Plate II), a species of Lower Miocene (and therefore Aquitanian) age. *S. blanckenhorni ornata* is associated at its type locality with *Miogypsina (s.s.)*.

In Algeria, Drooger & Magné (1959, pp. 277, 278) record *Miogypsina cushmani* and *Miogypsinella complanata* in natural association at their locality 9, and *Miogypsina mediterranea* and *Miogypsinoides* (they place *complanata* under *Miogypsinoides*) in association with Lower Miocene mollusca at locality 10 immediately below an *insueta* Zone planktonic fauna. They date these as Helvetian on the principle of nepionic acceleration applied in such a way that no overlap of species is permissible; we agree with Colom's remarks (1958) that such fossil assemblages are not of Helvetian age, and we consider them as a natural Aquitanian association, there being no need at all to consider any of the faunas as reworked.

In Persia, *Miogypsinella complanata* occurs practically at the top of the Aquitanian Middle Asmari limestone throughout which *Archaias hensoni*, *Meandropsina anahensis* and *Austrotrillina howchini* are of common occurrence; in East and North Africa *Archaias hensoni* occurs only in beds of Lower Miocene (Aquitanian) age and *Meandropsina anahensis* occurs in beds of both Aquitanian and Burdigalian age but not below.

In Tanganyika, in one survey area alone (in the Lindi area) we have ten records of *Miogypsinella complanata* occurring together with *Miogypsina (s.s.)*. At a depth of 37–41 feet in an information hole on Mafia Island, the low *f* stage *Flosculinella bontangensis* occurs in association with *Orbulina*.

The above observations concerning the ranges of miogypsinids and *Orbulina* are fully confirmed in the Far East where *Miogypsina (s.s.)* first appears at a low level in the *e* stage and extends to high levels in the *f* stage, *Miogypsinella* (not *M. complanata*) may first appear in the *d* stage and certainly extends up to a horizon almost at the top of the *e* stage, and *Miogypsinoides dehaarti* extends from at least low *e* stage horizons up into f_1 substage. These occurrences have been linked by LeRoy (1948, 1952) with the appearance of *Orbulina* at the top of the *e* stage.

Drooger (1952), in his clear and detailed account of the succession of miogypsinid species in the Central American region, recorded *Miogypsina* (*s.s.*) from the so-called 'Oligocene' Tabera formation of the Dominican Republic, the Culebra formation of Panama, the '*Globigerina*' *dissimilis* Zone of the Cipero formation of Trinidad, the San Luis formation of Venezuela, the San Sebastian formation of Porto Rico, the Jaruco formation of Cuba, the Suwannee formation of Florida, the Paso Real formation of Cuba, the Emperador limestone of Panama and from other unnamed 'Oligocene' beds; all these occurrences we consider to be broadly correlatable (not on the evidence of the *Miogypsina* alone) and to be not older than Miocene; Akers & Drooger (1957) agree that all Gulf Coast *Miogypsina* are Miocene. In the 'Oligocene' Montpelier limestone of Jamaica, for example, *Orbulina universa* occurs together with *Miogypsina bracuensis*, and also with and below a suite of *Lepidocyclina* species ubiquitous in the so-called 'Oligocene' of America (see Schuchert, 1935, p. 424); it has been suggested to us that one or both of these species may be incorrectly identified or redeposited, but from our stratigraphic table (Fig. 5) it will be seen that there is no reason why the two should not occur together at this horizon.

Turning now to other forms, the genus *Heterosteginoides* Cushman, 1918, is an acknowledged synonym of *Miolepidocyclina* Silvestri, 1907. Its type species, *H. panamensis* Cushman, comes from the Culebra formation of Panama, and Drooger (1952) recorded the species *mexicana* (Nuttall), under the synonymous subgeneric name *Miogypsinita*, from the Suwannee limestone of Florida. It appears to us to be highly significant that all records of this genus from the Old World are confined to the Lower Miocene.

We are aware that *Lepidosemicyclina* Rutten, 1911, has recently been re-instated by Mohan (1958) as a subgenus of *Miogypsina*, on the basis of the development of hexagonal chambers in equatorial section. This we consider to be due merely to the sections being slightly off the equatorial plane (cf. Mohan's plate 2, figs. 10 and 11), and

perhaps sometimes being a gerontic feature. Consequently, we regard *Lepidosemicyclina* as a synonym of *Miogypsina* (*s.s.*). No type species for *Lepidosemicyclina* appears to have been designated, so *Orbitoides* (*L.*) *thecidaeformis* Rutten, 1911, is here designated as type species. Also, *Flabelliporus dilatatus* Dervieux is here designated as type species of *Flabelliporus* Dervieux, 1894, so that, in conformity with current usage, the latter name is also a synonym of *Miogypsina* Sacco, 1893.

G. BRIEF REVIEW OF SIGNIFICANT FAUNAS FROM IMPORTANT PARTS OF THE CENTRAL AMERICAN REGION (see Fig. 5)

The close relationships of the so-called 'Oligocene' faunas of the Central American region with those of the Lower Miocene, both in that region and elsewhere, have become apparent to us and we hope that the following review and synthesis will make this clear. In mentioning species of *Lepidocyclina* we are aware of the synonymy published by Cole (1957*b*) but we feel that, although his illustrated specimens of some previously misidentified species may be correctly identified, we cannot yet accept the synonymy of many of the original species; this does not affect our conclusions in any way. Also, Cole (1958*a, c*) has indicated that he regards the nine species *vicksburgensis*, *semmesi*, *antiguensis*, *forresti*, *ellisorae*, *howei*, *muiri*, *palmarealensis* and *bullbrooki*, and the subspecies *semmesi ciperensis* as synonyms of the species we refer to as *Palaeonummulites dia*; this seems to us to be rather swinging the pendulum too much the other way, much as we are in favour of the recognition of intraspecific variation; here again, our usage of the original ten names does not affect our conclusions in any way. The correlations and dating shown on Fig. 5 have been arrived at by a cumulative synthesis and assessment of the evidence presented on the following pages, each case being checked against its predecessors individually and collectively, and finally checked as a whole. While there are bound to be personal opinions concerning the synonymy and taxonomy of some forms, it is the correlation and dating of the actual *assemblages* that matters.

We have had extensive discussions with Dr F. S. MacNeil (United States Geological Survey) concerning the intercorrelation of formations in the Gulf Coast of the United States. Fig. 3 was compiled by Dr MacNeil to illustrate his views on stratigraphical correlation within the region from Texas to Florida, basing his interpretation on the field evidence and the distribution of macrofossils.

question, and we feel that the anomalies can only be resolved by detailed examination of type-locality surface samples and their direct correlation with supposed subsurface equivalents. We suggest that the influence of facies upon some of the benthonic foraminifera, for example, has led to

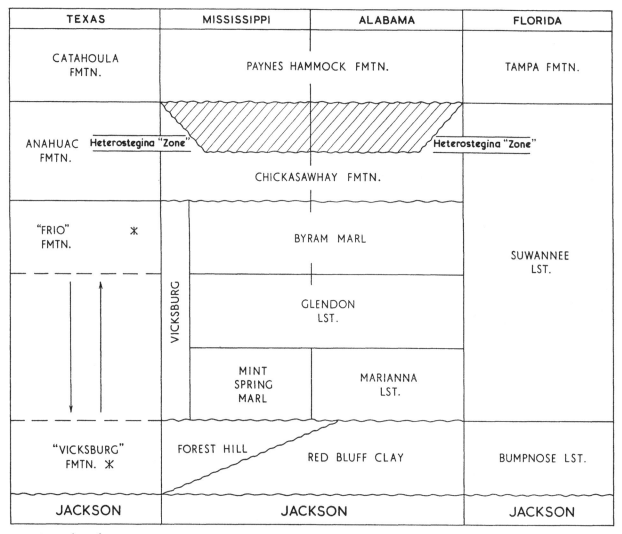

*Subsurface formations

Fig. 3. Correlation of some mid-Tertiary beds of the Gulf States based on outcrop evidence by F. S. MacNeil.

It can be seen that this interpretation differs in detail from that which is presented in Fig. 5, which has been compiled from evidence presented largely by Akers & Drooger (1957), Bolli (1957a) (for southern Louisiana) and other subsurface workers. These discrepancies do not affect our general conclusions as to the over-all dating of the beds in

some of the miscorrelations, and that more attention might profitably be paid to investigations into the distribution of the Globigerinaceae.

Dr MacNeil has pointed out to us the significant field-relationships often seen between the Red Bluff clay (Vicksburgian) and the sediments of the underlying Jacksonian. The uppermost three feet of the Jackson sediments show progressive leaching and disintegration; as the top of the Jackson is approached the fossils pass from a state of normal preservation, through varying degrees of rottenness, until they are represented only by phosphatized moulds. The top of the Jackson is clearly an eroded surface, for sand-filled pipes and borings occur in its highest part. The succeeding Red Bluff clay, constituting the basal bed of the Vicksburgian, contains at its base stringers of rolled, reworked and often phosphatized fossils and nodules derived from the Eocene. The nature and duration of this clearly observed hiatus is discussed in succeeding pages.

In FLORIDA (Akers & Drooger, 1957; Applin & Jordan, 1945; Cole, 1938, 1942, 1944, 1945; Cooke, 1945; Schuchert, 1935), *Lepidocyclina mantelli*, *Palaeonummulites dia*, *Pecten poulsoni* and *Clypeaster rogersi* have been recorded from the Marianna limestone, *C. rogersi* from the Glendon limestone, *Lepidocyclina supera*, *Anadara lesueuri*, *Pecten poulsoni* and *Clypeaster rogersi* from the Byram calcareous marls, *Heterostegina texana*, *Lepidocyclina (Eulepidina) favosa*, *L. (E.) gigas*, *L. supera*, *L. (E.) undosa*, *L. yurnagunensis*, *Miogypsina cushmani*, *M. gunteri*, *Miolepidocyclina mexicana*, *Palaeonummulites vicksburgensis*, *Orthaulax pugnax*, *Anadara* spp., *Chione bainbridgensis*, *Antiguastrea cellulosa* and *Diploastrea crassolamellata* from the Suwannee limestone, and *Lepidocyclina (Eulepidina)* cf. *favosa*, *Orthaulax pugnax* and *Clypeaster rogersi* from the Flint River formation (the littoral equivalent of the Suwannee limestone). It is also to be noted that, from the Suwannee limestone equivalent, the reworked Eocene forms *Coskinolina floridana* and *Dictyoconus cookei* occur in association with *Lepidocyclina (Eulepidina) favosa*, *L. (Nephrolepidina) suwanneensis*, *L. (E.) undosa* and *Palaeonummulites vicksburgensis*. From horizons in wells in north-west Florida, referred to merely as 'Oligocene', have been recorded *Heterostegina texana*, *Lepidocyclina parvula*, *L. (Eulepidina) undosa*, *Miogypsina gunteri*, *M. hawkinsi*, *M. venezuelana* and *Palaeonummulites forresti*, which fauna is the equivalent of the *Heterostegina* Zone of Texas. From subsurface beds reputed possibly to be younger than the Suwannee limestone, have

been recorded forms alleged to be intermediate between *Miogypsinella thalmanni* and *Miolepidocyclina panamensis*. All these beds we date as Aquitanian.

In SOUTH CAROLINA (Cooke & MacNeil, 1952), the microforaminiferal fauna of the Cooper marl (only 3 to 8 feet thick) has, we think rightly, been correlated with that of the Red Bluff clay and it, therefore, appears to be of low Aquitanian age. Cooke & MacNeil pointed out that the type of *Chlamys cocoana* (Dall) apparently came from the Red Bluff clay and not from the Cocoa sand, so that it appears to be inappropriately named. Not only does the Cooper marl contain abundant specimens of it, but also several representatives of primitive toothed whales, these latter being unknown in Eocene deposits elsewhere and, evidently, not being so old as was thought; Kellogg (1924) pointed out that the European middle Oligocene pelagic mammals are not known in American so-called 'Oligocene' deposits.

In NORTH CAROLINA, McLean (1947) recorded unnamed marls (now called Trenton marls, and believed to be equivalent to the Suwannee limestone) which contain *Angulogerina byramensis*, *Eponides byramensis*, *Virgulina vicksburgensis* and many other Vicksburgian benthonic microforaminifera which we date as Aquitanian.

In GEORGIA (Cooke, 1945, Schuchert, 1935, cum. bibl.), Vaughan has described, from the Bainbridge limestone, a rich coral fauna many of the species of which (e.g. *Antiguastrea cellulosa*) occur also in the Antigua limestone. *Lepidocyclina mantelli*, *Orthaulax pugnax* and *Pecten poulsoni* have been recorded from the Flint River formation; the old records of *Orbitoides papyracea* and *Orbitolites complanatus* are quite unreliable since both are Eocene species, but it seems likely that the latter may actually refer to '*Meandropsina*(?)' [*Orbitolites*] *americana*, originally described from the Culebra formation and Emperador limestone of Panama. These Georgia beds we date as Aquitanian.

In ALABAMA (Bandy, 1949; Cooke, 1935; MacNeil, 1944; Schuchert, 1935), *Ostrea* '*vicksburgensis*' has been recorded from the Red Bluff clay, *Lepidocyclina mantelli*, *Planorbulina mediterranensis*, *Ostrea* '*vicksburgensis*', *Pecten poulsoni* and *Clypeaster rogersi* from the Marianna lime-stone, *Lepidocyclina supera, L. (Eulepidina) undosa, Orthaulax pugnax, Chione bainbridgensis* and *Clypeaster rogersi* from the Bainbridge limestone, *Lepidocyclina supera, Anadara lesueuri* and *Chione bainbridgensis* from the Byram calcareous marl (including the Bucatunna clay), and *Lepidocyclina (Eulepidina) favosa, L. (E.) undosa, Anadara* aff. *lesueuri* and *Ostrea 'vicksburgensis'* from the Chickasawhay marl. The upper part of the Chicka-sawhay formation has been distinguished as the Paynes Hammock formation, and subsurface beds which have been correlated with this latter forma-tion contain *Miogypsinella bermudezi* and *M. complanata* (Akers & Drooger, 1957). All these beds we date as Aquitanian. The succeeding Tampa stage contains *Globorotalia lobata lobata* and is therefore Burdigalian.

In MISSISSIPPI (Cooke, 1935; Mornhinveg, 1941; Todd, M. R., 1947; Todd, R., 1952; Vaughan & Cole, 1936), Todd, M. R. (1947) stated that the microforaminiferal fauna of the Vicksburg group constituted a single unit; we agree that this is true except that certain forms such as the Eocene *Globorotalia (Turborotalia) cerroazulensis, Hant-kenina* spp. and *Pseudohastigerina micra* and Cretaceous microforaminifera must obviously have been derived from beds unconformably underlying the Red Bluff clay (see Jones, 1958). Further, Cushman (1935) remarked upon the close relation-ship of these microforaminiferal faunas with those of the Miocene of Australia. *Planorbulina medi-terranensis* has been recorded from the Mint Spring marl, *Lepidocyclina mantelli* and *Pecten poulsoni* from the Marianna limestone, *Lepido-cyclina mantelli, Planorbulina mediterranensis, Ostrea 'vicksburgensis', Pecten byramensis* and *Clypeaster rogersi* from the Glendon limestone, *Lepidocyclina supera, Palaeonummulites vicksburg-ensis, Planorbulina mediterranensis, Anadara le-sueuri, Ostrea 'vicksburgensis'* and *Pecten byra-mensis* from the Byram calcareous marl, and *Lepidocyclina (Eulepidina) favosa, L. (E.) undosa, Anadara lesueuri, Chione* cf. *bainbridgensis, C.* cf. *spenceri* (of the Antigua limestone) and *Ostrea 'vicksburgensis'* from the Chickasawhay marl. All these beds we date as Aquitanian.

In LOUISIANA (Akers, 1955; Gravell & Hanna, 1937; Howe, 1933), *Anadara lesueuri* and *Pecten poulsoni*, together with a large Byram calcareous marl fauna, have been recorded from Vicksburgian beds in the north. The *Heterostegina* 'Zone' of Texas, containing also *Lepidocyclina texana*, has been recorded as well. These beds we date as Aquitanian. In the delta region a thicker and more complete succession is found in subsurface sections, and this correlates more fully with the successions in Mississippi and Texas (Akers, 1955); *Globoro-talia (Turborotalia) fohsi barisanensis* has been observed near the base of the Fleming formation by Blow (unpublished), which indicates that this part of the succession is highest Aquitanian (see below, under Venezuela and Trinidad).

In TEXAS (Applin, Ellisor & Kniker, 1925; Ellisor, 1944; Gravell & Hanna, 1937; Schuchert, 1935), there are no marine 'Oligocene' outcrops; subsurface formations containing marine faunas have been correlated with Vicksburgian beds. *Ostrea georgiana* occurs in beds believed to be equivalent to the Frio formation, and the *Marginu-lina* 'Zone' contains *Amphistegina lessoni* and *Globigerinoides trilobus sacculifer*. The overlying beds (which have been equated to the Anahuac formation and which have been rightly correlated with the *Heterostegina* 'Zone' of Florida) contain *Globigerinoides trilobus sacculifer, Heterostegina israelskyi, H. texana, Lepidocyclina colei, L. texana, Palaeonummulites ellisorae* and *P. howei*; the record of *Orbulina universa* we have been given to under-stand is incorrect. All these beds we date as Aquitanian. The equivalent *Heterostegina* Zone in Florida contains three species of *Miogypsina*.

Akers & Drooger (1957) have recorded *Mio-gypsina gunteri* and *M. tani* from the *Heterostegina* Zone of the so-called Anahuac formation from several subsurface samples in the Gulf Coast.

In CALIFORNIA (Atwill, 1935; Graham & Drooger, 1952), *Miolepidocyclina ecuadorensis* has been recorded from a calcareous grit correctly dated in 1938 as Lower Miocene; its occurrence in Ecuador is evidently of the same general age. The Tumey formation, with the Eocene *Discocyclina clarki*, does not appear to be of Oligocene age as recorded.

Samples from the lower Relizian (*Siphogenerina hughesi* Zone), kindly sent to us by Professor M. N. Bramlette (University of California) have been examined. These samples, from the Adelaide and Bradley Quadrangles, Salinas Valley, yielded a fauna consisting very largely of *Globigerina ciperoensis ciperoensis* and *G. praebulloides* (*s.l.*) but without any species referable to the genus *Globigerinoides* Cushman, the Miocene species of which are descended from *Globigerina praebulloides* in the *Globorotalia kugleri* Zone (see Part 2 of this work). Accordingly, the *Siphogenerina hughesi* Zone (i.e. the lower Relizian) of California can be correlated with the *Globigerina ciperoensis ciperoensis* Zone of the Caribbean and south-eastern Sicily. Bolli (1957, p. 107) has correlated the upper part of the Chickasawhay formation of the Gulf Coast with the *G. ciperoensis ciperoensis* Zone, and, whilst the Chickasawhay has, previous to this study, been considered to be Oligocene, the Relizian of California has always been considered as Miocene; Kleinpell (1938) has even equated it with the Burdigalian. It now appears that the Relizian should be considered as low in the Aquitanian and approximately equivalent to the Chickasawhay formation of the Gulf Coast.

In MEXICO, in THE RIO GRANDE EMBAYMENT (Schuchert, 1935), the Gueydan formation, with large *Ostrea* and a basal reef limestone at Mendez, has been dated by the United States Geological Survey as ?Lower Miocene; it rests unconformably on the Eocene.

In the TAMPICO EMBAYMENT (Barker, 1939; Schuchert, 1935; Vaughan, 1919a, 1933) the Alazan formation contains *Lepidocyclina asterodisca, L. (Eulepidina) favosa, L. mantelli, L. supera, Miolepidocyclina mexicana, Palaeonummulites muiri* and *Ostrea 'vicksburgensis'*. In this formation, Nuttall (1932, 1933) has shown the presence of *Globigerina ampliapertura* (*G. apertura* of Nuttall), *G. ciperoensis* (*G. concinna* of Nuttall), *G. ouachitaensis* (*G. bulloides* of Nuttall) and *Globorotalia menardii* (of Nuttall, more likely the subspecies *archeomenardii* Bolli), a collection of forms which indicates the presence within the Alazan formation of beds equivalent to several zones of the lower Cipero formation at least; in fact, these three

species of *Globigerina* (but not the *Globorotalia*) would indicate either an Oligocene or basal Aquitanian age. However, the *Globorotalia* (a carinate form) shows, with the larger foraminifera listed above, that part at least (and probably the greater part) of the Alazan formation belongs to the Lower Miocene (Aquitanian). The lower part of the Alazan formation can now be confirmed as one of the few true Central American representatives of the Oligocene, for Drs Bolli and Bermúdez have, after seeing our specimens of *Globigerina oligocaenica* from the Rupelian of East Africa (see Part 2, p. 71), kindly forwarded to us authentic specimens of this species from the lower Alazan, indicating that this part of the formation is older than the Cipero formation, or the Vicksburgian of the Gulf States, or their equivalents, which lack this species but which contain the true Aquitanian (Lower Miocene) fauna. The Meson formation, which overlies the Alazan, contains *Heterostegina antillea, Lepidocyclina canellei, L. (Nephrolepidina) crassata, L. (Eulepidina) favosa, L. (E.) gigas mexicana, L. parvula, L. (N.) tournoueri, L. (E.) undosa, L. waylandvaughani, Miogypsinella sanjosensis* (*Miogypsina complanata* of Nuttall, renamed and dated as 'Aquitanian (Lower Miocene)' by Hanzawa in 1940), *Palaeonummulites antiguensis, P. palmarealensis* and *P. semmesi*. According to Schuchert (1935), Böse has reported that the upper part of the overlying San Rafael formation contains faunas that are related to those of the acknowledged Miocene Gatun formation; in its lower part occur *Clypeaster*, and at its base is an extensive coral reef containing *Antiguastrea cellulosa* and two other forms common to the Antigua limestone. The Meson and the lower part of the San Rafael formations we date as Aquitanian.

In NICARAGUA (Schuchert, 1935), the equivalent of the Aquitanian part of the Alazan formation of the Tampico Embayment occurs.

In COSTA RICA (Cole, 1953c; Goudkoff & Porter, 1942; Schuchert, 1935), *Pliolepidina tobleri* has been recorded from an unnamed limestone 'which definitely correlates with...the San Fernando formation of Trinidad'; we regard this limestone as of Aquitanian age, and consider the associate Eocene foraminifera (*Asterocyclina georgiana,*

A. mariannensis, A. minima, Heterostegina ocalana, Lepidocyclina macdonaldi and *Palaeonummulites ocalanus*) to be derived from underlying Eocene beds, just as well-preserved Eocene faunas (*Coskinolina floridana, Dictyoconus cookei*) are acknowledged to be reworked from Eocene beds into Vicksburgian beds in Florida (Cole, 1945). Although the supposed 'Upper Oligocene' Nicoyan series is reported to contain small foraminifera and orbitoidal forms like those of the Manzanilla series, and although the latter contains *Asterocyclina* and *Discocyclina*, in view of the above interpretations we do not attempt to date the Nicoyan series on present evidence. The Guallava beds (dated as 'Lower Oligocene') contain *Lepidocyclina hilli, L. mantelli* and *Phos.* The Amoura shale, with *Orbulina universa*, is evidently highest Aquitanian and Burdigalian, and the overlying Uscari shale (itself unconformably overlain by the Gatun formation) is evidently also Burdigalian.

In PANAMA (Cole, 1952, 1953b, 1957a; Cushman, 1919a; Schuchert, 1935; Vaughan, 1919a; Woodring, 1957, 1958, 1959; Woodring & Thompson, 1949), the basal complex is overlain in some areas by the supposedly Eocene Gatuncillo formation and in others by the conglomeratic 'Oligocene' Bohio formation. The Bohio formation contains *Cyclorbiculina* [*Archaias*] *compressa* (see Smout & Eames, 1958), *Heterostegina antillea* and/or *panamensis, Lepidocyclina canellei, L. (Eulepidina) favosa, L. (E.) gigas, L. giraudi, L. parvula, L. (Nephrolepidina) vaughani, L. waylandvaughani, L. yurnagunensis, Miogypsina antillea, M. gunteri, Miolepidocyclina panamaensis, Pliolepidina tobleri, Turritella* cf. *altilira, T.* cf. *caleta* and the plant *Taenioxylon multiradiatum*, most of them from Barro Colorado Island (locality 42d), an assemblage which is typical of the Vicksburgian, the Alazan formation, the Antigua limestone, etc., and which we regard as Aquitanian; Dall (1890–1903) also recorded the Miocene Mollusca '*Corbula* (*Aloidis*) *heterogenea*', '*C.* (*Bothrocorbula*) *viminea*' and *Clementia dariena.* From the upper part of the formation Woodring (1958, 1959) records *Orthaulax* cf. *pugnax*; the 'strong representation of genera and subgenera' (e.g. *Anadara, Anomalocardia, Chione, Naticarius*) which Woodring (1958, p. 18)

claims to have first appeared here 'in late Oligocene time' and to have 'reached their fullest development in late Tertiary and modern seas', clearly belong to the Miocene surge. The supposed earliest known *Morum* ('*Oniscidia*') of the Gatuncillo formation, and *Anomalocardia* from the Bohio formation, are not really as old as has been thought. In some localities (e.g. Trinidad Island in Gatun Lake) the Bohio formation contains Aquitanian *Pliolepidina tobleri* in association with reworked Middle and Upper Eocene larger foraminifera (Woodring's localities 42 and 42b, with the Upper Eocene *Lepidocyclina macdonaldi, L. pustulosa, Nummulites striatoreticulatus, Palaeonummulites jacksonensis, P. kugleri* and *P. trinitatensis* and the Middle Eocene *Fabiania cubensis*). Some deposits thought to belong to the Gatuncillo formation contain almost identical assemblages of Aquitanian *Pliolepidina tobleri* with mixed Middle and Upper Eocene larger foraminifera (Woodring's localities 1a, 2, 10, 19, 22, 37, S.L. 84, with the Upper Eocene *Asterocyclina georgiana, A. minima, Helicolepidina spiralis, Lepidocyclina montgomeriensis, Palaeonummulites floridensis, P. moodybranchensis, P. ocalanus, P. vaughani* and *Pseudophragmina* (*Proporocyclina*) *flintensis* and the Middle Eocene *Yaberinella jamaicensis* together with the Miocene *Lepidocyclina* (*Nephrolepidina*) *chaperi*). These latter deposits we believe to have been formed during the same major Miocene transgression and contemporaneously with the Bohio formation. At Woodring's locality 39 the Bohio formation yielded smaller foraminifera (e.g. *Bolivina alazanensis, B. byramensis, B. plicatella mera, B. tectiformis, Bulimina alazanensis, Chiloguembelina cubensis, Cibicides mexicanus, C. perlucidus* (see below under Colombia), *Eponides umbonatus multisepta, Gaudryina alazanensis, Globigerina ciperoensis, G. ouachitaensis, Karreriella mexicana, Nonion pompilioides, Planulina marialana, Plectofrondicularia alazanensis, P. vaughani, Pleurostomella alternans, Siphonina tenuicarinata, Spiroloculina texana, Uvigerina gardnerae nuttalliana, U. spinicostata* and *Vaginulinopsis alazanensis*—identified by Renz and Bermúdez) which confirm the correlation of the sample with horizons in the Alazan formation, the lower part of the Cipero formation and the rest of the Vicksburgian equivalents, again, in our opinion, Aquitanian. A sample (Woodring's locality 21) supposedly from the Gatuncillo formation of the Rio Gatuncillo area, dated as 'Eocene' by Bermúdez and Renz, was found by them to contain a fauna of smaller foraminifera which had much in common with that of the Bohio formation fauna from locality 39, including *Globigerina ciperoensis* and *G. ouachitaensis*, at least eight further Neogene benthonic species, but no planktonic species known to be restricted to the Eocene. The remaining Gatuncillo formation microforaminiferal samples (Woodring's localities 17, 24, 31 and 35) contain mixed Middle Eocene (e.g. *Porticulasphaera mexicana*) and Upper Eocene (e.g. *Hantkenina suprasuturalis*) species together with (in the same sample) many of the above-mentioned forms we regard as Lower Miocene. These points indicate that many of the deposits referred to the Gatuncillo formation should be dated as Aquitanian. The geological map accompanying Woodring's paper (1957) is admittedly lacking in detail in some areas (e.g. the Madden Lake district) owing to the nature of the terrain, but at least two interpretations are possible: first, the Gatuncillo formation is in part, at least, unconformably overlain by the Bohio formation, in which case the areas to which Gatuncillo age has been assigned may be complicated tectonically, as suggested by Woodring & Thompson (1949, p. 227), and outliers of Bohio formation may not have been recognized; or secondly, the Gatuncillo formation is, in part *at least*, the lateral facies equivalent of the Bohio formation.

From the lower part of the Caimito formation have been recorded *Cyclorbiculina [Archaias] compressa, Heterostegina antillea, H. israelskyi, H. panamensis, Lepidocyclina asterodisca, L. canellei, L. (Nephrolepidina) dartoni, L. giraudi, L. parvula, L. (N.) tournoueri, L. (Eulepidina) undosa, L. (N.) vaughani, L. waylandvaughani, L. yurnagunensis, Miogypsina antillea, Miolepidocyclina panamensis* and *Palaeonummulites panamensis*; the formation has also yielded *Globularia (Ampullinopsis) spenceri* and *Orthaulax* cf. *pugnax*. On Barro Colorado Island, Gatun Lake area, the Caimito formation rests directly upon the Bohio formation, and, in its

marine facies, contains a similar assemblage of larger foraminifera; nearby samples, at a closely similar stratigraphical level, yielded a rich planktonic foraminiferal fauna characteristic of the *Globorotalia kugleri* Zone (Bolli, in Woodring, 1958, p. 22). Of the Tonosi limestone corals, from near the base of the Lower Culebra formation, four out of five are common to the Antigua limestone; also recorded from the Culebra formation are *Lepidocyclina pancanalis, Miogypsina intermedia, Pliolepidina duplicata, P. panamensis, P. tobleri, Orthaulax gabbi* and *Turritella* cf. *altilira*. The Upper Culebra formation contains *Lepidocyclina canellei, L. (Nephrolepidina) chaperi, L. (N.) vaughani, L. waylandvaughani, 'Meandropsina(?)' americana, Miogypsina cushmani, Miolepidocyclina panamensis, Palaeonummulites panamensis, Orthaulax gabbi* and six species of corals four of which are common to the Antigua limestone and four to the Anguilla formation. The Emperador limestone member has yielded *Heterostegina panamensis, Lepidocyclina miraflorensis, L. (Nephrolepidina) vaughani, Miolepidocyclina panamensis, Pliolepidina duplicata, P. panamensis, P. tobleri, Orthaulax* cf. *aguadillensis, Chione (Lirophora), Clypeaster, Holoporella albirostris* and twenty-six species of corals nine of which are common to the Anguilla formation and six to the Antigua limestone. The Cucuracha formation contains *Anadara*, large *Ostrea* and the plant *Taenioxylon multiradiatum*. The Panama formation contains *Lepidocyclina miraflorensis, L. parvula* and *Miolepidocyclina panamensis*. It will be obvious how closely related are all these faunas from the Bohio formation up to the Panama formation, inclusive; all formations up as far as a horizon near the top of the Culebra, except possibly for the uppermost part of the Caimito formation, in some localities, we date as Aquitanian.

In COLOMBIA (Becker & Dusenbury, 1958; Petters & Sarmiento, 1956; Schuchert, 1935), Olsson's 'Upper Oligocene' is correctly correlated with the Uscari shales of Costa Rica, but these are of Burdigalian age, underlie the Middle Miocene Gatun formation, and are equivalent to the Alum Bluff. That part of Olsson's 'Middle Oligocene' containing *Globularia (Ampullinopsis)* cf. *spenceri* is evidently correctly correlated with the Antigua limestone and would be Aquitanian; that part containing the *Hannatoma* fauna is probably, in conformity with the general consensus of opinion (Durham *et al.*, 1949), of Eocene age. The marine deposits at Toluviejo, with *Helicolepidina*, are of course of Upper Eocene, not Lower Oligocene, age. We accept the Antiguan affinities of the 'Middle Oligocene' non-marine Mugrosa series molluscan fauna as indicated by Pilsbry and Olsson; since these beds would therefore be of Aquitanian age, the overlying 'Upper Oligocene' non-marine Colorado series molluscan faunas would also be Miocene.

Of the foraminifera listed by Petters & Sarmiento, it is evident from the distribution of the species *Catapsydrax dissimilis* (not above Aquitanian) and *Globoquadrina altispira* (s.l., Aquitanian to Lower Pliocene) that the so-called 'Middle Oligocene' ('*Globigerina*' *dissimilis* Zone) of the Carmen-Zambrano area cannot be older than Aquitanian; their 'Lower Oligocene' (*Cibicides perlucidus* Zone) is equivalent to the *Globorotalia opima* Zone–lower *Globorotalia kugleri* Zone interval of Trinidad and Venezuela (see below), which on regional grounds we consider to belong to the Aquitanian (see also under Panama Bohio formation above). The occurrences of *Orbulina universa* (uppermost Aquitanian to Recent), *Globorotalia* (*Turborotalia*) *fohsi* and *G. praemenardii* from the top almost to the base of their so-called 'Upper Oligocene' indicate that it is of Burdigalian age; also, the appearance of *G. menardii* (uppermost Burdigalian to Recent) in the lower part of the *Bulimina carmenensis* Zone (lower portion of so-called 'Lower Miocene') indicates that the upper part of the *carmenensis* Zone marks the beginning of the Vindobonian. This is confirmed by the relationships of the molluscan faunas, in the higher part of the section, to those of the Gatun and Bowden formations (e.g. *Drillia gatunensis*, *Terebra bowdenensis*, *T. gatunensis*, *Turritella gatunensis*).

Becker & Dusenbury (1958) have recorded acknowledged Aquitanian foraminifera (e.g. *Bolivina byramensis*, *Bulimina alazanensis*, *Plectofrondicularia vaughani*, etc.), as well as *Miolepidocyclina*

panamensis—referred to incorrectly as *Heterosteginoides ecuadorensis*, since the embryonic spiral of the megalospheric form is far too long for that species) from beds in the Goajira Peninsula area.

In ECUADOR (Barker, 1932; Stainforth, 1948 a), the so-called 'Lower Oligocene' with *Lepidocyclina* (*Eulepidina*) *undosa*, *L. yurnagunensis* and *Pliolepidina tobleri* (Stainforth, 1948 a, p. 134) is, as in the case of the Culebra formation of Panama, of Aquitanian age; these beds contain common reworked (or misidentified) Eocene species such as *Globorotalia* (*Turborotalia*) *centralis*. Affinity with the Aquitanian Alazan formation of Mexico is indicated by the presence of such forms as *Cibicides mexicanus*, *Rotalia mexicana mecatepecensis* and *Uvigerina mexicana*. The so-called 'Middle Oligocene' contains *Globigerinatella insueta*, *Globigerinoides trilobus* and *Globorotalia* (*Turborotalia*) *fohsi barisanensis* which are placed by Bolli (1957 a) as not lower than Miocene, an assemblage which we would refer to the upper part of the Aquitanian. The so-called 'Upper Oligocene', in addition to the above three species, contains *Orbulina universa* (throughout) and (from the San Pedro sandstones) *Lepidocyclina* (*Nephrolepidina*) *verbeeki* (Upper Aquitanian to Burdigalian), *Miogypsina* cf. *staufferi* and *Miolepidocyclina ecuadorensis*, an assemblage which shows that much of this so-called 'Upper Oligocene' is of Burdigalian age. The Burdigalian age of the succeeding 'Lower Miocene' is confirmed by the presence of *Globorotalia menardii*.

In PERU (Drooger, 1955; Olsson, 1931; Stainforth, 1955; Stainforth & Rüegg, 1953; Travis, 1953; Weiss, 1955), much confusion was caused (e.g. in the case of the interpretation of Olsson's molluscan faunas) by miscorrelation of beds (see Stainforth, 1955, p. 2069), the clarification of which has resolved many of the anomalous ranges of molluscan species; further, some of the illustrations of Mollusca indicate that more than one species has occasionally been included under one specific name (e.g. Olsson's plate 14, figs. 8, 12 and 13, and plate 16, figs. 4 and 6, each include more than one species). The molluscan fauna of the Chira shale seems to be closely related to that of the underlying Eocene beds, and, amongst other forms, includes the Eocene genus *Cypraedia* in association with

Upper Eocene foraminifera such as *Clavigerinella eocanica*, *Hantkenina* and *Stichocassidulina*. The Mirador formation also contains Upper Eocene foraminifera such as *Clavigerinella eocanica*, *Hantkenina alabamensis* and *Stichocassidulina thalmanni*. The Cone Hill shale, with *Hantkenina primitiva*, forms the top of the Upper Eocene. The Mancora formation, with *Globularia (Ampullinopsis) spenceri* (of the Antigua limestone) and *Agriopoma*, is believed to be the equivalent of the horizons in southern Peru and southern Ecuador at which *Miogypsina gunteri* and *Miolepidocyclina ecuadorensis* occur, and this formation is to be dated as Aquitanian. The overlying Heath formation contains a number of Mollusca of modern type occurring also in the Mancora formation, and includes the genus *Chione*. The Heath formation also contains *Catapsydrax dissimilis*, *Globigerinoides trilobus sacculifer*, *Globorotalia (Turborotalia) mayeri* and *Hastigerina* cf. *aequilateralis*; since these planktonic foraminifera allow ready correlation with Venezuela and Trinidad (see below), this formation contains beds of both high Aquitanian and low Burdigalian age. From undifferentiated 'Oligo-Miocene' beds in the Sechura Desert, which occur at about the same horizon as the Heath formation, *Globorotalia (Turborotalia) fohsi barisanensis*, *G. (T.) fohsi fohsi* and *Hastigerina* cf. *aequilateralis* have been recorded, indicating that these beds also are equivalent to the highest Aquitanian and lowest Burdigalian deposits of Venezuela and Trinidad. Shales above the acknowledged Miocene Montera formation contain *Globorotalia menardii* (*s.s.*) and *Sphaeroidinellopsis subdehiscens* (recorded as *Sphaeroidinella dehiscens*), indicating a Vindobonian and/ or, at oldest, highest Burdigalian age. When the different horizons from which Olsson's Mollusca were obtained are all satisfactorily correlated, we feel it likely that, apart from a few possible reworked specimens in the basal Miocene, there will be discovered a very marked palaeontological break between the Miocene and Eocene assemblages from above and below the Miocene/Eocene boundary.

In VENEZUELA (Blow, 1959; Cushman & Renz, 1941; Franklin, 1944; Gorter & van der Vlerk, 1932; Gravell, 1933; Hedberg, 1937; Kugler, 1957; Liddle, 1946; Renz, 1948, 1957; Schuchert, 1935; Senn, 1935), Upper Eocene beds such as the Jarillal shales, with *Raetomya*, the Pauji shales, with *Hantkenina*, and the Mene Grande formation, with *Asterocyclina georgiana*, *Helicolepidina spiralis*, *Lepidocyclina pustulosa*, *Pseudophragmina (Proporocyclina) flintensis*, etc., are directly overlain by beds of Miocene age. The Tinajitas member, constituting the basal part of the Merecure formation, contains the Aquitanian *Pliolepidina tobleri* and *Anadara*(?), *Clypeaster* cf. *concavus* (of Antigua and Anguilla) and *C.* cf. *cotteaui* (of Antigua), associated with *Asterocyclina*, *Discocyclina* and *Lepidocyclina pustulosa* which must be reworked from the underlying Eocene. The Churuguara series contains *Amphistegina lessoni*, *Heterostegina* cf. *antillea*, *Lepidocyclina canellei*, *L. falconensis*, *L. (Eulepidina) gigas*, *L. (E.) senni*, *L. (E.) undosa*, *Miogypsina hawkinsi*, *Miogypsinella sanjosensis* and *Palaeonummulites* and is also of Aquitanian age. The Lower San Luis formation contains *Heterostegina*, *Lepidocyclina (Eulepidina) favosa*, *L. (E.) undosa*, *Miogypsina*, *Palaeonummulites* and *Spiroclypeus*, which fauna must be Aquitanian. The Middle San Luis formation (San Luis limestone) contains *Amphistegina lessoni*, *Heterostegina* cf. *antillea*, *H. panamensis*, *Lepidocyclina canellei*, *L. (Eulepidina) favosa*, *L. forresti*, *L. (E.) gigas*, *L. sanluisensis*, *L. (E.) undosa*, *Miogypsina bramlettei*, *M. hawkinsi*, *Palaeonummulites*, *Pliolepidina duplicata* and *Turritella altilira*, again an Aquitanian fauna. The Agua Clara shales and the El Mene (*vel* El Salto) Sand member contain *Amphistegina lessoni*, *Lepidocyclina* cf. *canellei*, *Miogypsina hawkinsi*, *M. staufferi*, *M. venezuelana*, *Palaeonummulites*, *Spiroclypeus*, *Orthaulax*, *Phos costatus*, *Turris albida*, *Turritella* cf. *altilira*, *Anadara latidentata*, *Chione*, *Clementia dariena* and *Raeta*, also of Aquitanian age.

The upper Agua Clara and overlying Cerro Pelado formations have been correlated by Cushman & Renz (1941) with the lower Agua Salada group of east Falcón, comprising the '*Marginulina wallacei*' and the '*Rectuvigerina multicostata*' Zones; these can be correlated with the middle and upper parts of the Acostian of Renz (1948), which comprise the '*Robulus wallacei*' Zone of the San

Lorenzo (*vel* Tocuyu) formation and the '*Rectuvigerina transversa*' Zone of the upper San Lorenzo (*vel* Tocuyo) and lower Pozón formations, including the Policarpio 'Greensand' member. Blow (1959) has observed that the base of the *Rectuvigerina transversa* Zone occurs above the highest occurrence of *Catapsydrax dissimilis* and before the occurrence of *Globigerinoides bisphericus*. However, the base of the *Rectuvigerina transversa* Zone occurs in sediments with *Globigerinatella insueta* and *Globigerinoides trilobus* (*s.s.*).

Globigerinoides bisphericus and the intermediate stages in the evolution of *Orbulina* (i.e. various subspecies of *Porticulasphaera glomerosa*) occur in the topmost part of the San Lorenzo (Tocuyo) formation immediately below the base of the Policarpio 'Greensand' member, which is the basal member of the overlying Pozón formation. Hence the uppermost part of the San Lorenzo (Tocuyo) formation and the basal part of the Pozón formation with *Globigerinatella insueta* and *Globigerinoides bisphericus* are high Aquitanian. The value of *G. bisphericus* cannot be over-emphasized as a marker for the upper half of the *Globigerinatella insueta* Zone in the absence of *G. insueta* itself within the area of northern South America.

Blow (1959) also found that the top of the *Rectuvigerina transversa* Zone occurred in the middle part of his *Globorotalia (Turborotalia) fohsi fohsi* Zone where a fauna consisting of *G. (T.) fohsi* (*s.s.*), *G. menardii praemenardii* and *Orbulina* has been observed. The top of the *Rectuvigerina transversa* Zone occurs shortly after the extinction of *Globorotalia (Turborotalia) fohsi barisanensis* and *G. menardii archeomenardii*, and this horizon is considered to be within the Burdigalian on evidence seen in the Mediterranean area (Blow, 1957) and that reported by Ruscelli from Italy (see Blow, 1959).

The lower part of the San Lorenzo (Tocuyo) formation (i.e. lower to middle part of the *Robulus wallacei* Zone and upper part of the underlying '*Uvigerinella*' *sparsicostata* Zone) contains a planktonic fauna equivalent to that of the *Catapsydrax dissimilis* and *C. stainforthi* Zones of the south Trinidad succession. Indeed, *C. dissimilis* does not quite range to the top of the *Robulus wallacei* Zone. The lower San Lorenzo (Tocuyo) formation also

contains *Heterostegina* cf. *antillea*, *Lepidocyclina canellei* and *Miogypsina*, the last probably ranging upwards into the lower Pozón formation.

The Guacharaca formation, which underlies the San Lorenzo (Tocuyo) formation, has been correlated by Blow (1959, chart 4) with the Trinidad Zones of *Globorotalia (Turborotalia) kugleri*, *Globigerina ciperoensis* (*s.s.*), *Globorotalia (Turborotalia) opima* (*s.s.*) and *Globigerina ampliapertura*. The benthonic foraminiferal fauna of the Guacharaca formation below the '*Uvigerinella*' *sparsicostata* Zone is characterized by such forms as *Bolivina mexicana aliformis*, *Bulimina sculptilis* and *Uvigerina mexicana*; these forms are known in the type Vicksburgian of the Gulf Coast States, and Renz (1948, p. 30) suggested a correlation of this interval of the Guacharaca with the Vicksburgian which is, however, dated here as Aquitanian.

The upper Agua Clara, most of the Cerro Pelado (of Central Falcón), the San Lorenzo (Tocuyo) and the lowermost Pozón formations (comprising the lower part of the Agua Salada group and the lower two-thirds of the Acostian of Renz in eastern Falcón) are therefore correlated with the Anahuac formation of Mexico and Texas and with the *Heterostegina* Zone of the Gulf Coast which contains the highest occurrence of *Lepidocyclina* (*Eulepidina*); all are dated as highest Aquitanian.

The Carapita formation contains *Rectuvigerina transversa* and several species of benthonic microforaminifera common to the Alazan and Byram formations (e.g. *Anomalina alazanensis*, *Bolivina alazanensis*, *Ceratobulimina alazanensis*, *Eponides byramensis* and *Siphonina tenuicarinata*); apart from reworked Eocene forms (e.g. *Globorotalia spinulosa*), the presence of *Globigerina ciperoensis* (*s.s.*), and *Hastigerina* aff. *aequilateralis* indicates that the Carapita formation includes beds of both Aquitanian and Burdigalian age.

In the Eastern Venezuela Basin, the upper part of the Oficina formation contains *Globorotalia (Turborotalia) fohsi* (*s.s.*) and can therefore be correlated either with the uppermost part of Renz's *Rectuvigerina transversa* Zone or the lower part of his '*Globorotalia fohsi* Zone' of the Pozón formation; this is lower Burdigalian. Blow (1959) has subsequently taken the highest occurrence of *G. fohsi robusta* (*vel G. lobata robusta*) as being the top of the Burdigalian portion of the Pozón formation; this occurs approximately half-way up the '*Valvulineria*' *herricki* Zone of Renz, and may be broadly correlated with a horizon near the middle of the Socorro formation of Central Falcón. The overlying Damsite formation, with *Heterostegina*, *Miogypsina*, *Clementia dariena*, *Turris albida*, *Turritella altilira* and *T. gatunensis*, is correlated with the Middle Miocene Gatun formation.

Finally, there have been a few records of mixed Miocene (*Pliolepidina*) and Eocene (*Helicolepidina*, etc.) faunas which have only very tentatively been placed stratigraphically, although they were dated as Eocene; we have no doubt that these samples are really of low Aquitanian age. In the Maracaibo Basin it is well known that the Miocene rests directly on the Eocene with a distinct break.

In TRINIDAD (van den Bold, 1958, 1960; Bolli, 1957*a*, *b*; Bolli *et al.*, 1957; Brönnimann, 1950; Cushman & Renz, 1947; Drooger, 1952; Kugler, 1953, 1954; Nuttall, 1928; Renz, 1942; Schuchert, 1935; Stainforth, 1948; Tobler, 1926; Vaughan & Cole, 1941), the 'San Fernando conglomerate' or its equivalents (not the true underlying San Fernando formation) rest with every evidence of important unconformity on Eocene rocks of various ages (Palaeocene to Upper Eocene). For example, on Soldado Island, Vaughan & Cole (1941) and Kugler (1953, p. 45; 1954) have given an excellent account of the succession, although it is evident that the importance of the unconformity involved was not fully realized. The acknowledged Palaeocene Soldado formation (beds 1 to 3) is immediately overlain by a conglomerate consisting of rubble derived from the underlying Palaeocene limestone; plate 2 of Vaughan and Cole's paper shows the outcrop of the conglomerate forming a 'Y' around what can only have been an Eocene island or a slipped mass, the conglomerate being the basal member of a new transgression. In fact, reworked material in the form of limestone blocks and boulders (up to 24 × 12 × 12 feet) and *Pholas*-bored pebbles and brecciated rubble are recorded as high as bed 10. In our opinion this is an excellent example of a major Tertiary unconformity; the transgressive deposits contain abundant fossils and rock material derived from the Eocene rocks exposed at the time, many of these fossils being derived from soft rocks and being very well preserved. Furthermore, it is acknowledged that the 'San Fernando conglomerate' grades up into the Cipero formation and that reworked boulders occur as high as the *dissimilis* Zone which we have shown to be high Aquitanian. Some of these boulders are of soft material from which 'exceptionally well preserved' delicate small foraminifera have been obtained by washing in the laboratory (Bolli, 1957*b*, p. 158) and consequently many such small forms must be expected as reworked material in the enclosing rocks. There is no lithological distinction between the matrix of the 'San Fernando conglomerate' and the overlying Cipero formation, the only differences being that the former contains a very much larger proportion of reworked Eocene large and small foraminifera and that *Cassigerinella chipolensis* and *Globigerina ciperoensis* (*s.s.*, but not *s.l.*) are absent from it. Furthermore, the 'San Fernando conglomerate' is notorious for the quantity of Eocene and older reworked planktonic foraminifera that it contains, and, in fact, we have seen, in a single sample from the 'Mount Moriah silts' collected near San Fernando, the upper Campanian–lower Maestrichtian *Globotruncana contusa patelliformis*, the Eocene *Globigerina linaperta*, the Middle Eocene *Globorotalia* aff. *crassata* and *Truncorotaloides rohri*, the Middle–Upper Eocene *Globigerinatheka barri* and *Globorotalia* (*Turborotalia*) *centralis*, as well as the Upper Eocene *Globigerapsis semi-involuta* and *Hantkenina alabamensis*. These forms are present in a good state of preservation. Blow (whilst in Trinidad) has also noted the occurrence of the Lower–Middle Eocene *Globorotalia aragonensis* and the Palaeocene *G. velascoensis* in the 'Mount Moriah silts' in an equally good, if not better, stage of preservation as the autochthonous fauna. Vaughan & Cole (1941, p. 10) record that bed 7 of the 'San Fernando formation' (*vel* 'San Fernando conglomerate') on Soldado Rock contains the Eocene *Hantkenina primitiva* in association with the (Aquitanian) Alazan forms *Plectofrondicularia vaughani* and *Vaginulina elegans mexicana*, which show an Aquitanian affinity.

The molluscan faunas of Soldado Rock have been described by Maury (1912, 1929). Above the Palaeocene lower beds, the 'Boca de Serpiente formation' (bed 8) contained a 'rich molluscan fauna, but the species were entirely different. The characteristically Lower Eocene forms of Bed No. 2...are all absent' (Maury, 1929, p. 180). Most of the species listed by Maury (1912) from bed 8 were new, and only four known species were firmly identified; these were believed by Maury to show affinity with the Eocene of the Gulf Coast. These four species were all poorly preserved, but careful comparison of her figures with those given by the other workers quoted in her synonymic lists shows that there was no firm basis for her identifications and that some of them were definitely wrong (e.g. *Corbula subengonata* Dall and *Dentalium microstria* Heilprin we believe to have been misidentified). Further, Maury herself recanted in one case (1929), reducing the identification of *Fusoficula juvenis* to '*Fusoficula* like *juvenis*', which we think still to be optimistic. Lastly, she included *Venerupis* in her list from bed 8; this genus is only doubtfully known beneath the Miocene, and her poorly preserved specimen probably belongs to the genus *Irus* Oken, which is known to be restricted to the interval Miocene to Recent. We consider that the mollusca of bed 8 show as much a Miocene as an Eocene affinity. The specimens of *Tubulostium*, a genus not known above the Eocene, recorded by Rutsch (1940) from bed 11 on Soldado Island, are the only recorded ones which are not undoubtedly of Eocene age, and it is evident that these small but thick-shelled forms are to be regarded as reworked.

Vaughan & Cole (1941) recorded that bed 9 contains the Eocene forms *Asterocyclina asterisca*, *Discocyclina cubensis*, *D. vaughani*, *Helicolepidina soldadensis*, *H. spiralis*, *Lepidocyclina pustulosa*, *Palaeonummulites ocalanus*, *P. trinitatensis* and *Pseudophragmina* (*Proporocyclina*) *flintensis* in association with *Pliolepidina tobleri* which we regard as being of Miocene age. We would mention in passing that exactly parallel circumstances have been encountered by us in the Mediterranean area, where *Lepidocyclina* (*Eulepidina*) *dilatata* occurs in association with an abundant fauna of reworked

Eocene material consisting of *Alveolina*, *Discocyclina*, etc. Stainforth (1948, p. 1309) has stated that until recently the whole San Fernando 'formation' was regarded as Upper Eocene, but that B. Caudri has shown that the Upper Eocene larger foraminifera do not reach the top of the formation, whereas several 'Oligocene' species (e.g. *Lepidocyclina* (*Eulepidina*) *favosa*, *L.* (*E.*) *gigas*, *L.* (*E.*) *undosa*, *L. supera*, *L. yurnagunensis* (*s.l.*) and *Palaeonummulites semmesi ciperensis*) appear in the topmost beds of the 'San Fernando formation' at Point Bontour and Vista Bella. The apparently anomalous association of Eocene and 'Oligocene' index fossils within this one formation is easily explained by H. G. Kugler (*verb.*), who believes that lower Cipero beds were confused with the 'San Fernando conglomerate' (and consequently with the 'San Fernando formation') by Stainforth, with the result that Stainforth and Caudri placed the upper limit of the San Fernando formation too high. This illustrates the confused nature of the field-relationships of the sediments of this part of the stratigraphical section, which is so common in Trinidad. The San Fernando conglomerate and its local equivalents, with the 'Mount Moriah silts', we consider to be of Aquitanian age, and to be closely analogous to, and broadly correlatable with, the Red Bluff clay and the Bohio formation.

From the lowest beds of the Cipero formation a rich orbitoidal fauna with *Lepidocyclina* (*Eulepidina*) *favosa*, *L.* (*E.*) *gigas*, *L. supera*, *L.* (*E.*) *undosa* and *L. yurnagunensis* as well as *Palaeonummulites semmesi ciperensis* has been recorded. The *Globigerina* cf. *concinna* Zone of Kugler (1953, 1954) and the 'Zone I' of Stainforth (1948) has been subdivided into three further zones; the lowest of these three zones is the *G. ampliapertura* Zone (Bolli, 1957a). Bolli (1957a, p. 103) also referred the displaced wedge of the 'Bamboo clay' to the *G. ampliapertura* Zone and this contains, as well as those species listed above, *Lepidocyclina* cf. *canellei*, *L. parvula*, *L. subglobosa*, *L. waylandvaughani* and *Parvamussium bronni pennyi*. The first occurrence of *Miogypsina* (*s.s.*) occurs in the *Globorotalia opima opima* Zone which is the middle zone of the new subdivision of the old 'Zone I' of Stainforth (see also Part 2, §IV, of this work). The supposed *Catapsy-*

drax dissimilis Zone of Kapur Quarry contains *Miolepidocyclina ecuadorensis*, and the same quarry has yielded *Lepidocyclina (Eulepidina) favosa, L. (E.) gigas, L. (Nephrolepidina) tempanii, L. (E.) undosa, L. (N.)* cf. *verbeeki, Miogypsina gunteri, M. tani* and '*M. tani-bronnimanni*'. Brönnimann (1950) has shown that the Ste. Croix formation falls within the Zone of *Globigerinatella insueta*; it contains *Ceratobulimina alazanensis, Guttulina byramensis, Lepidocyclina (Eulepidina) favosa, L. (E.) gigas, L. (Nephrolepidina) tempanii, L. (E.) undosa, L. (N.)* cf. *verbeeki, Miogypsina gunteri, Pseudoclavulina alazanensis, Spiroloculina texana, Uvigerina mexicana*, etc., and *Turris* aff. *albida* and *Parvamussium bronni pennyi*, an assemblage which is of Aquitanian age. Other *Lithothamnium* limestones within the Cipero formation (Morne Diablo, Mejias, Quinam, etc.), all probably slipped masses from approximately the same horizon within the Cipero formation but not yet precisely placed stratigraphically, contain *Archaias, Heterostegina antillea, Lepidocyclina asterodisca, L. canellei, L. (Eulepidina) favosa, L. forresti, L. giraudi, L. pancanalis, L. parvula, L. subglobosa, L. supera, L. (E.) undosa, L. yurnagunensis, Miogypsina basraensis, M. bramlettei, Miolepidocyclina mexicana, Palaeonummulites bullbrooki* (not an *Amphistegina* as it possesses a marginal cord), *P. semmesi, Sorites, Turritella* aff. *altilira, Parvamussium bronni pennyi* and pteropods of European Burdigalian and Helvetian(?) affinities, which are typical Lower Miocene assemblages. Tobler (1926) found at Erin Point *Lepidocyclina (Eulepidina) dilatata, L. (Nephrolepidina) marginata, L. (N.) persimilis, L. (N.) tournoueri* and *Miogypsina*, to which he attributed an Aquitanian age. The detailed zonation of the Lower Miocene Cipero and the Middle Miocene Lengua formations developed by Bolli using planktonic foraminifera has been found to have very wide application. By comparison with the succession of planktonic faunas as recognized in Venezuela, Mexico and the Gulf Coast States, we place the top of the Aquitanian at the top of the *Globorotalia (Turborotalia) fohsi barisanensis* Zone of Bolli (not at the top of the range of the species), the succeeding Zones up to *G. fohsi robusta (vel G. lobata robusta)* constituting the Burdigalian and

comprising the upper part of the Cipero formation, which contains the *Pleurophopsis* fauna. The succeeding zones of *G. (Turborotalia) mayeri* and *G. menardii (s.s.)* (Lengua formation) are shown by Blow (1959) to be of Vindobonian age.

Van den Bold (1958) indicates that the Brasso formation can be readily correlated with the higher zones of the Cipero formation; it is therefore of upper Aquitanian and Burdigalian age. Van den Bold (1960) also shows on his Charts 1–4 that there is a very noticeable palaeontological break separating the ostracod faunas above and below the top of the Hospital Hill marl.

In CARRIACOU (Cole, 1958a; Martin-Kaye, 1958; Senn, 1940; Trechmann, 1935), the Hillsborough Rectory limestone, which forms the lower part of the Lower Tuffs, has been found to contain abundant *Pliolepidina tobleri*, associated with rare *Heterostegina ocalana, Lepidocyclina macdonaldi* and *L. pustulosa* and 'a few fragments of *Asterocyclina minima*', an assemblage closely comparable to those of the Bohio and San Fernando formations and which we consider to be of Aquitanian age, with reworked Eocene material. The higher part of the Lower Tuffs contains fossiliferous limestone lenses and foraminiferal ashes. These limestone lenses were found by Cole to contain *Heterostegina antillea, Lepidocyclina canellei, L. (Nephrolepidina) tournoueri, L. (N.) vaughani, L. waylandvaughani, Miogypsina antillea, Miolepidocyclina panamensis* and *Palaeonummulites dia*, an Aquitanian assemblage common throughout the whole Caribbean area. The lower Aquitanian age of these beds is confirmed by the presence of the *Globorotalia opima opima* Zone fauna in the associated foraminiferal ashes (Bolli, in Martin-Kaye, 1958, p. 398).

The succeeding Carriacou limestone series commences with limestone lenses and the Belmont beds, which are disconformable upon the underlying Lower Tuffs. The Belmont beds contain, according to Bolli (in Martin-Kaye, 1958, p. 399), a planktonic fauna with *Catapsydrax stainforthi*, which is considered by him to be an assemblage belonging to either the *C. dissimilis* or the *C. stainforthi* Zone. We believe that the *C. dissimilis* Zone is more likely, because Bolli did not record the presence of *Globigerinatella insueta*. The overlying

Calcareous Tuffs (of Lehner, see Martin-Kaye, 1958) do contain *G. insueta*, and could be referable to either or both of the *C. stainforthi* and *G. insueta* Zones. If the record of the presence of *Globorotalia fohsi fohsi* elsewhere within these Calcareous Tuffs is correct, then these beds must be diachronous, and may be partly equivalent to the Grand Bay beds. The Carriacou limestone immediately follows the Calcareous Tuffs; it has been found to contain indigenous *Miogypsina staufferi* and *Palaeonummulites cojimarensis*, which indicate a faunal affinity with the Cojimar limestone of Cuba, and which suggest a high Aquitanian or low Burdigalian age (rare Eocene *Coskinolina floridana* present in the Carriacou limestone are admitted by Cole to be reworked, as in Florida).

The Grand Bay beds and their associated basal limestone lenses are separated from the underlying Carriacou limestone series by a hiatus which probably comprises the *Globorotalia fohsi barisanensis* Zone and part of the *G. fohsi fohsi* Zone. Bolli (in Martin-Kaye, 1958) records a rich planktonic foraminiferal fauna which is of *G. fohsi fohsi* Zone age. The Burdigalian age is confirmed by Blow's (unpublished) observation of *G. lobata lobata* at a high level within the Grand Bay beds, immediately below the succeeding (Sarmatian?) Upper Tuffs. The Grand Bay beds contain a rich molluscan fauna, including such forms as *Architectonica gatunense*, *Phos costatus* and *P. semicostatus*, which range from Burdigalian to Vindobonian.

In BARBADOS (Beckmann, 1953; Bolli, 1957a; Senn, 1940, 1948), Bolli has shown that the upper part of the richly fossiliferous Oceanic formation (the Codrington College beds) can readily be correlated with the lower part of the Cipero formation of Trinidad, comprising all his Zones as high as the Zone of *Globorotalia (Turborotalia) kugleri*; it contains the same planktonic foraminifera together with *Cibicides mexicanus*, *Plectofrondicularia vaughani*, *Uvigerina spinicostata*, etc. This upper part is therefore of Aquitanian age; the lower part is, of course, Eocene. The Bissex Hill formation is broadly equivalent to the *Globigerinatella insueta* Zone of Bolli (late Aquitanian). The lower part of the succeeding *Globigerina* marls is

correctly dated by Bolli as equivalent to the Zone of *Globorotalia (Turborotalia) fohsi barisanensis* (highest Aquitanian), but the higher part (unpublished observations of Blow) is equivalent to part of the Trinidad Zone of *G. (T.) fohsi fohsi* (early Burdigalian).

In MARTINIQUE (Schuchert, 1935; Senn, 1940; Trechmann, 1935), the Ste. Marie limestones, with *Lepidocyclina* and *Spiroclypeus*, are evidently of Aquitanian age. Near La Trinité beds with *Lepidocyclina giraudi*, *Phos*, *Turritella gatunensis*, and *T. tornata* are believed to be of Burdigalian age. The Macabao limestone, with abundant small *Lepidocyclina*, and succeeding Morne Vent limestone, with the last rare and small *Lepidocyclina*, are probably also of Burdigalian age.

In ANTIGUA (Canu & Bassler, 1919; Cooke, 1919; Cushman, 1919b; Schuchert, 1935; Thomas, 1942; Trechmann, 1941; Vaughan, 1919a, 1933; Vaughan & Cole, 1936), the Seaforth limestone contains the Aquitanian *Lepidocyclina (Eulepidina) favosa*, *L. (E.) undosa* and *Orthaulax seaforthensis*, together with the corals *Antiguastrea cellulosa*, *Astrocoenia decaturensis* (also Antigua limestone and 'Middle Oligocene' of Georgia and Cuba), *A. guantanamensis* (also Antigua limestone and 'Middle Oligocene' of Cuba and Panama), *Cladocora recrescens* (also Antigua limestone and Chattahoochee formation of Georgia), *Diploastrea crassolamellata*, *Goniopora regularis* (also Antigua limestone) and *Stylophora imperatoris* (also Emperador limestone of Panama and 'Upper Oligocene' of Anguilla). Overlying cherty beds contain *Lepidocyclina mantelli*. The Antigua limestone above contains *Heterostegina antillea*, *Lepidocyclina antiguensis*, *L. (Eulepidina) favosa*, *L. forresti*, *L. (E.) gigas*, *L. hodgensis*, *L. mantelli*, *L. pancanalis*, *L. parvula*, *L. tempanii*, *L. (E.) undosa*, *L. undulata*, *L. (Nephrolepidina) vaughani*, *L. waylandvaughani*, *L. wetherellensis*, *Palaeonummulites antiguensis*, *P. forresti*, common *Clypeaster* (including *C. concavus* and *C. cotteaui*), a molluscan fauna which Cooke found to be closely related to acknowledged Lower Miocene faunas, six bryozoan species of which one was new and five were known only from the Neogene, a large coral fauna which we have discussed previously (p. 26), and *Taenioxylon multiradiatum*, which fauna is typical of the Aquitanian; further-more, Bolli (1957a) has found *Globigerina cipero-ensis* (*s.l.*), and correlates the Antigua limestone with the *Globigerina ciperoensis* Zone or below in Trinidad. We ourselves have found, in samples of Antigua limestone, a species of *Sporadotrema* which occurs (abundantly) elsewhere in the Indo-Pacific region only in the Neogene.

In ST BARTHOLOMEW ISLAND (Schuchert, 1935), the supposed 'Middle Oligocene', with *Clypeaster antillarum* and *C. concavus*, would be of Aquitanian age; underlying beds, attributed to the 'Lower Oligocene', and overlying the Upper Eocene, seem likely to be of Aquitanian age also.

In ANGUILLA (Cushman, 1919b; Drooger, 1952; Schuchert, 1935; Senn, 1940; Vaughan, 1919a), the Anguilla formation contains *Miolepidocyclina antillea*, seventeen species of corals (including *Antiguastrea cellulosa*) twelve of which are common to the combined Culebra formation and Emperador limestone faunas, *Clypeaster antillarum*, *C. con-cavus*, *Echinolampas semiorbis* (also in the Empera-dor limestone) and *Orthaulax pugnax*; these faunas are evidently of Burdigalian age, and have correctly been correlated with the Emperador limestone. Senn (1940, p. 1596) records that the Anguilla formation contains the last, rare, small *Lepido-cyclina*, which would confirm the dating in this region.

On ST CROIX (VIRGIN ISLANDS) (Schuchert, 1935), the supposed 'Middle Oligocene' contains *Carpen-teria americana*, *Lepidocyclina yurnagunensis mor-ganopsis* and four of the Antigua limestone coral species, and is of Aquitanian age. The so-called 'Upper Oligocene' contains *Heterostegina antillea*, *Miolepidocyclina* and *Orthaulax aguadillensis*, and would appear to be of high Aquitanian and/or lower Burdigalian age.

In PUERTO RICO (Drooger, 1952; Hubbard, 1923; Petters & Sarmiento, 1956; Schuchert, 1935; Vaughan, 1919a), on the north shore, the San Sebastian shales contain *Lepidocyclina*, '*Mean-dropsina*(?)' cf. *americana*, '*Miogypsina gunteri-tani*', *Miolepidocyclina ecuadorensis*, *Anadara*, *Cle-mentia dariena* and *Spondylus bostrychites*; the Pepino limestone (upper part of San Sebastian shales) has yielded twelve species of corals eight of

which (e.g. *Antiguastrea cellulosa*) are common to the Antigua limestone. The Lares limestone con-tains *Lepidocyclina* cf. *mantelli*, *Anadara*, *Chione hendersoni*, *C. woodwardi* and *Pecten* spp. The Cibao limestone contains *Lepidocyclina*, '*Meandrop-sina*(?)', *Orthaulax aguadillensis*, *Chione hender-soni*, *Spondylus bostrychites*, and *Echinolampas* cf. *aldrichi*. The Los Puertos limestone contains *Lepi-docyclina*, '*Meandropsina*(?)', *Orthaulax aguadil-lensis*, *O. portoricoensis*, *Chione hendersoni* and *C. woodwardi*. The Quebradillas limestone contains *Globorotalia menardii* and abundant Mollusca (including *Orthaulax aguadillensis*, *O. portori-coensis*, *Phos costatus*, *Chione woodwardi* and *Ostrea antiguensis*) many of which are of Bowden type. We agree with Vaughan that the San Sebastian shales and the Lares limestone are the approximate equivalent of the Antigua limestone, but date them as Aquitanian; the Cibao limestone and Los Puertos limestone appear likely to be of Burdi-galian age, and the Quebradillas limestone of Middle Miocene age.

On the south shore, the Juana Diaz formation contains *Lepidocyclina mantelli* and the sirenian *Halitherium antillense*, and is of Aquitanian age. The overlying Ponce formation contains '*Mean-dropsina*(?)', *Palaeonummulites* and *Pecten chipo-lensis*, and is probably of Burdigalian and Vindo-bonian age.

In THE DOMINICAN REPUBLIC (Bermúdez, 1949; Schuchert, 1935), the Tabera formation, which rests unconformably and with a basal conglo-merate on basement, contains *Heterostegina antil-lea*, *Lepidocyclina* (*Nephrolepidina*) *crassata*, *L.* (*N.*) *marginata*, *L. schlumbergeri*, *L. yurnagunensis morganopsis*, *Miogypsinella thalmanni*, *Palaeo-nummulites* and twenty-four species of corals (including *Antiguastrea cellulosa*) together with reworked Eocene *Lepidocyclina pustulosa*, and we agree that it is the equivalent of the Antigua lime-stone, being of Aquitanian age. The Sombrerito formation contains *Biorbulina bilobata*, *Catapsy-drax dissimilis*, *Globorotalia menardii archeo-menardii* (recorded as *G. menardii*), *G.* (*Turboro-talia*) *fohsi* (*s.l.*), *G.* (*T.*) *mayeri*, *Hastigerinella bermudezi* (recorded as *H. eocanica*), *Heterostegina antillea*, *Lepidocyclina subraulini*, *Miogypsina*, *Or-*

bulina universa and *Uvigerina mexicana*, which correlate it with the highest Aquitanian Zones of *Catapsydrax stainforthi*, *Globigerinatella insueta* and *Globorotalia (Turborotalia) fohsi barisanensis* (of Bolli) of the Cipero formation of Trinidad.

The Cervicos limestone contains *Orthaulax aguadillensis* and *Clypeaster concavus*. The Trinchera limestone, which is the lowest member of the Florentino formation, contains *Globigerinoides ruber*, *Globorotalia lobata* and *Sphaeroidinellopsis seminulina* (recorded as *Sphaeroidinella grimsdalei* and *S. rutschi*), a fauna which indicates a Burdigalian age; later levels of the Florentino formation contain a rich peneroplid fauna which is evidently also of Burdigalian age; the Lemba formation (equivalent to the lower part of the Florentino formation) contains *Cibicides mexicanus*, *Planulina marialana* and *Rotalia mexicana* of the Lower Miocene; the Baitoa conglomerate contains *Orthaulax inornata*, *Phos costatus*, *P. semicostatus* and *Anadara hispaniola*; the Cercado beds contain *Phos gabbi*, *Anadara* and *Corbula (Bothrocorbula) viminea*—these are equivalent to the upper Cervicos limestone and upper Florentino formation, and we date them as high Burdigalian. The succeeding Gurabo formation has a rich molluscan fauna of Middle Miocene (Bowden) type.

In HAITI (Schuchert, 1935), the so-called 'Middle Oligocene' contains *Heterostegina*, *Lepidocyclina (Eulepidina) favosa*, *L. (E.) gigas*, *L. (E.) undosa*, *L. undulata* and *L. yurnagunensis*; we agree that it is the equivalent of the Antigua limestone, being of Aquitanian age. The Madame Joie formation contains *Lepidocyclina giraudi*, '*Meandropsina(?)* americana', *Miogypsina antillea* and several corals; it is correlated with the upper part of the Sombrerito formation of the Dominican Republic by Bermúdez, and may therefore be of high Aquitanian and/or low Burdigalian age. The Thomonde formation contains *Phos semicostatus*, *Anadara*, *Cavolina* and corals, and, being correlated with the Baitoa conglomerate, would be of Burdigalian age. The Las Cahobas formation contains a number of corals, and the fauna of the overlying Port-au-Prince shales is the acknowledged equivalent of the Middle Miocene Bowden fauna.

In JAMAICA (Cole, 1956; Cushman & Jarvis,

1930; L. M. Davies, 1952; Matley, 1951; Schuchert, 1935; Vaughan, 1928; Vaughan & Cole, 1936), the lower part of the White limestone is of Middle and Upper Eocene age, and it is succeeded, after an acknowledged hiatus, by the Montpelier member of the upper part of the White limestone which contains *Lepidocyclina canellei*, *L. (Eulepidina) favosa*, *L. (E.) undosa*, *Miogypsina bracuensis*, *Orbulina universa* and *Clypeaster cotteaui*, indicating a high Aquitanian age. The succeeding Moneague member contains *Lepidocyclina canellei*, *L. (Eulepidina) favosa*, *L. (E.) gigas*, *L. miraflorensis*, *L. parvula*, *L. supera*, *L. (E.) undosa* and *L. yurnagunensis*, an assemblage which must also be of high Aquitanian age. From so-called 'Oligocene' localities listed by L. M. Davies and by Cole are recorded *Heterostegina antillea*, *H. panamensis*, *H.* cf. *texana*, *Lepidocyclina canellei*, *L. (Eulepidina) favosa*, *L. (E.) gigas*, *L. mantelli*, *L. parvula*, *L. supera*, *L. (E.) undosa*, *L. yurnagunensis*, *Miogypsina antillea*, *Miolepidocyclina panamensis*, *Palaeonummulites antiguensis*, *P.* cf. *panamensis*, *P.* cf. *semmesi* and *P. vicksburgensis*, together with the reworked Eocene *Lepidocyclina ocalana*, *Lockhartia* and *Palaeonummulites* cf. *floridensis*; these samples belong to the upper part of the White limestone, are of Aquitanian age, and obviously correlate with the '*Heterostegina* Zone' of the Gulf Coast States; as in Florida, reworked Eocene fossils are present in the Aquitanian. One further example of reworking is that of Cole's record of Vaughan's sample V.128 which contains abundant *Pliolepidina tobleri* in association with rare derived *Dictyoconus cookei*, *Fabiania cubensis* and *Lepidocyclina macdonaldi*, the sample being of Aquitanian (not Eocene) age. The Bowden formation, overlying the White limestone, contains a rich molluscan fauna agreed to be of Middle Miocene age, together with *Globigerinoides trilobus sacculifer*, *Globorotalia menardii miocenica* and '*Sphaeroidinella dehiscens*' (probably *Sphaeroidinellopsis subdehiscens*). From the 'Miocene' have also been recorded *Amphisorus matleyi*, *Miogypsina* cf. *hawkinsi*, *Palaeonummulites* aff. *nassauensis* and *P.* cf. *panamensis*, some of which may belong to the Burdigalian.

In THE CAYMAN ISLANDS (Schuchert, 1935), the Bluff limestone contains *Carpenteria americana*,

Lepidocyclina (*Eulepidina*) *gigas*, *L.* (*E.*) *undosa* and *L. yurnagunensis*; it is agreed to be equivalent to part of the Antigua limestone, and is of Aquitanian age.

In CUBA (De Albear, 1947; Beckmann, 1957; Cushman & Bermúdez, 1949; Drooger, 1952; Hadley, 1934; Kugler, 1938; D. K. Palmer, 1934, 1940*b*, *c*; D. K. Palmer & Bermúdez, 1936; R. H. Palmer, 1945; Petters & Sarmiento, 1956; Rutten, 1935; Schuchert, 1935; Thiadens, 1937; Vaughan, 1933), the Middle and Upper Eocene are overlain by the Tinguaro marls which contain *Lepidocyclina* (*Eulepidina*) *favosa*, *L.* (*E.*) *undosa* and reworked Eocene *Globorotalia* (*Turborotalia*) *cerroazulensis*, which we date as Aquitanian. Radiolarian marls which have yielded *Lepidocyclina* (*Eulepidina*) *gigas* are also of Aquitanian age. The (Finca) Adelina marls have yielded a rich benthonic microforaminiferal fauna including *Cibicides perlucidus* (see Venezuela), *Angulogerina vicksburgensis*, *Anomalina alazanensis*, *Bolivina alazanensis*, *Bulimina alazanensis*, *Cassidulina subglobosa*, *Chiloguembelina cubensis* (probably reworked), *Planulina marialana*, *P. mexicana*, *Plectofrondicularia mexicana*, *P. vaughani*, *Siphonina tenuicarinata* and *Uvigerina spinicostata*, an Aquitanian assemblage correlating well with the Alazan formation of Mexico and with the Vicksburgian of the Gulf Coast States. The Jaruco limestone contains *Miogypsina globulina* (recorded under the invalid name *M. irregularis*); tentative records of *Globorotalia praemenardii* and *G. fohsi* from this limestone probably refer to *G. menardii archeomenardii* and *G.* (*Turborotalia*) *fohsi barisanensis*. The Jaruco limestone is evidently equivalent to the Paso Real limestone which contains *Miogypsina globulina*, *M. intermedia* and '*M. intermedia-cushmani*', and we regard the fauna as of high Aquitanian age. The marls at the base of the Yumuri limestone contain *Globorotalia menardii* (*vel G. menardii archeomenardii*) and *Orbulina universa*, and are of high Aquitanian age. The Yumuri limestone contains *Lepidocyclina canellei*, *L.* (*Nephrolepidina*) *crassata*, *L.* (*N.*) *dartoni*, *L.* (*Eulepidina*) cf. *dilatata*, *L.* (*E.*) *favosa*, *L.* (*N.*) *marginata*, *L. schumbergeri*, *L.* (*N.*) *sumatrensis*, *L. yurnagunensis*, *L. yurnagunensis morganopsis* and *Turritella altilira*, and must therefore be of highest

Aquitanian age. The Cojimar formation contains *Anomalina alazanensis*, *Bigenerina floridana* (of the Alum Bluff), *Biorbulina bilobata*, *Bolivina alazanensis*, *B. floridana*, *Cassidulina subglobosa*, *Chiloguembelina cubensis* (small, rare, and probably reworked—see Beckmann, 1957, p. 89), *Globoquadrina altispira*, *Globorotalia* (*Turborotalia*) *fohsi fohsi*, *G.* (*T.*) *mayeri*, *G. menardii praemenardii*, *Globotruncana* spp. (reworked), *Hastigerina* aff. *aequilateralis*, *Orbulina universa*, *Palaeonummulites cojimarensis*, *Paraspiroclypeus chawneri*, *Rectuvigerina transversa*, *Sphaeroidinella seminulina* (*vel Sphaeroidinellopsis seminulina kochi*) and *Tretomphalus altanticus*, a lower Burdigalian assemblage correlating well with Bolli's *G.* (*T.*) *fohsi fohsi* Zone in Trinidad. The Güines limestone has *Lepidocyclina* and *Miogypsina* near its base, and also contains *Globorotalia menardii* (*s.s.*), indicating an upper Burdigalian age. Other Burdigalian limestones (believed to be Anguilla equivalent) contain *Lepidocyclina* (*Nephrolepidina*) *perundosa*, *L. subraulini* (also Sombrerito formation of the Dominican Republic), *Orthaulax aguadillensis* and several Anguillan echinoids including *Clypeaster concavus*, *C. cotteaui* and *Echinolampas semiorbis*. The Canimar formation, with *Globorotalia menardii multicamerata*, is of post-Burdigalian age, and is succeeded by the La Cruz marl which has a rich Vindobonian Bowden fauna.

De Albear (1947) has recorded a sample (A-220), from the Camagüey District, which contains the Palaeocene *Ranikothalia antillea* together with the Middle Eocene *Econuloides wellsi* and the Upper Eocene *Dictyoconus americanus*; since these are in association with *Pliolepidina tobleri*, all the Eocene forms must be derived, as the sample must be of Aquitanian age.

Vaughan (1933), Rutten (1935) and Thiadens (1937) have recorded several Miocene species of *Lepidocyclina* and other larger foraminifera from Cuban beds; since, however, no formation names or stratigraphical information of any sort has been given, the records are of little value except in so far as they provide no evidence of Oligocene age, and all post-Eocene forms fit into the succession described in the previous paragraphs.

47

H. PALAEOGEOGRAPHY AND CONCLUSIONS

That important earth movements took place at the close of Eocene times in the Central American region is well known and well documented (e.g. Schuchert, 1935); the following extract from Vaughan (1919*b*, p. 607) is a typical example of the importance attached to them:

There was between the Upper Eocene and the Middle Oligocene deposition periods great deformation in the Antilles. The folding in the principal mountains of Jamaica, the Sierra Maestra of Cuba, and apparently those of Haiti, Porto Rico, the Virgin Islands, and St Croix appears to have taken place at this time. Diastrophism seems also to have been active in Chiapas, Tabasco, Petén, Guatemala, Nicaragua, Costa Rica, and Panama.

While agreeing that important earth movements of the type described did take place, we feel sure that their real importance has still not been realized. We consider that they were even more important and *lasted longer* than has yet been thought to be the case. The evidence we have given indicates that all the southern part of the United States, most of the Central American region and the northern part of South America was uplifted during the whole of Oligocene times. Whereas there was no orogeny, and erosion does not appear to have been intensive in the area of the Gulf Coast States, both were extremely active in the Central American region (including the unstable archipelagic Antilles), and it is in this latter region (e.g. Panama, Trinidad, Venezuela, Cuba), where reworking of Eocene fossils into Miocene deposits occurred on the largest scale.

We feel that further consideration of the palaeogeographical history of the southern part of the United States, on a truly large and continental scale, is necessary. We do not suggest an orogeny there, except in a very minor way in Florida. We believe that during the important Oligocene orogeny in the archipelagic Caribbean region, the whole of the southern portion of the United States was uplifted *without* orogeny as a gigantic coastal plain (much as it is now in the present period of unconformity there, but even more extensive), resulting, not in unconformity, but in non-sequence and disconformity. That is why reworked fossils are uncommon or small (e.g. the *Hantkenina* of the

Red Bluff clay), there being no topography worth mentioning. The horizon of this disconformity we contend to be between the Vicksburgian or younger beds above and Jacksonian or older beds below, depending upon local circumstances and location. The last planktonic zones found below this datum and the first planktonic zones found above it, in the whole Gulf Coast–Caribbean–northern South American region, are known in the Tethyan region. Blow and Banner discuss this in detail in Part 2, where they describe two completely new and well-defined Zones of planktonic foraminifera, the lower of Uppermost Eocene age (confirmed by the presence of *Discocyclina*) and the higher of Oligocene age (confirmed by the presence of the well-known *Nummulites intermedius-fichteli*), both these being controlled above and below; these two zones are missing in the Gulf States, the Caribbean archipelago and northern South America precisely at the above-mentioned datum line. In the Gulf Coast States and on the continental shelf we feel sure that it is above and below this datum line that (quite apart from the effects of other associated minor unconformities and their manifestations) new beds come in seawards, but we know of no published evidence that real Oligocene has appeared from within this disconformity—we call it disconformity rather than unconformity since the angle of discordance is so low. Cyclical deposition, as recognized, occurred both above and below the datum. The new information (p. 35) on the occurrence of true Oligocene foraminifera in the lower part of the Alazan formation of the Tampico embayment of Mexico demonstrates the possibility of their presence in the region as a whole, and confirms that their absence elsewhere in the region is due to a diastem at the Jacksonian–Vicksburgian boundary.

In our endeavours to resolve the anomalous evidence from the middle Tertiaries of the Central American region we have reviewed what appears to us to be all the pertinent evidence. It may be that some references have not been quoted, or that some age allocations we have given on the stratigraphical table (Fig. 5) may need slight adjustment, but we feel fully convinced that our basic conclusions are logical and sound, and will stand the test of time in so far as the world-wide evolutionary succession of Tertiary faunas is concerned. Our investigations have covered a very wide area, and have included all kinds of marine fossils (Foraminifera, Gastropoda, Lamellibranchia, Pteropoda, Echinoidea, Anthozoa, Bryozoa, Decapoda, etc.) as well as plants, but nowhere in the vast region under consideration can we find any published evidence at all for the presence of marine Oligocene. This involves a radical revision of the conception of Tertiary palaeogeography in the region.

Woodring (1927, p. 996) stated that 'all American records of *Pliolepidina* (*Lepidocyclina duplicata*, *L. tobleri*, *L. panamensis*) are Eocene so perhaps the assignment of Sumatra beds carrying *luxurians* to the Aquitanian stage needs confirmation'; we have shown that the Miocene dating we give is in accord with the known range of all members of this group in extra-American localities. Again, Vaughan has more than once (e.g. 1919 *b*, p. 585) regarded the Aquitanian as equivalent to the Chattian since he thought that American faunas he had under consideration were of Oligocene age, in spite of the fact that European palaeontologists (e.g. Douvillé) regarded the Aquitanian as Lower Miocene; our conclusions that the beds in question are all of Lower Miocene age disposes of all possible confusion.

We conclude that, in the whole region we have considered (and even as far north as the state of Washington), there are no published records of stratigraphical sections of fossiliferous marine beds which can be dated as Oligocene. Almost all those beds previously dated as Oligocene, and even some previously referred to the Upper Eocene, we regard as being undoubtedly Miocene (mainly Lower Miocene). Bolli has recently informed us (letter dated 27 August 1959) that he and Bermúdez have recognized an occurrence of the Oligocene index species *Globigerina oligocaenica* Blow & Banner sp. nov. (see Part 2, p. 71) in the Dominican Republic; this could represent the existence of post-Jacksonian and pre-Vicksburgian (pre-Cipero age) deposits in the more peripheral areas of the Central American province. The presence of typical *G. oligocaenica* in the lower Alazan formation of Mexico (see p. 35) similarly indicates that its

apparently general absence in those beds previously dated as Oligocene elsewhere is due to geological factors rather than to the merely geographical factors hitherto postulated.

The evolutionary succession of all kinds of fossils within the Tertiary of America, on this basis, is closely correlatable with, and not out of phase with, that in other parts of the world—compare, for example, the dating of miogypsinids given by Hanzawa (1940) and by Drooger (1952).

To sum up:

(*a*) The geological uniformity of the Vicksburgian, its equivalents, and acknowledged Miocene becomes very apparent.

(*b*) The palaeontological uniformity of the same beds is likewise very apparent.

(*c*) The single isochronous major unconformity at the base of these beds demonstrates the uniformity of the geological history throughout this region and the Tethys as a whole.

(*d*) Reworking and redeposition of faunas, previously recognized to have taken place locally, can now be seen to have been more extensive than was thought and to be a common feature associated with this important horizon of unconformity, as it is elsewhere.

(*e*) There is a resulting isochronous distribution of all kinds of faunas in relation to the type stages in Europe and elsewhere also.

(*f*) It is apparent that previously there have been no adequate published records of Oligocene planktonic foraminifera available and that there have been no published species that could be recognized as characterizing the Oligocene; it is significant that, while Emiliani (1954) was able to identify his Eocene planktonic species, all five of his Oligocene species were recorded as '*Globigerina* sp.nov.', and it is evident that research (which has now been carried out) on reliably dated Oligocene planktonic faunas is essential in order to fill this gap; some new Oligocene planktonic forms have been discovered by us, and are described in Part 2.

(*g*) As a result of our conclusions, hundreds of species and many genera recorded from America, belonging to numerous different groups of fossils, must have their ages and ranges revised.

IX. General Conclusions

We now find that the dating and age-determination of the planktonic foraminifera is in complete accord with that of other groups of fossils in the Central American region and in other parts of the world, the evolutionary succession of faunas being the same and not being out of phase in any part. It is on this basis that the dating and ranges of the species of the Globigerinaceae, examined in detail and assessed in Part 2 of this study which follows, will be proposed.

Appendix 1. (Recognition of the genus *Palaeonummulites* Schubert, 1908)

We have re-examined the specimens of *Nummulina pristina* Brady, 1874, the monotype of *Palaeonummulites*. These specimens were originally supposed to have come from the Carboniferous limestone of a quarry near Namur in Belgium, but van den Broeck, who originally sent the material to Brady, subsequently (1899) indicated that they were inadvertently mixed with some Carboniferous samples, and were probably of Tertiary origin. Our re-examination of the type material confirms that it must be of Tertiary age. A small proportion of this type material, which was unfigured by Brady, consists of small rotalids which we hereby expressly exclude from the species and the genus. We here designate the British Museum (Natural History) specimen P. 35413 (plate 12, fig. 1 of Brady, 1874) as the lectotype of *Nummulina pristina* Brady, 1874, the type species of *Palaeonummulites* Schubert, 1908. Brady's plate 12, fig. 3 (specimen no. P. 35503) and specimen no. P. 35504 (here figured—plate I, fig. F) are topotypic syntypes. The synonymy of these forms is therefore:

Genus: *Palaeonummulites* Schubert, 1908

Type species: *Nummulina pristina* Brady, 1874; monotypy (related to *Nummulites wemmelensis* de la Harpe and van den Broeck, 1881; Bartonian); ? Bartonian.

Synonyms: *Operculinella* Yabe, 1918 (type species: *Amphistegina cumingi* Carpenter, 1859; Recent); *Operculinoides* Hanzawa, 1935 (type species: *Nummulites (Nummulina) willcoxi* Heilprin, 1883; Eocene).

Stratigraphical distribution: Eocene to Recent; world wide.

The examination of the type material of *Nummulina pristina* and of typical Belgian *Nummulites wemmelensis* leaves us in no doubt as to their very close relationship. In view of the fact that the names *Operculinella* and *Operculinoides* have been approximately equally used in the Eastern and Western Hemispheres for forms which are obviously congeneric, we have no hesitation in replacing both by the name *Palaeonummulites*.

Appendix 2. (Revision of the genus *Pliolepidina* H. Douvillé, 1915)

Owing to the confused nature of the records of forms referred to *Lepidocyclina* and *Pliolepidina*, both from the point of view of generic and subgeneric criteria and of their stratigraphical record in the Central American region, we have been obliged to reassess their value. *Lepidocyclina (s.s.)* ranges from Eocene to Lower Miocene, and *Pliolepidina* has been regarded as being restricted to the Eocene; however, the latter is correctly reported from the Emperador limestone of Panama which current American opinion (again, we think, rightly) dates as Miocene.

The name *Pliolepidina* was proposed in 1915 by H. Douvillé for an unnamed species to which he in 1917 gave the name *P. tobleri* which becomes the type by monotypy. Subsequent references to the type as being *Lepidocyclina pustulosa* by Vaughan & Cole (1941) or '*L. (P.) pustulosa forma tobleri* H. Douvillé *forma teratologica*' by Cole, in Cushman (1948) are incorrect and misleading, being based (as we show below) upon a false interpretation of the synonymy of these and certain other lepidocyclines. H. Douvillé distinguished his *Pliolepidina* from *Lepidocyclina (s.s.)* because of its large, multicellular embryont. In 1924–25 he suggested that this large embryont might be teratological, even though he still regarded it as being of subgeneric value. Vaughan & Cole (1941) redefined the subgenus and unjustifiably asserted that *L. tobleri* was merely a teratological form of *Isolepidina pustulosa* H. Douvillé—which has all the characters of *Lepidocyclina (s.s.)*. However, similar large, multicellular embryonts characterize *Multilepidina* Hanzawa, 1932, forms of which are fairly widespread in the Lower Miocene only (see van der Vlerk, 1950, 1955) of the Indo-Pacific and Mediterranean regions and are not accompanied by forms which could by any stretch of imagination be regarded as their normal non-teratological counterparts. We entirely agree with Rutten's opinion (1941) that not even *Pliolepidina* is to be regarded as teratological, and we can unequivocally assert that *Simplorbites*, to which he also referred. is also not a teratological form.

The type species of *Multilepidina* is *Lepidocyclina* (*M.*) *irregularis* Hanzawa, 1932 (by monotypy), from the Burdigalian of Formosa, which Hanzawa distinguished from *Pliolepidina* solely on account of the smaller chambers of the embryont being inside, not outside, the common thick wall; we agree with Rutten (1941, figs. 20, 21) that the embryonts of these two forms are indistinguishable and that the smaller chambers occur within the common thick wall in both cases, and even a brief inspection of illustrations of the embryont of *tobleri* would confirm this point. Rutten also indicated that the only remaining point of difference was that *Multilepidina* had a 'six-stolon' system (see pp. 18–19), and *Pliolepidina* a 'four-stolon' system. However, Brönnimann (1947) showed that both these 'stolon systems' can be found in one and the same specimen of *tobleri*. Hence, we consider that *Multilepidina* must be regarded as a synonym of *Pliolepidina* and a brief definition and synonymy of the genus follows:

Genus *Pliolepidina* H. Douvillé, 1915

Type species: *Lepidocyclina* (*Pliolepidina*) sp. H. Douvillé, 1915 = *Pliolepidina tobleri* H. Douvillé, 1917; near San Fernando, Trinidad (collected at Point Bontour, *teste* H. G. Kugler, *verb.*); 'Mount Moriah silts' (*teste* Kugler), dated by us as Aquitanian—see pp. 41–42; monotypy.

Synonyms: *Multicyclina* Cushman, 1919; type species *Lepidocyclina* (*M.*) *duplicata* Cushman, 1919; 2 miles north of David, Panama; originally given as Oligocene but dated by us as Aquitanian (see p. 36); original designation and monotypy. *Multilepidina* Hanzawa, 1932; type species *Lepidocyclina* (*M.*) *irregularis* Hanzawa, 1932; Formosa; Kaizan beds (Burdigalian); monotypy. *Cyclolepidina* Whipple, 1934; type species *Lepidocyclina* (*C.*) *suvaensis* Whipple, 1934; Fiji; Suva formation (Burdigalian); original designation. *Pliorbitoina* van de Geyn and van der Vlerk, 1935; type species *Pliolepidina tobleri* H. Douvillé, 1917; near San Fernando, Trinidad; Mount Moriah (= San Fernando) formation (dated by us as Aquitanian); by substitution and monotypy.

DIAGNOSIS

Lenticular Lepidocyclininae of medium to large size which are characterized by a large megalospheric embryont surrounded by a thick wall enclosing one large chamber and from four to ten smaller peripheral chambers the walls between which are relatively thin, there being no spirality as in *Polylepidina*; four to six intercameral foramina.

REMARKS

Of other names which have been associated with these forms, *Orbitoina* (van de Geyn & van der Vlerk, 1935, *nom.nud.*) Schenck & Frizzell, 1936 (type species *Lepidocyclina trinitatis* H. Douvillé, 1924; first valid designation), with its objective typonym *Isorbitoina* (van de Geyn & van der Vlerk, 1935, *nom.nud.*) Schenck & Frizzell, 1936, and *Neolepidina* Brönniman, 1947 (type species *Isolepidina pustulosa* H. Douvillé, 1917; original designation) are synonyms of *Lepidocyclina* (*s.s.*). *Lepidocyclina* (*Pliolepidina*) *tobleri* (H. Douvillé), 1917, was incorrectly quoted as the type of *Neolepidina* by Thalmann, 1948. The differences between the embryonts of *Lepidocyclina*, *Triplalepidina*, *Nephrolepidina* and *Eulepidina*, on the one hand, and *Pliolepidina* and *Polylepidina*, on the other, are so marked that we regard the latter two as full genera.

GEOLOGICAL RANGE

Aquitanian to Burdigalian.

ASSIGNMENT OF SPECIES

Species now referred to *Pliolepidina*
(1) Lower Miocene species

Lepidocyclina (*Pliolepidina*) *amoentai* Zuffardi-Comerci, 1929; Borneo; Lower Miocene.

L. (*Multilepidina*) *fijiensis* Cole, 1945; Fiji; *f* stage (Burdigalian).

L. (*Multilepidina*) *irregularis* Hanzawa, 1932; Formosa; Burdigalian. (*N.B.*, *L. omphalus* Tan, 1935, from the Burdigalian of Java, is probably the microspheric form of this species.)

L. (*Pliolepidina?*) *luxurians* Tobler, 1925; Sumatra; middle Aquitanian.

L. stigteri van der Vlerk, 1925; Dutch Borneo; post-Eocene (*vel* Lower Miocene).

L. subdilatata R. Douvillé, 1908; Italy; upper Burdigalian.

L. (*Cyclolepidina*) *suvaensis* Whipple, 1934; Fiji; Suva formation (Burdigalian). (*N.B.*, *L.* (*Eulepidina*) *dilatata* var. *laddi* Whipple, 1934, also from the Suva formation of Fiji, is probably the microspheric form of this species.)

L. (*Multilepidina*) *wanneri* van den Abeele, 1949; Java; Rembang beds (Burdigalian).

(2) Other species now assigned to the Lower Miocene

Lepidocyclina (*Multilepidina*) *palustris* Azzaroli, 1958; Somaliland; Aquitanian *e* stage (not Chattian).

L. panamensis Cushman, 1919; type locality near mouth of the Tonosi River, Panama; 'Oligocene'. Also recorded from the Culebra formation and doubtfully from the Emperador limestone, as well as from the '?Oligocene' of David, Panama. (These, and all other occurrences of the species, we now date as Lower Miocene—see p. 36.)

L. duplicata Cushman, 1919; type locality 2 miles north of David, Panama; 'Oligocene'. Also from the 'Oligocene; near the mouth of the Tonosi River, Panama'. (These, and all other occurrences of the species, we now date as Lower Miocene—see p. 36.)

Pliolepidina tobleri H. Douvillé, 1917; type locality near San Fernando, Trinidad (H. G. Kugler has informed us, *verb.*, that Douvillé obtained his specimens from Point Bontour, from the 'Mount Moriah silts'); 'Oligocene (lower Stampian)'. (This and other records of the species have subsequently been regarded as of Upper Eocene age, but we now regard them all as of Lower Miocene age—see pp. 41–42.) (*N.B. Lepidocyclina curasavica* Koch, 1928, from the 'Oligocene' of Venezuela and Curaçao, is, we agree with Cole (1952), the microspheric form of *Pliolepidina tobleri*. We would also point out the close specific relationship of *P. tobleri* to the Aquitanian *P. luxurians*.)

Species incorrectly referred to *Pliolepidina*

Lepidocyclina (Pliolepidina) ariana Cole & Ponton, of Cole (1945); this is a *Lepidocyclina (s.s.)*.

L. (Pliolepidina) cedarkeysensis Cole, 1942; this is a *Lepidocyclina (s.s.)*.

L. (Pliolepidina) gubernacula Cole, 1952; this is a *Lepidocyclina (s.s.)*.

L. hubbardi aurarensis Hodson, 1926, as *L. (Pliolepidina) aurarensis* (Hodson), in Vaughan & Cole (1941); this is a *Lepidocyclina (s.s.)*.

L. (Pliolepidina) kinlossensis Vaughan, 1928; this is a *Lepidocyclina (s.s.)*, possibly also including a *Polylepidina*.

L. (Pliolepidina) macdonaldi Cushman, of Cole (1945, 1952); this is a *Lepidocyclina (s.s.)*.

L. (Polylepidina) mirandana Hodson, 1926, placed in synonymy of *L. (Pliolepidina) pustulosa tobleri* (H. Douvillé) by Cole (1952); appears to be a true Eocene *Polylepidina*.

L. (Pliolepidina) peruviana Cushman, of Cole (1944); this is a *Lepidocyclina (s.s.)*.

Isolepidina pustulosa H. Douvillé, 1917, as *Lepidocyclina (Pliolepidina) pustulosa* (H. Douvillé), in Vaughan & Cole (1941); this is a *Lepidocyclina (s.s.)*.

Lepidocyclina (Pliolepidina) r. douvillei (Lisson), of Cole (1944); this is a *Lepidocyclina (s.s.)*.

L. (Pliolepidina?) subglobosa Nuttall, of Vaughan & Cole (1941); this is a *Lepidocyclina (s.s.)*.

FURTHER REMARKS

Dr J. P. Beckmann has recently (1958) recorded, but not illustrated, the genus *Pliolepidina* from the Middle Eocene and Upper Eocene of Cuba. Since we felt that he was using the name in the lax sense

of Cole, we wrote to him, and he kindly sent us a representative lepidocycline-bearing sample from both the Middle and Upper Eocene beds concerned. Preparations (equatorial sections and polished surfaces at the level of the equatorial plane) were made of a large proportion of the lepidocyclines, twelve preparations from the Middle Eocene sample, and seventeen preparations from the Upper Eocene sample, covering all observed variants of the lepidocyclines present. Not a single embryont of *Pliolepidina* type was found, every preparation that was not that of a microspheric specimen having an embryont typical of *Lepidocyclina (s.s.)*. These findings confirm our conclusions given above, and we have communicated them to Dr Beckmann.

Cole (1960) has recently tentatively suggested that *Nephrolepidina* and *Multilepidina* both be placed in the synonymy of *Eulepidina*, since he considered that embryonts of all these types occurred in one species. This is contrary to the practical experience, over many years, of ourselves and many colleagues, and we consider the suggestion to be quite unwarranted. It is evident that speciation has not been correct, and that in some cases it has not been realized that species of both Eocene and Miocene age were present in the samples considered. (See Eames, Banner, Blow, Clarke & Smout, *Micropaleontology*, **8**, (1962)—in the press).

Selected References

Accordi, B. (1956). Stratigrafia e Paleontologia delle formazioni Oligo-Mioceniche del Trevigiano Orientale. *Mem. Ist. Geol. Min. Univ. Padova*, **19**, 1–64, 5 pls.

AGIP Mineraria (1957). *Foraminiferi padani (Terziario e Quaternario); Atlante iconografico e distribuzione stratigrafica.* 52 pls. Milano.

Akers, W. H. (1955). Some planktonic foraminifera of the American Gulf Coast and suggested correlations with the Caribbean Tertiary. *J. Paleont.* **29**, 647–664, pl. 65.

Akers, W. H. & Drooger, C. W. (1957). Miogypsinids, planktonic foraminifera, and Gulf Coast Oligocene-Miocene correlations. *Bull. Amer. Ass. Petrol. Geol.* **41**, 656–678.

Albear, J. F. de (1947). Stratigraphic paleontology of Camagüey District, Cuba. *Bull. Amer. Ass. Petrol. Geol.* **31**, 71–91.

Applin, E. R., Ellisor, A. E. & Kniker, H. T. (1925). Subsurface stratigraphy of the coastal plain of Texas and Louisiana. *Bull. Amer. Ass. Petrol. Geol.* **9**, 79–122, pl. 3.

Applin, E. R. & Jordan, L. (1945). Diagnostic foraminifera

REFERENCES

from subsurface formations in Florida. *J. Paleont.* **19**, 129–148, pls. 18–21.

Atwill, E. R. (1935). Oligocene Tumey formation of California. *Bull. Amer. Ass. Petrol. Geol.* **19**, 1192–1204.

Azzaroli, A. (1958). L'Oligocene e il Miocene della Somalia. Stratigrafia, Tettonica, Paleontologia (Macroforaminiferi, Coralli, Molluschi). *Palaeontogr. ital.* **52**, (N.S., **22**), 1–142, pls. 1–36.

Bandy, O. L. (1949). Eocene and Oligocene foraminifera from Little Stave Creek, Clarke County, Alabama. *Bull. Amer. Paleont.* **32**, no. 131, 1–210, pls. 1–27.

Banner, F. T. & Blow, W. H. (1959). The classification and stratigraphical distribution of the Globigerinaceae. *Palaeontology*, **2**, part 1, 1–27, pls. 1–3.

Barker, R. W. (1932). Three species of larger Tertiary foraminifera from S.W. Ecuador. *Geol. Mag.* **69**, 277–281, pl. 16.

Barker, R. W. (1939). Species of the foraminiferal family Camerinidae in the Tertiary and Cretaceous of Mexico. *Proc. U.S. Nat. Mus.* **86**, 305–330, pls. 11–22.

Becker, L. E. & Dusenbury, A. N. (1958). Mio-Oligocene (Aquitanian) foraminifera from the Goajira Peninsula, Colombia. *Spec. Publ. Cushman Found. Foram. Res.* no. 4, 1–48, pls. 1–7.

Beckmann, J. P. (1953). Die Foraminiferen der Oceanic Formation (Eocaen-Oligocaen) von Barbados, Kl. Antillen. *Ecl. geol. Helv.* **46**, 301–412, pls. 16–30.

Beckmann, J. P. (1957). *Chiloguembelina* Loeblich & Tappan and related foraminifera from the Lower Tertiary of Trinidad, B.W.I. *Bull. U.S. Nat. Mus.* **215**, 83–95, pl. 21.

Beckmann, J. P. (1958). Correlation of pelagic and reefal faunas from the Eocene and Paleocene of Cuba. *Ecl. geol. Helv.* **51**, no. 2, 416–422.

Bellen, R. C. van (1956). The stratigraphy of the 'Main Limestone' of the Kirkuk, Bai Hassan, and Qarah Chauq Dagh structures in North Iraq. *J. Inst. Petrol.* **42**, no. 393, 233–263.

Bellen, R. C. van (1959). Tertiary. (In: *Lexique Stratigraphique Internationale*, 3, fasc. 10a, *Iraq*) Congr. Geol. Internat., Comm. Strat., Centre Nat. de la Recherche Sci., Paris.

Bermúdez, P. J. (1949). Tertiary smaller foraminifera of the Dominican Republic. *Spec. Publ. Cushman Lab. Foram. Res.* **25**, 1–322, pls. 1–26.

Beyrich, E. (1854). Über die Stellung der Hessischen Tertiärbildungen. *Ber. Verh. preuss. Akad. Wiss. Berl.* pp. 640–666.

Blow, W. H. (1956). Origin and evolution of the foraminiferal genus *Orbulina* d'Orbigny. *Micropaleontology*, **2**, 57–70.

Blow, W. H. (1957). Transatlantic correlation of Miocene sediments. *Micropaleontology*, **3**, 77–79.

Blow, W. H. (1959). Age, correlation and biostratigraphy of the Upper Tocuyo (San Lorenzo) and Pozón Formations, Eastern Falcón, Venezuela. *Bull. Amer. Paleont.* **39**, no. 178, 59–251, pls. 6–19.

Bold, W. A. van den (1958). Ostracoda of the Brasso formation of Trinidad. *Micropaleontology*, **4**, 391–418, pls. 1–5.

Bold, W. A. van den (1960). Eocene and Oligocene

Ostracoda of Trinidad. *Micropaleontology*, **6**, no. 2, 145–196, pls. 1–8.

Bolli, H. M. (1957a). Planktonic foraminifera from the Oligocene-Miocene Cipero and Lengua formations of Trinidad, B.W.I. *Bull. U.S. Nat. Mus.* **215**, 97–125, pls. 22–29.

Bolli, H. M. (1957b). Planktonic foraminifera from the Eocene Navet and San Fernando Formations of Trinidad, B.W.I. *Bull. U.S. Nat. Mus.* **215**, 155–172, pls. 35–39.

Bolli, H. M. (1959) (1960). Planktonic foraminifera as index fossils in Trinidad, West Indies, and their value for worldwide stratigraphic correlation. *Ecl. geol. Helv.* **52**, no. 2, 627–637, 1 table.

Bolli, H. M., Loeblich, A. R. & Tappan, H. (1957). Planktonic foraminiferal families Hantkeninidae, Orbulinidae, Globorotaliidae, and Globotruncanidae. *Bull. U.S. Nat. Mus.* **215**, 1–50, pls. 1–11.

Boussac, J. (1911). Études Paléontologiques sur le Nummulitique Alpin. *Mém. Explic. Carte géol. Détail. France*, i–vii, 1–439, Atlas (22 pls.).

Boussac, J. (1912). Études Stratigraphiques sur le Nummulitique Alpin. *Mém. Carte géol. Détail. Fr.* i–xxx, 1–662, 20 pls.

Bowen, R. N. C. (1955). The stratigraphical range of the foraminiferal genus *Orbulina* d'Orbigny 1839. *Geol. Mag.* **92**, 162–167.

Brady, H. B. (1874). On a true Carboniferous Nummulite. *Ann. Mag. Nat. Hist.* (4), **13**, 222–231, pl. 12, figs. 1–5.

Broeck, E. van den (1899). Petites notes rhizopodiques. *Ann. Soc. Malac. Belg.* (1898), **33**, 35–38.

Brongniart, A. (1823). *Mémoire sur les terrains de sédiment supérieurs calcaréo-trappéens du Vicentin*, i–vi, 1–86, pls. 1–6. Paris.

Brönnimann, P. (1947). Zur Neue-Definition von *Pliolepidina* H. Douvillé, 1915. *Ecl. geol. Helv.* **39**, 373–379.

Brönnimann, P. (1950). Occurrence and ontogeny of *Globigerinatella insueta* Cushman and Stainforth from the Oligocene of Trinidad, B.W.I. *Contr. Cushman Found. Foram. Res.* **1**, 80–82, pls. 13–14.

Canestrelli, G. (1908). Revisione della fauna Oligocenica di Laverda nel Vicentino. *Atti Soc. ligust. Sci. nat. geogr.* **19**, 27–79, 97–150, pls. 1–2.

Canu, F. & Bassler, R. S. (1919). Fossil Bryozoa from the West Indies. *Publ. Carneg. Inst*, **291**, 73–102, pls. 1–7.

Carter, A. N. (1958a). Pelagic foraminifera in the Tertiary of Victoria. *Geol. Mag.* **95**, no. 4, 297–304.

Carter, A. N. (1958b). Pelagic foraminifera in Southern Victoria and the Murray Basin. *Australia and New Zealand Association for the Advancement of Science, Murray Basin Symposium*, 2 pls.

Carter, A. N. (1958c). Tertiary foraminifera from the Aire District, Victoria. *Bull. geol. Surv. Vict.* **55**, 1–76.

Carter, A. N. (1959). Guide foraminifera in the Tertiary stages in Victoria. *J. Vict. Min. Geol.* **6**, no. 3, 48.

Chapman, F. (1914). Description of a limestone of Lower Miocene Age from Bootless Inlet, Papua. *J. roy. Soc. N.S.W.* **48**, 281–301, pls. 7–9.

Cole, W. S. (1938). Stratigraphy and micropaleontology of two deep wells in Florida. *Bull. Fla geol. Surv.* **16**, 1–48, pls. 1–12.

Cole, W. S. (1942). Stratigraphic and paleontologic studies of wells in Florida. No. 2. *Bull. Fla geol. Surv.* **20**, 1–55*b*, pls. 1–16.

Cole, W. S. (1944). Stratigraphic and paleontologic studies of wells in Florida. No. 3. *Bull. Fla geol. Surv.* **26**, 1–168, pls. 1–29.

Cole, W. S. (1945). Stratigraphic and paleontologic studies of wells in Florida. No. 4. *Bull. Fla geol. Surv.* **28**, 1–160, pls. 1–22.

Cole, W. S. (1949). Upper Eocene larger foraminifera from the Panama Canal Zone. *J. Paleont.* **23**, 267–275, pls. 52–55.

Cole, W. S. (1952). Eocene and Oligocene larger foraminifera from the Panama Canal zone and vicinity. *U.S. Geol. Surv., Prof. Paper*, **244**, 1–41, pls. 1–28.

Cole, W. S. (1953*a*). Criteria for the recognition of certain assumed Camerinid genera. *Bull. Amer. Paleont.* **35**, no. 147, 1–22, pls. 1–3.

Cole, W. S. (1953*b*). Some late Oligocene larger foraminifera from Panama. *J. Paleont.* **27**, 332–337, pls. 43–44.

Cole, W. S. (1953*c*). Larger foraminifera from the Upper Eocene of Costa Rica. *J. Paleont.* **27**, 748–749.

Cole, W. S. (1954). Larger foraminifera and smaller diagnostic foraminifera from Bikini Drill Holes. *U.S. Geol. Surv., Prof. Paper*, **260**–O, 569–608, pls. 204–222.

Cole, W. S. (1956). Jamaican larger foraminifera. *Bull. Amer. Paleont.* **36**, no. 158, 205–225, pls. 24–31.

Cole, W. S. (1957*a*). Late Oligocene larger foraminifera from Barro Colorado Island, Panama Canal Zone. *Bull. Amer. Paleont.* **37**, no. 163, 313–338, pls. 24–30.

Cole, W. S. (1957*b*). Variation in American Oligocene species of *Lepidocyclina*. *Bull. Amer. Paleont.* **38**, no. 166, 31–51, pls. 1–6.

Cole, W. S. (1957*c*). Larger foraminifera of Saipan Island. *U.S. Geol. Surv., Prof. Paper*, **280**–I, 321–360.

Cole, W. S. (1957*d*). Larger foraminifera from Eniwetok Atoll drill holes. *U.S. Geol. Surv., Prof. Paper*, **260**–V, 743–784, pls. 231–249.

Cole, W. S. (1958*a*). Names of and variation in certain American larger foraminifera. No. 1. *Bull. Amer. Paleont.* **38**, no. 170, 179–204, pls. 18–25.

Cole, W. S. (1958*b*). Larger foraminifera from Carriacou, British West Indies. *Bull. Amer. Paleont.* **38**, no. 171, 219–233, pls. 26–29.

Cole, W. S. (1958*c*). Names of and variation in certain American larger foraminifera, particularly the Camerinids. No. 2. *Bull. Amer. Paleont.* **38**, no. 173.

Cole, W. S. (1958*d*). Names of and variation in certain American larger foraminifera, particularly the Discocyclinids. No. 3. *Bull. Amer. Paleont.* **38**, no. 176.

Cole, W. S. (1958*e*). Names of and variations in certain Indo-Pacific Camerinids. *Bull. Amer. Paleont.* **39**, no. 181.

Cole, W. S. (1960). Variability in embryonic chambers of *Lepidocyclina*. *Micropaleontology*, **6**, no. 2, 133–144, pls. 1–4.

Cole, W. S., Todd, R. & Johnson, C. G. (1960). Conflicting age determinations suggested by foraminifera on Yap, Caroline Islands. *Bull. Amer. Paleont.* **41**, no. 186, 77–112, pls. 11–13.

Colom, G. (1952). Aquitanian-Burdigalian diatom deposits of the North Betic Strait, Spain. *J. Paleont.* **26**, no. 6, 867–885.

Colom, G. (1958). The age of the beds with *Miogypsina mediterranea* Brönnimann on the island of Majorca. *Micropaleontology*, **4**, 347–362.

Comité des travaux historiques et scientifiques (1958). 'Colloque sur le Miocène.' *Comptes rendus du Congrès des Sociétés Savantes de Paris et des départements*; *83e Congrès*; *Section des Sciences, sous-section de Géologie*, 1–421. Paris, Gauthier-Villars.

Cooke, C. W. (1919). Tertiary mollusks from the Leeward Islands and Cuba. *Publ. Carneg. Instn*, **291**, 103–156, pls. 1–16.

Cooke, C. W. (1923). The correlation of the Vicksburg group. *U.S. Geol. Surv., Prof. Paper*, **133**, 1–9.

Cooke, C. W. (1935). Notes on the Vicksburg group. *Bull. Amer. Ass. Petrol. Geol.* **19**, 1162–1172.

Cooke, C. W. (1945). Geology of Florida. *Bull. Fla Geol. Surv.* **29**, 1–339.

Cooke, C. W. & MacNeil, F. S. (1952). Tertiary stratigraphy of South Carolina. *U.S. Geol. Surv., Prof. Paper*, **243**–B, 19–29.

Cossmann, M. (1919). Monographie illustrée des mollusques oligocèniques des environs de Rennes. *J. Conchyliol.* **64**, 133–199, pls. 4–7.

Cossmann, M. (1921–22). Synopsis illustré des Mollusques de l'Éocène et de l'Oligocène en Aquitaine. *Mém. Soc. géol. Fr. Paléont.* no. 55 (= Vol. **23**, fasc. 3–4, 1–112, pls. 1–8; Vol. **24**, fasc. 1–2, 113–220, pls. 9–15).

Cossmann, M. & Peyrot, A. (1909–35). Conchologie Néogènique de l'Aquitaine. *Actes Soc. Linn. Bordeaux*, **63**–**86**.

Crespin, I., Kicinski, F. M., Patterson, S. J. & Belford, D. J. (1956). Papers on Tertiary Micropalaeontology. *Rep. Bur. Miner. Resour. Aust.* **25**, 1–77.

Cushman, J. A. (1919*a*). The larger fossil foraminifera of the Panama Canal Zone. *Bull. U.S. Nat. Mus.* **103**, 89–102, pls. 34–35.

Cushman, J. A. (1919*b*). Fossil foraminifera from the West Indies. *Publ. Carneg. Instn*, **291**, 21–71, pls. 1–15.

Cushman, J. A. (1922). The Byram calcareous marl of Mississippi and its foraminifera. *U.S. Geol. Surv., Prof. Paper*, **129**–E, 87–105, pls. 14–28.

Cushman, J. A. (1935). New species of foraminifera from the Lower Oligocene of Mississippi. *Contr. Cushman Lab. Foram. Res.* **11**, 25–37, pls. 4–5.

Cushman, J. A. (1948). *Foraminifera, their Classification and Economic Use.* Harvard University Press.

Cushman, J. A. & Bermúdez, P. J. (1949). Some Cuban species of *Globorotalia*. *Contr. Cushman Lab. Foram. Res.* **25**, 26–48, pls. 5–8.

Cushman, J. A. & Jarvis, P. W. (1930). Miocene foraminifera from Buff Bay, Jamaica. *J. Paleont.* **4**, 353–368, pls. 32–34.

Cushman, J. A. & Renz, H. H. (1941). New Oligocene-Miocene foraminifera from Venezuela. *Contr. Cushman Lab. Foram. Res.* **17**, 1–27, pls. 1–7.

Cushman, J. A. & Renz, H. H. (1947). The foraminiferal fauna of the Oligocene, Ste. Croix Formation, of

REFERENCES

Trinidad, B.W.I. *Spec. Publ. Cushman Lab. Foram. Res.* no. 22, 1–46, pls. 1–8.

Cushman, J. A. & Todd, R. (1948). Foraminifera from the Red Bluff-Yazoo section at Red Bluff, Mississippi. *Contr. Cushman Lab. Foram. Res.* **24**, 1–12, pls. 1–2.

Cuvillier, J. & Sacal, V. (1956). *Stratigraphic Correlations by Microfacies in Western Aquitaine*, pp. 1–33, pls. 1–100. Leiden: E. J. Brill.

Daguin, F. (1948). L'Aquitaine Occidentale. *Géologie régionale Fr.* **5**, 1–232.

Dall, W. H. (1890–1903). Contributions to the Tertiary fauna of Florida, with especial reference to the Miocene Silex beds of Tampa and the Pliocene beds of the Caloosahatchie River. *Trans. Wagner Inst. Sci. Philad.* **3** (1–6), 1–1654, pls. 1–60.

David-Sylvain, É. (1937). Étude sur quelques grands Foraminifères tertiaires. *Mém. Soc. géol. Fr. N.S.* **15**, Fasc. 1, Mém. no. 33, 1–44, pls. 1–4.

Davies, A. M. (1934). *Tertiary Faunas*, **2**. Thomas Murby and Co.

Davies, L. M. (1952). Foraminifera of the White Limestone of the Kingston District, Jamaica. *Trans. Edinb. Geol. Soc.* **15**, 121–132.

Deninger, K. (1901). Beitrag zur Kenntniss der Molluskenfauna der Tertiärbildungen von Reit im Winkel und Reichenhall. *Geogr. Jber.* **14**, 221–245, pls. 7–8.

Depéret, C. & Roman, F. (1902). Monographie des Pectinidés néogènes de l'Europe et des régions voisines. Genre *Pecten*. *Mém. Soc. géol. Fr. Paléont.* no. 26 (= Vol. **10**, fasc. 1, 1–73, 8 pls.).

Depéret, C. & Roman, F. (1905). Monographie des Pectinidés néogènes de l'Europe et des régions voisines. Genre *Pecten*. *Mém. Soc. géol. Fr. Paléont.* no. 26 (= Vol. **13**, fasc. 2, 75–104, pls. 6–8).

Dollfus, G. F. (1918). L'Oligocène supérieur marin dans le bassin de l'Adour. *Bull. Soc. géol. Fr.* (4), **17**, 89–102, pl. 8.

Douvillé, H. (1917). Sur l'âge des couches à Lépidocyclines de l'Aquitaine. *C.R. Somm. Soc. géol. Fr.* no. 11, 144–146.

Douvillé, H. (1924–25). Révision des Lépidocyclines. *Mém. Soc. géol. Fr.*, N.S., Mém. no. 2, 1–115, pls. 1–6.

Dreger, J. (1892). Die Gastropoden von Häring bei Kirchbichl in Tirol. *Ann. naturh. (Mus.) Hofmus., Wien*, **7**, 11–34, pls. 1–4.

Dreger, J. (1903). Die Lamellibranchiaten von Häring bei Kirchbichl in Tirol. *Jb. geol. Reichsanst. (Bundesanst.), Wien*, **53**, 253–294, pls. 11–13.

Drooger, C. W. (1952). *Study of American Miogypsinidae.* Zeist: Vonk and Co.

Drooger, C. W. (1954*a*). *Miogypsina* in Northern Italy. *Proc. K. Akad. Wet. Amst.* B, **57**, 227–249, pls. 1–2.

Drooger, C. W. (1954*b*). The Oligocene-Miocene boundary on both sides of the Atlantic. *Geol. Mag.* **91**, 514–518.

Drooger, C. W. (1955). Two species of *Miogypsina* from Southern Peru. *Bull. Soc. Geol. Peru*, **26**, 9–16.

Drooger, C. W. (1956). Transatlantic correlation of the Oligo-Miocene by means of foraminifera. *Micropaleontology*, **2**, no. 2, pp. 183–192, pl. 1.

Drooger, C. W. (1958). Das Älter der Miogypsinengesteine von Kephallinia. *Ann. Geol. Pays Helléniques*, **9**, 115–118.

Drooger, C. W. (1960). *Miogypsina* in north-western Germany. *Proc. K. Akad. Wet. Amst.* B, **63**, no. 1, 38–50, pls. 1–2.

Drooger, C. W. & Magné, J. (1959). Miogypsinids and planktonic foraminifera of the Algerian Oligocene and Miocene. *Micropaleontology*, **5**, no. 3, 273–284, pls. 1–2.

Drooger, C. W. & Socin, C. (1959). Miocene foraminifera from Rosignano, northern Italy. *Micropaleontology*, **5**, no. 4, 415–426, pls. 1–2.

Dubertret, L., Vautrin, H., Keller, A. & David, É. (1933). Contribution à l'Étude géologique de la Syrie septentrionale: le Miocène en Syrie et au Liban. *Haut-Comm. Republ. franc. en Syrie et au Liban, Notes et Mém.* **1**, 1–208, pls. 1–12.

Durham, J. W. (1955). Classification of Clypeasteroid Echinoids. *Univ. Calif. Bull. Dep. Geol.* **31**, no. 4, 73–198, pls. 3–4.

Durham, J. W. *et al.* (1949). The age of the *Hannatoma* mollusk fauna of South America. *J. Paleont.* **23**, 145–160.

Eames, F. E. (1950). The Pegu system of Central Burma. *Rec. geol. Surv. India*, **81** (2), 377–388, pl. 4.

Eames, F. E. (1953). The Miocene/Oligocene boundary and the use of the term Aquitanian. *Geol. Mag.* **90**, 388–392.

Eames, F. E., Banner, F. T., Blow, W. H. & Clarke, W. J. (1960). Mid-Tertiary stratigraphical Palaeontology. *Nature, Lond.*, **185**, no. 4711, 447–448.

Eames, F. E. & Clarke, W. J. (1957). The ages of some Miocene and Oligocene foraminifera. *Micropaleontology*, **3**, 80.

Ellisor, A. C. (1944). Anahuac formation. *Bull. Amer. Ass. Petrol. Geol.* **28**, 1355–1375, pls. 1–7.

Emiliani, C. (1954). The Oligocene microfaunas of the central part of the northern Apennines. *Palaeontogr. ital.* **48** (N.S., **18**), 77–184, pls. 21–25.

Evernden, J. F. (1959). First results of research on the dating of Tertiary and Pleistocene rocks by the potassium/argon method. *Proc. Geol. Soc. Lond.*, no. 1565, 17–19.

Fabiani, R. (1915). Il Paleogene del Veneto. *Mem. Ist. Geol. Univ. Padova*, **3**, 1–337, pls. 1–9.

Flandrin, J. (1938). Contribution à l'Étude paléontologique du Nummulitique algérien. *Mat. Carte géol. Algérie*, 1re Ser., Pal., no. 8, 1–158, pls. 1–15.

Fontannes, F. & Depéret, C. (1889–92). *Études stratigraphiques et paléontologiques pour servir à l'histoire de la période tertiaire dans le bassin du Rhône. IX and X. Les terrains tertiaires de la côte de Provence.* Pt. 9, 1–116, 2 pls., Bâle, Genève et Lyon; Pt. 10, 1–79, Paris.

Franklin, E. S. (1944). Microfauna from the Carapita formation of Venezuela. *J. Paleont.* **18**, 301–319, pls. 44–48.

Fuchs, T. (1870). Beitrag zur Kenntniss der Conchylienfauna des Vicentinischen Tertiärgebirges. *Denkschr. Akad. Wiss. Wien*, **30**, 137–216, pls. 1–11.

Glaessner, M. F. (1953). Time-stratigraphy and the Miocene Epoch. *Bull. Geol. Soc. Amer.* **64**, 647–658.

Glaessner, M. F. (1959). Tertiary stratigraphic correlation in the Indo-Pacific region and Australia. *J. Geol. Soc. India*, **1**, 51–67, table 1.

Glaessner, M. F. (1960). West-Pacific stratigraphic correlation. *Nature, Lond.*, **186**, no. 4730, 1039–1040.

Glaessner, M. F. & Wade, M. (1959). Revision of the foraminiferal family Victoriellidae. *Micropaleontology*, **5**, no. 2, 193–212, pls. 1–3 (*cum. bibl.*).

Glibert, M. (1957). Pélécypodes et gastropodes du Rupélien supérieur et du Chattien de la Belgique. *Mém. Inst. Sci. nat. Belg.*, no. 137, 1–98, 6 pls.

Glibert, M. & Heinzelin, I. de (1954). L'Oligocène inférieur belge. *Inst. Roy. Sci. Nat. Belg.*, Vol. *jubilaire Van Straelen*, pp. 282–438, 7 pls.

Görges, J. (1952). Die Lamellibranchiaten und Gastropoden des oberoligozänen Meeressandes von Kassel. *Abh. hess. Landesamt. Bodenforsch.* **4**, 1–134, pls. 1–3.

Gorter, N. E. & Vlerk, I. M. van der (1932). Larger foraminifera from Central Falcón (Venezuela). *Leid. geol. Meded.* **4**, 94–122, pls. 11–17.

Goudkoff, P. P. & Porter, W. W. (1942). Amoura Shale, Costa Rica. *Bull. Amer. Ass. Petrol. Geol.* **26**, 1647–1655.

Graham, J. J. & Drooger, C. W. (1952). An occurrence of *Miogypsina* in California. *Contr. Cushman Found. Foram. Res.* **3**, 21–22.

Grateloup, J. P. S. de (1847–48). *Conchyliologie fossile des terrains tertiaires du bassin de l'Adour (environs de Dax)*, i–xx, 1–12, 48 pls. and explan. Bordeaux.

Gravell, D. W. (1933). Tertiary larger foraminifera of Venezuela. *Smithson. misc. Coll.* **89**, no. 11, 1–44, pls. 1–6.

Gravell, D. W. & Hanna, M. A. (1937). The *Lepidocyclina texana* horizon in the *Heterostegina* Zone, Upper Oligocene, of Texas and Louisiana. *J. Paleont.* **11**, 517–529, pls. 60–65.

Gravell, D. W. & Hanna, M. A. (1938). Subsurface Tertiary Zones of correlation through Mississippi, Alabama and Florida. *Bull. Amer. Ass. Petrol. Geol.* **22**, no. 8, 984–1013, pls. 1–7.

Gregory, J. W. (1891). The Maltese fossil Echinoidea, and their evidence on the correlation of the Maltese Rocks. *Trans. Roy. Soc. Edinb.* **36**, 585–639, pls. 1–2.

Grekoff, N. & Gubler, Y. (1951). Données complémentaires sur les terrains tertiaires de la Nouvelle-Calédonie. *Rev. Inst. franç. Pétrole*, **6**, 283–293.

Grimsdale, T. F. (1959). Evolution in the American Lepidocyclinidae. *Proc. K. Akad. Wet. Amst.* B, **62**, (1), 8–33.

Gripp, K. (1916). Über das marine Altmiocän im Nordseebecken. *Neues Jb. Min. Geol. Paläont.*, Beil. Band **41**, 1–59, pls. 1–2.

Hadley, W. H. (1934). Some Tertiary foraminifera from the North Coast of Cuba. *Bull. Amer. Paleont.* **20**, no. 70A, 1–40, pls. 1–5.

Hanzawa, S. (1940). Micropalaeontological studies of drill cores from a deep well in Kita-Daito-Zima (North Borodino Island). *Jubilee Publ. Commemm. Prof. Yabe's Sixtieth Birthday*, **2**, 766–775. Tohoku Imp. Univ.

Haug, É. (1908–11). *Traité de Géologie*, **1–2**. Paris: Librairie Armand Colin.

Hedberg, H. D. (1937). Foraminifera of the Middle Tertiary Carapita Formation of North-eastern Venezuela. *J. Paleont.* **11**, 661–697, pls. 90–92.

Hornibrook, N. de B. (1958). New Zealand Upper Cretaceous and Tertiary foraminiferal zones and some overseas correlations. *Micropaleontology*, **4**, no. 1, 25–38, pl. 1 (*cum bibl.*).

Howe, H. V. (1933). Review of Tertiary Stratigraphy of Louisiana. *Bull. Amer. Ass. Petrol. Geol.* **17**, 613–655.

Hubbard, B. (1923). The Geology of the Lares District, Porto Rico. *N.Y. Acad. Sci., Sci. Serv. Porto Rico*, **2**, 1–115.

Jenkins, D. G. (1958). Pelagic foraminifera in the Tertiary of Victoria. *Geol. Mag.* **95**, 438–439.

Jones, D. J. (1958). Displacement of Microfossils. *J. sediment. Petrol.* **28**, 453–467.

Kautsky, F. (1925). Das Miocän von Hemmoor und Basbeck-Osten. *Abh. geol. Landesamt., Berl.*, N.F. **97**, 1–255, pls. 1–12.

Kellogg, R. (1924). Tertiary pelagic mammals of Eastern North America. *Bull. Geol. Soc. Amer.* **35**, 755–766.

Kent, P. E., Slinger, F. C. & Thomas, A. N. (1951). Stratigraphical exploration surveys in S.W. Persia. *Proc. 3rd World Petrol. Congr.*, Section I.

Kleinpell, R. M. (1938). Miocene stratigraphy of California. *Spec. Publ. Amer. Ass. Petrol. Geol.* London: Murby.

Koenen, A. von (1867–68). Das marine Mitteloligocän Norddeutschlands und seine Mollusken-Fauna. *Palaeontographica*, **16**, 53–128, pls. 6–7; 223–296, pls. 26–30.

Koenen, A. von (1893–94). Das norddeutsche Unter-Oligocän und seine Mollusken-Fauna. 5, Pelecypoda. *Abh. Geol. Spec. Preussen*, **10**, Heft 5, 1005–1248, pls. 63–86; Heft 6, 1249–1338, pls. 87–95.

Kranz, W. (1910). Das Tertiär zwischen Castelgomberto, Montecchio Maggiore, Creazzo und Monteviale im Vicentin. *Neues Jb. Min. Geol. Paläont.*, Beil. Band **29**, 180–268, pls. 4–6.

Kranz, W. (1914). Das Tertiär zwischen Castelgomberto, Montecchio Maggiore, Creazzo und Monteviale im Vicentin. *Neues Jb. Min. Geol. Paläont.*, Beil. Band **38**, 273–324, pl. 6.

Kugler, H. G. (1938). El Eoceno de la Roca del Soldado cerca de Trinidad. *Bol. Geol. Min. Estados Unidos Venezuela*, **2** (2–4), 202–225.

Kugler, H. G. (1953). Jurassic to recent sedimentary environments in Trinidad. *Bull. Ass. Suisse Géol. Ing. Pétrole*, **20**, no. 59, 27–60.

Kugler, H. G. (1954). The Miocene/Oligocene boundary in the Caribbean region. *Geol. Mag.* **91**, 410–414.

Kugler, H. G. (1957). Contribution to the geology of the Islands Margarita and Cubagua, Venezuela. *Bull. Geol. Soc. Amer.* **68**, 555–566, pl. 1.

Lemoine, P. & Douvillé, R. (1904). Sur le Genre *Lepidocyclina* Gümbel. *Mém. Soc. géol. Fr., Pal.*, Mém. no. 32, 1–41, pls. 1–3.

LeRoy, L. W. (1948). The foraminifer *Orbulina universa* d'Orbigny, a suggested Middle Tertiary time indicator. *J. Paleont.* **22**, 500–508.

LeRoy, L. W. (1952). *Orbulina universa* d'Orbigny in Central Sumatra. *J. Paleont.* **26**, 576–584.

Leupold, W. & Vlerk, I. M. van der (1931). Stratigraphie van Nederlandsch Oost-Indie. *Leid. geol. Meded.* **5**, 611–648.

REFERENCES

Liddle, R. A. (1946). *The Geology of Venezuela and Trinidad.* Pal. Research Inst., Ithaca.

MacNeil, F. S. (1944). Oligocene stratigraphy of southern United States. *Bull. Amer. Ass. Petrol. Geol.* **28**, 1313–1354.

Marks, P. (1951). A revision of the smaller foraminifera from the Miocene of the Vienna Basin. *Contr. Cushman Found. Foram. Res.* **2** (2), 33–73, pls. 5–8.

Martin-Kaye, P. H. (1958). The geology of Carriacou. *Bull. Amer. Paleont.* **38**, no. 175, 395–406, 1 map (*cum bibl.*).

Matley, C. A. (1951). *Geology and Physiography of the Kingston District, Jamaica.* Crown Agents of the Colonies, London.

Maury, C. J. (1902). A comparison of the Oligocene of western Europe and the southern United States. *Bull. Amer. Paleont.* **3**, 313–404, pls. 20–29.

Maury, C. J. (1912). A contribution to the paleontology of Trinidad. *J. Acad. Nat. Sci. Philadel.* Ser. 2, **15**, 25–112, pls. 1–13.

Maury, C. J. (1929). The Soldado Rock Type Section of Eocene. *J. Geol.* **37**, 177–181.

McLean, J. (1947). Oligocene and Lower Miocene microfossils from Onslow County, N. Carolina. *Acad. Nat. Sci. Philadel. Not. Nat.* no. 200, 1–9.

Mohan, K. (1958). Miogypsinidae from western India. *Micropaleontology*, **4**, 373–390, pls. 1–3.

Mornhinveg, A. R. (1941). The Foraminifera of Red Bluff. *J. Paleont.* **15**, 431–435.

Nuttall, W. L. F. (1928). Tertiary foraminifera from the Naparima Region of Trinidad (British West Indies). *Quart. J. geol. Soc. Lond.* **84**, 57–115, pls. 3–8.

Nuttall, W. L. F. (1932). Lower Oligocene foraminifera from Mexico. *J. Paleont.* **6**, 3–35, pls. 1–9.

Nuttall, W. L. F. (1933). Two species of *Miogypsina* from the Oligocene of Mexico. *J. Paleont.* **7**, 175–177, pl. 24.

Olsson, A. A. (1931). Contributions to the Tertiary Paleontology of Northern Peru. Part 4, The Peruvian Oligocene. *Bull. Amer. Paleont.* **17**, no. 63, 1–264, pls. 1–21.

Oppenheim, P. (1900). Paläontologische Miscellaneen. III (2). Beiträge zur Kenntniss des Oligocän und seiner Fauna in den venetianischen Voralpen. *Z. dtsch. geol. Ges.* **52**, 243–326, pls. 9–11.

Oppenheim, P. (1903). Ueber die Ueberkippung von S. Orso, des Tertiär des Tretto und Fauna wie Stellung der Schioschichten. *Z. dtsch. geol. Ges.* **55**, 98–235, pls. 8–11.

Palmer, D. K. (1934). Some large fossil foraminifera from Cuba. *Mem. Soc. Cubaña Hist. Nat.* **8**, 235–264, pls. 12–16.

Palmer, D. K. (1940*a*). Foraminifera of the Upper Oligocene Cohimar Formation of Cuba. *Mem. Soc. Cubaña Hist. Nat.* **14**, 113–132, pls. 17–18.

Palmer, D. K. (1940*b*). Foraminifera of the Upper Oligocene Cohimar Formation of Cuba. *Mem. Soc. Cubaña Hist. Nat.* **14**, 277–304, pls. 51–53.

Palmer, D. K. (1940*c*). Foraminifera of the Upper Oligocene Cohimar Formation of Cuba. *Mem. Soc. Cubaña Hist. Nat.* **15**, 281–306, pls. 28–31.

Palmer, D. K. & Bermúdez, P. J. (1936). An Oligocene foraminiferal fauna from Cuba. *Mem. Soc. Cubaña Hist. Nat.* **10**, 227–316, pls. 13–20.

Palmer, R. H. (1945). Outline of the Geology of Cuba. *J. Geol.* **53**, 1–34.

Petters, V. & Sarmiento, S. R. (1956). Oligocene and Lower Miocene biostratigraphy of the Carmen-Zambrano area, Colombia. *Micropaleontology*, **2**, 7–35, pl. 1.

Peyrot, A. (1933). Conchologie néogènique de l'Aquitaine. Conclusions. *Act. Soc. linn. Bordeaux*, **85**, 38–68.

Piggott, C. D. (1959). Genetics and the origin of species. *Nature, Lond.*, **184**, no. 4686, 587–588.

Raulin, V. (1891). Sur quelques faluns bleus inconnus du Département des Landes. *Bull. Soc. géol. Fr.* (3), **19**, 8–14.

Raulin, V. (1896). Sur la faune Oligocène de Gaas (Landes). *Bull. Soc. géol. Fr.* (3), **23**, 546–555.

Renz, H. H. (1942). Stratigraphy of Northern South America, Trinidad and Barbados. *8th Amer. Sci. Congr. Proc.* **4**, Geol. Sci., 513–571.

Renz, H. H. (1948). Stratigraphy and fauna of the Agua Salada group, State of Falcón, Venezuela. *Mem. Geol. Soc. Amer.* **32**, 1–219, pls. 1–12.

Renz, H. H. (1957). Stratigraphy and geological history of Eastern Venezuela. *Geol. Rdsch.* **45**, (3), 728–759, pls. 8–13.

Renz, O. & Küpper, H. (1947). Über morphogenetische Untersuchungen an Grossforaminiferen. *Ecl. geol. Helv.* **39**, no. 2 (1946), 317–342, pl. 18.

Roger, J. (1944). Révision des pectinidés de l'Oligocène du domaine nordique. *Mém. Soc. géol. Fr.*, N.S., **23**, fasc. 1; Mém. no. 50, 1–57, 2 pls.

Rovereto, G. (1900). Illustrazione dei Molluschi Fossili Tongriani posseduti dal Museo Geologico della R. Universita di Genova. *Atti Univ. Genova*, **15**, 31–210, pls. 1–9.

Rutsch, R. (1940). Die gattung *Tubulostium* im Eocaen der Antillen. *Ecl. geol. Helv.* **32**, no. 2, 231–244, pl. 12.

Rutten, M. G. (1935). Larger Foraminifera of Northern Santa Clara Province, Cuba. *J. Paleont.* **9**, 527–545, pls. 59–62.

Rutten, M. G. (1941). Synopsis of the Orbitoididae. *Geol. Mijnbouw.*, 3e jahrg., no. 2, 34–62, pls. 1–2.

Sacco, F. (1906). Les Étages et les faunes du bassin tertiare du Piémont. *Bull. Soc. géol. Fr.* (4), **5**, 893–916, pls. 30–31, maps.

Schenck, H. G. & Frizzell, D. L. (1936). Subgeneric nomenclature in foraminifera. *Amer. J. Sci.* (5), **31**, 464–466.

Schlosser, M. (1922). Revision der Unteroligocänfauna von Häring und Reit im Winkel. *Neues Jb. Min. Geol. Pal., Stuttgart*, Beil. Band 47, 254–294.

Schubert, R. J. (1911). Die fossilen Foraminiferen des Bismarckarchipels und einiger angrenzender Inseln. *Geol. Reichsanst., Abh., Wien*, **20**, Heft 4, 1–130, pls. 1–6.

Schuchert, C. (1935). *Historical Geology of the Antillean-Caribbean Region.* John Wiley and Sons (*cum. bibl.*).

Senn, A. (1935). Die stratigraphische Verbreitung der Tertiären Orbitoididen. *Ecl. geol. Helv.* **28**, 51–113, 369–373, pls. 8–9.

Senn, A. (1940). Paleogene of Barbados and its bearing on history and structure of Antillean-Caribbean Region. *Bull. Amer. Ass. Petr. Geol.* **24**, 1548–1610.

Senn, A. (1948). Die Geologie der Insel Barbados B.W.I. (Kleine Antillen) und die Morphogenese der umliegenden marinen Grossformen. *Ecl. geol. Helv.* **40**, 199–222.

Smout, A. H. & Eames, F. E. (1958). The genus *Archaias* (foraminifera) and its stratigraphical distribution. *Palaeontology*, **1**, 207–225, pls. 39–42.

Socin, C. 1959 (1958). Una proposta per il limite cronologico Elveziano-Tortoniano. *Boll. Soc. Geol. Ital.* **77**, fasc. 1, 27–38.

Stainforth, R. M. (1948 a). Applied Micropaleontology in Coastal Ecuador. *J. Paleont.* **22**, 113–151, pls. 24–26.

Stainforth, R. M. (1948 b). Description, correlation, and paleoecology of Tertiary Cipero Marl Formation, Trinidad, B.W.I. *Bull. Amer. Ass. Petrol. Geol.* **32**, 1292–1330.

Stainforth, R. M. (1955). Ages of Tertiary formations in northwest Peru. *Bull. Amer. Ass. Petrol. Geol.* **39**, 2068–2077.

Stainforth, R. M. (1960). Current status of Transatlantic Oligo-Miocene correlation by means of planktonic foraminifera. *Rev. Micropaléontologie*, **2**, no. 4, 219–230, text-figs. (*cum bibl.*).

Stainforth, R. M. & Rüegg, W. (1953). Mid-Oligocene transgression in Southern Peru. *Bull. Amer. Ass. Petrol. Geol.* **37**, 568–569.

Stratigraphical Lexicon of Venezuela (English Edition) (1956). Dirección de Geología, Min. de Minas e Hidrocarburos. *Bol. Geol.* (Spec. Publ.), no. 1.

Szöts, E. (1956). La Limite entre le Paléogène et le Néogène et le problème des Étages Chattien et Aquitanien. *Acta Geol. Acad. Sci. Hung.* **4**, (2), 209–219.

Thalmann, H. (1948). Bibliography and index to the genera, species and varieties of foraminifera for the year 1946. *J. Paleont.* **22**, 193–221.

Thiadens, A. A. (1937). Cretaceous and tertiary foraminifera from Southern Santa Clara Province, Cuba. *J. Paleont.* **11**, 91–109, pls. 15–19.

Thomas, H. D. (1942). On fossils from Antigua, and the age of the Seaforth Limestone. *Geol. Mag.* **79**, 49–61, pls. 3–4.

Tobler, A. (1926). *Miogypsina* in untersten Neogen von Trinidad und Borneo. *Ecl. geol. Helv.* **19**, 719–722.

Todd, M. R. (1947). Vicksburg Oligocene foraminifera from Mississippi (Abstract). *Bull. Geol. Soc. Amer.* **58**, no. 12, pt. 2, 1233.

Todd, R. (1952). Vicksburg (Oligocene) smaller foraminifera from Mississippi. *U.S. Geol. Surv., Prof. Paper*, **241**, 1–47, pls. 1–6.

Todd, R. (1957). Geology of Saipan, Mariana Islands. Smaller foraminifera. *U.S. Geol. Surv., Prof. Paper*, **280**–H, 265–320, pls. 64–93.

Todd, R., Cloud, P. E., Low, D. & Schmidt, R. G. (1954). Probable occurrence of Oligocene on Saipan. *Amer. J. Sci.* **252**, 673–682, pl. 1.

Todd, R. & Post, R. (1954). Smaller foraminifera from Bikini drill holes. *U.S. Geol. Surv., Prof. Paper*, **260**–N, 547–568, pls. 198–203.

Travis, R. B. (1953). La Brea-Pariñas Oil Field, northwestern Peru. *Bull. Amer. Ass. Petrol. Geol.* **37**, 2093–2118.

Trechmann, C. T. (1935). The geology and fossils of Carriacou, West Indies. *Geol. Mag.* **72**, 529–555, pls. 20–22.

Trechmann, C. T. (1941). Some observations on the geology of Antigua, West Indies. *Geol. Mag.* **78**, 113–124.

Vaughan, T. W. (1919a). Fossil corals from Central America, Cuba, and Porto Rico, with an account of the American Tertiary, Pleistocene and Recent Coral Reefs. *U.S. Nat. Mus. Bull.* **103**, 189–507, pls. 68–152.

Vaughan, T. W. (1919b). The biologic character and geologic correlation of the sedimentary formations of Panama in their relation to the Geologic history of Central America and the West Indies. *U.S. Nat. Mus. Bull.* **103**, 547–612.

Vaughan, T. W. (1924a). Criteria and status of correlation and classification of Tertiary Deposits. *Bull. Geol. Soc. Amer.* **35**, 677–742.

Vaughan, T. W. (1924b). American and European Tertiary larger foraminifera. *Bull. Geol. Soc. Amer.* **35**, 785–822, pls. 30–36.

Vaughan, T. W. (1928). Species of large arenaceous and orbitoidal foraminifera from the Tertiary Deposits of Jamaica. *J. Paleont.* **1**, 277–298, pls. 43–50.

Vaughan, T. W. (1933). Studies of American species of foraminifera of the genus *Lepidocyclina*. *Smithson. Misc. Coll.* **89**, no. 10, 1–53, pls. 1–32.

Vaughan, T. W. & Cole, W. S. (1936). New Tertiary foraminifera of the genera *Operculina* and *Operculinoides* from North America and the West Indies. *Proc. U.S. Nat. Mus.* **83**, no. 2996, 487–496, pls. 35–38.

Vaughan, T. W. & Cole, W. S. (1941). Preliminary report on the Cretaceous and Tertiary larger foraminifera of Trinidad, B.W.I. *Geol. Soc. Amer., Special Papers*, no. 30, 1–137, pls. 1–45.

Venzo, S. (1937). La fauna Cattiana delle glauconie Bellunesi (Oligocene). *Mem. Ist. Geol. Univ. Padova*, **13**, 1–207, pls. 1–12.

Vlerk, I. M. van der (1950). Stratigraphy of the Caenozoic of the East Indies based on Foraminifera. *Proc. Int. Pal. Union. Rep. 18th Sess., Int. Geol. Congr.* 1948, pt. xv, 61.

Vlerk, I. M. van der (1955). Correlation of the Tertiary of the Far East and Europe. *Micropaleontology*, **1**, 72–75.

Vlerk, I. M. van der (1959). Problems and principles of Tertiary and Quaternary Stratigraphy. *Quart. J. Geol. Soc. Lond.* **115** (1), 49–63.

Vlerk, I. M. van der & Umbgrove, J. H. F. (1927). Tertiaire gidsforaminiferen van Nederlandsch Oost-Indië. *Dutch East Indies, Dienst Mijnb. Wetensch. Meded.* no. 6.

Wade, M. & Carter, A. N. (1957). The foraminiferal genus *Sherbornina* in southeastern Australia. *Micropaleontology*, **3**, no. 2, 155–164, pls. 1–3.

Weiss, L. (1955). Planktonic index foraminifera of northwestern Peru. *Micropaleontology*, **1**, 301–319, pls. 1–3.

Whipple, G. L. (1934). Larger foraminifera from Vitilevu, Fiji. *Bull. Bishop Mus., Honolulu*, no. 119, 141–153, pls. 19–23.

Wolff, W. (1897). Die Fauna der südbayerischen Oligocaenmolasse. *Palaeontographica*, **43**, 223–311, pls. 20–28.

Woodring, W. P. (1927). Marine Eocene deposits on the east slopes of the Venezuelan Andes. *Bull. Amer. Ass. Petrol. Geol.* **11**, 992–996.

REFERENCES

Woodring, W. P. (1957). Geology and paleontology of Canal Zone and Adjoining Parts of Panama. *U.S. Geol. Surv., Prof. Paper*, **306**–A, 1–145, pls. 3–23.

Woodring, W. P. (1958). Geology of Barro Colorado Island, Canal Zone. *Smithson. Misc. Coll.* **135**, no. 3, 1–39, pls. 1–3.

Woodring, W. P. (1959). Geology and Paleontology of Canal Zone and adjoining parts of Panama; description of Tertiary Mollusks (Gastropods: Vermetidae to Thaididae). *U.S. Geol. Surv., Prof. Paper*, **306**–B, 147–239, pls. 24–38.

Woodring, W. P. & Thompson, T. F. (1949). Tertiary formations of Panama Canal Zone and adjoining parts of Panama. *Bull. Amer. Ass. Petrol. Geol.* **33**, 223–247.

EXPLANATION OF PLATES

PLATE I

Comparison of *Nummulites* and *Palaeonummulites*; all × 40.

A, B. *Nummulites vascus* Joly & Leymerie; Oligocene, Cyrenaica; specimens deposited in the British Museum (Natural History), registered numbers P. 44493 and P. 44494 respectively.

C–E. *Palaeonummulites cumingi* (Carpenter); Recent, Port Moresby, Papua; specimens deposited in the British Museum (Natural History), registered numbers 1960 6.1–2; E, young stage of specimen illustrated by D.

F–H. *Palaeonummulites pristinus* (Brady); ?Carrière du Fond d'Arquet, Namur; ?Wemmelian (Bartonian). F, equatorial section, BM(NH), registered number P. 35504; G, split specimen (after Brady, 1874), BM(NH), no. P. 35412; H, axial section (after Brady, 1874), BM(NH), no. P. 35503; all three specimens are syntypic.

All the specimens illustrated here are megalospheric forms.

PLATE II

The Aquitanian of Malta (A–D) and Aquitaine (E).

A, Natural association of *Eulepidina*, *Miogypsinella* and *Spiroclypeus*; Lower Coralline limestone, Ghar Lapsi, Malta; sample WACR 266; × 12·5.

B–D. Details of A, all × 30, showing, respectively, *Miogypsinella* (off-centre axial section), *Spiroclypeus* (skew section) and *Eulepidina* (axial section of flange), and a near-centre axial section of *Miogypsinella*.

E. *Miogypsinella complanata* (Schlumberger), equatorial section of nepionic spiral; Escornebéou, Aquitaine; sample FWA 66; × 30.

PLATE III

Natural associations of *Miogypsina* (*s.s.*) with *Miogypsinella* and *Miogypsinoides*; Aquitanian of Sicily and *e* stage (Kereruan) of Papua.

A. Ragusa limestone formation, Ragusa Platform area, south-east Sicily; sample HRW(S) 27; × 30. Compare Pl. *C*, fig. 1, Cuvillier & Sacal (1956).

B. Kaban limestone formation, Upper Sirebi River district, Papua; sample 156 KH; × 30; *Eulepidina* (unfigured) is also present in this sample.

PLATE IV

Natural association of *Eulepidina*, *Nephrolepidina*, *Miogypsina* (*s.s.*), *Miogypsinoides* and *Spiroclypeus*; Aquitanian of Tanganyika.

A. Sample FCRM 2013, from a stream south-west of Mtwero, Lindi district, south Tanganyika; × 12·5.

B, C. Details of the same thin section, both × 30. B, *Miogypsina* and *Miogypsinoides*; C, *Miogypsinoides*, *Eulepidina* and *Spiroclypeus*.

PLATE V

Natural association of *Eulepidina*, *Nephrolepidina*, *Miogypsinoides* and *Spiroclypeus*, and of *Miogypsina* (*s.s.*) and *Spiroclypeus*; from the *e* stage (Kereruan) of Papua.

A. Kaban limestone formation, Upper Sirebi River district, Papua; sample 158 KH, × 30; *Miogypsinoides*, *Spiroclypeus* and *Eulepidina* flanges.

B. Kaban limestone formation, northern Darai Hills, Kanau Creek area, Papua (6° 50′ S., 143° 5′ E.); sample 383 KO; × 30; axial sections of *Miogypsina* (*s.s.*) and *Spiroclypeus*.

C. Murufi limestone, Central Highlands, Papua; sample 360 HC; × 10; *Eulepidina* in axial section.

D. Detail of C, showing *Spiroclypeus* in association with *Eulepidina*; × 30.

PLATE VI

The Burdigalian of East Africa and the f_{1-2} stage (Taurian) of Papua.

A, B. Natural association of *Austrotrillina howchini* and *Taberina malabarica*; Burdigalian, Kenya, north of Sabaki River, 10 miles west of Mambrui; both figures from one thin section of sample REL 3166; × 30.

C. Natural association of *Flosculinella bontangensis* with *Miogypsina* (*s.s.*) and *Nephrolepidina*; Burdigalian, Zanzibar Protectorate, near Chake-Chake, Pemba Island; sample APT 309; × 30.

D, E. Natural association of *Alveolinella*, *Austrotrillina howchini* and *Miogypsina* (*s.s.*); Darai limestone formation, f_{1-2} stage (Taurian), Omati anticline, Darai Hills, Papua; both from one thin section of sample 687 KI; × 30.

PLATE VII

The Burdigalian of East Africa and the f_{1-2} stage (Taurian) of Papua.

A, B. Natural association of *Taberina malabarica* and *Miogypsina* (*s.s.*); Omati Well No. 1 (7° 25′ S., 143° 57′ E.), Darai Hills, Omati River, Papua; Darai limestone formation; orientated thin-sections from the same core-sample; × 30.

C, D. Natural association (parts of the same thin section) of *Miogypsina* (*s.s.*) and *Meandropsina anahensis*; Burdigalian limestone, near Mkakara, Mtwara-Mikindani district, Tanganyika; sample LOG 86; × 30.

E, F. Natural association (parts of the same thin-section) of *Flosculinella bontangensis* and *Borelis melo*; Burdigalian limestone, near Chanjani, Pemba Island, Zanzibar Protectorate; sample APT 406; × 30.

PLATE I

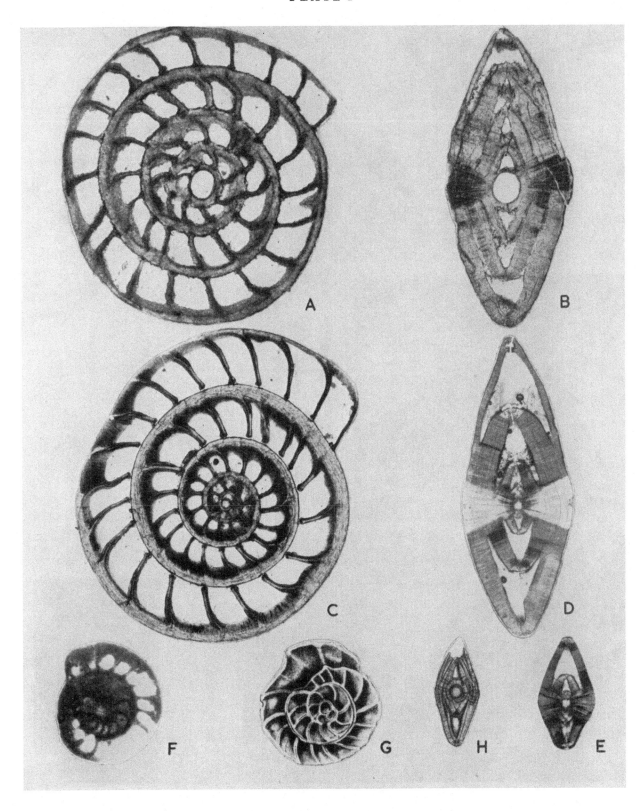

PLATE II

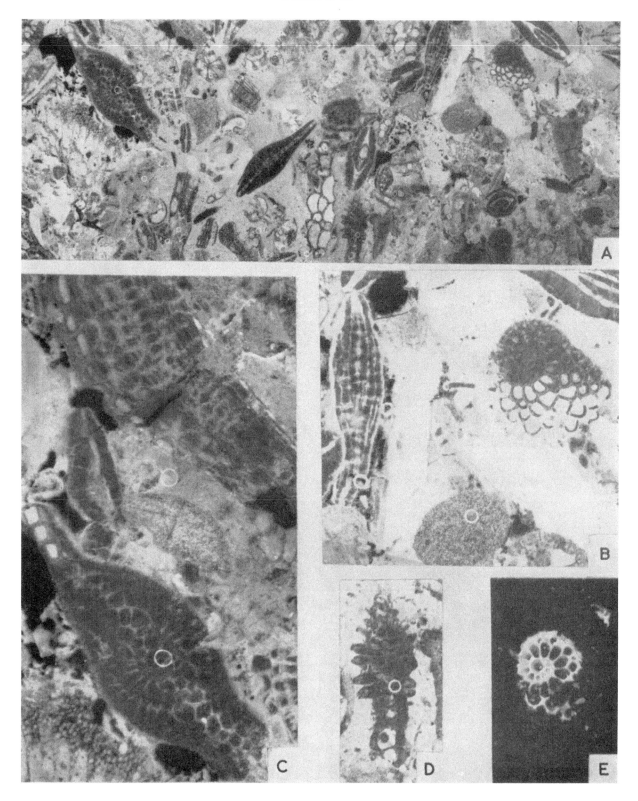

PLATE III

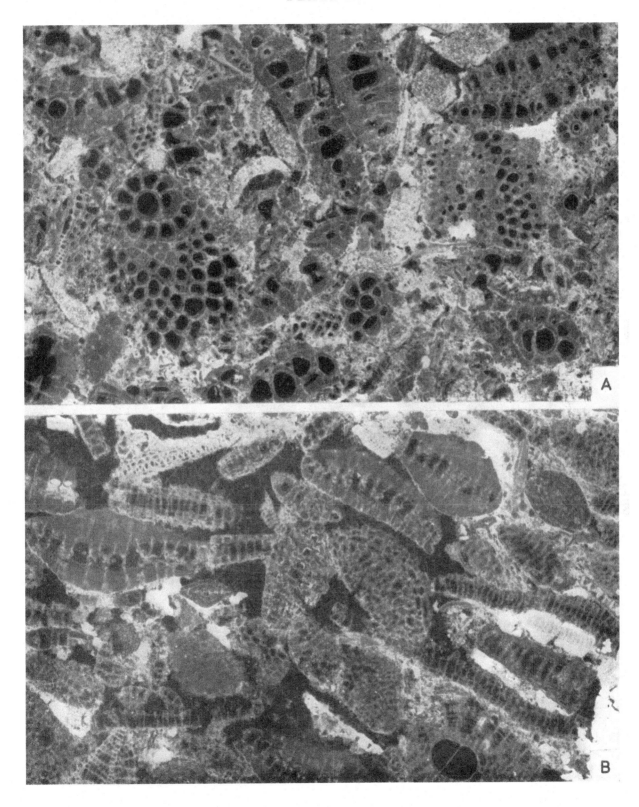

PLATE IV

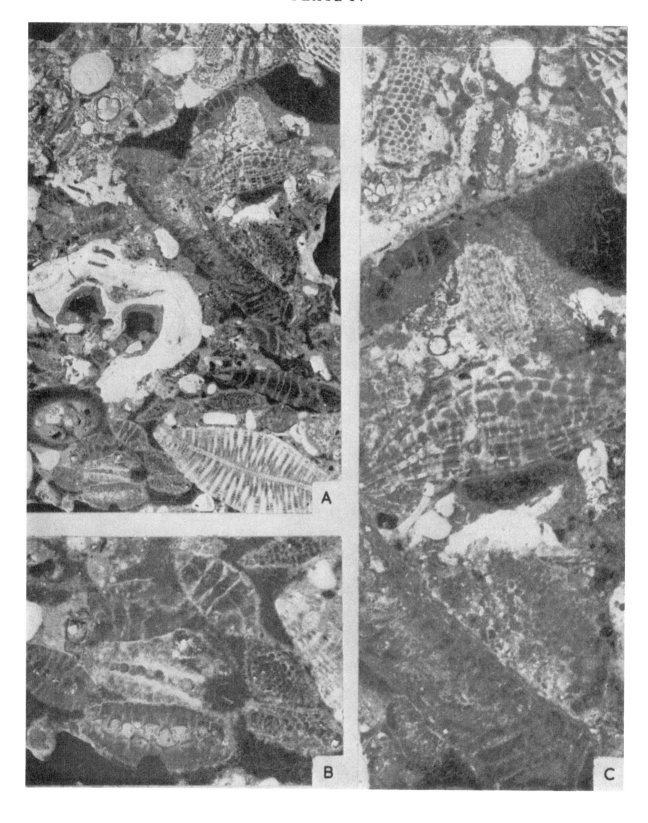

PLATE V

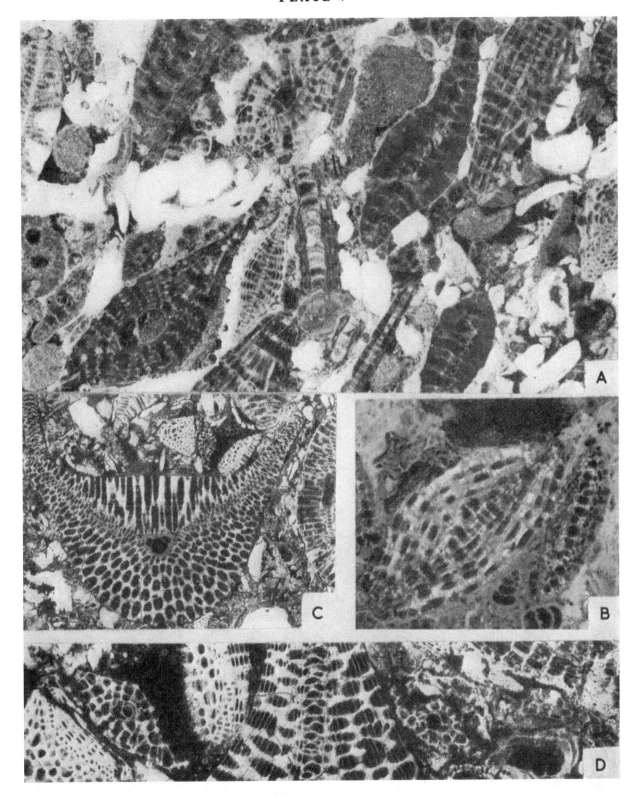

PLATE VI

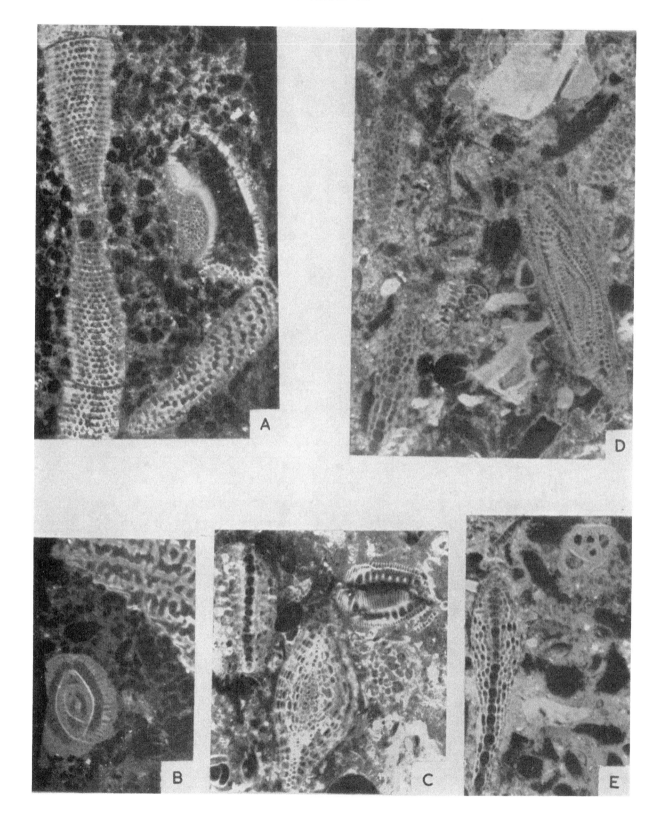

PLATE VII

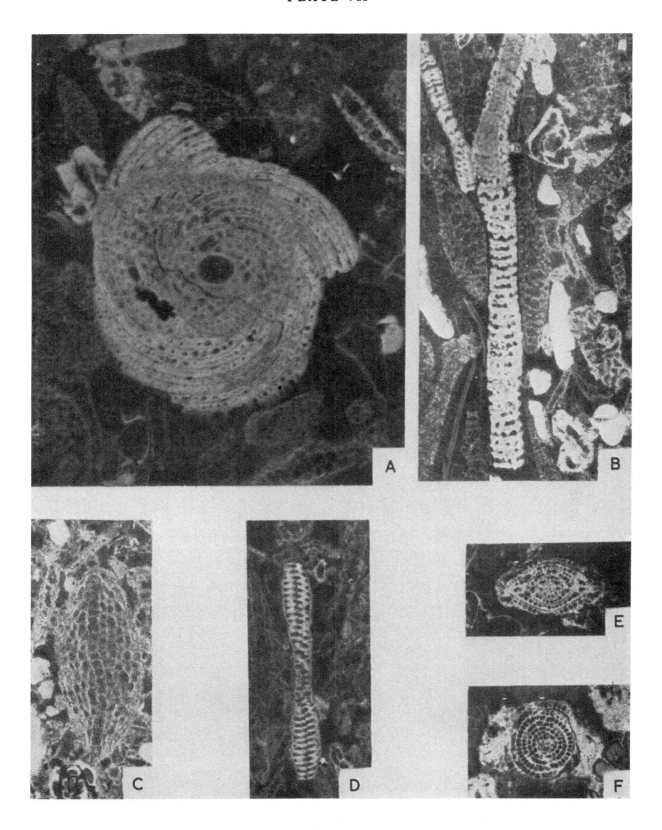

PART 2

THE MID-TERTIARY (UPPER EOCENE TO AQUITANIAN) GLOBIGERINACEAE

ABSTRACT. This part describes and discusses the important genera and species of the Globigerinaceae which are of great stratigraphical importance for the correlation of Mid-Tertiary sediments. In Tanganyika richly fossiliferous sediments in an area of simple geological structure enable a direct correlation to be made between abundant and distinctive assemblages of planktonic foraminifera and well-known stratigraphically important larger foraminifera. A series of biozones are proposed for the interval of the Upper Eocene, Oligocene and lower Aquitanian, and the species and assemblages are compared to and correlated with sediments comprising the same time-interval in Southern Europe, the Mediterranean and, in particular, the Caribbean Region. The results of this study confirm the conclusions reached by Eames, Banner, Blow and Clarke in Part 1 of this work which indicates the absence of the upper Bartonian and the whole of the Oligocene in the well-known stratigraphical sections of the Central American Region.

I. Introduction

Planktonic foraminifera, as applied to Tertiary biostratigraphy, have received their chief study from the outstanding researches of Bolli (1950, 1951, 1957a–c) and the major use of these most valuable stratigraphical indices has, so far, been mainly confined to work in the Caribbean Region (Bolli, 1950, 1951, 1957a–c; Beckmann, 1953; Blow, 1959), although some records from other areas exist, notably in New Zealand (Hornibrook, 1958). However, it is in Trinidad that the most detailed sequences have been documented in beds dating from Palaeocene to Middle Miocene. Indeed, owing to the detailed work of Bolli, Trinidad must, for the present, remain a standard of reference notwithstanding that this island is a far from ideal locality for detailed biostratigraphical studies. The problems resulting from both post-depositional tectonic complications and deposition in a sedimentary environment of high tectonic intensity are appreciated both by Bolli and the present writers. One of us (Blow, 1959) has already discussed some aspects of these problems for the Miocene sediments of Trinidad as seen in the light of evidence from eastern Falcón, Venezuela. Kugler (1953) has discussed many of the problems associated with sedimentation in an environment of high tectonic

intensity, such as reworking of foraminifera, 'wild-flysch' sedimentation, the occurrence of 'morros' and rapid change of facies. Kugler's work (1953) concerning these problems in the Cretaceous and Tertiary strata of southern Trinidad has been taken into full account when correlations between the Lindi area and southern Trinidad are discussed below (see also Bolli, 1957a–c).

This present Part concentrates its attention on the biostratigraphy of the Upper Eocene, Oligocene and lowermost Miocene (Aquitanian) sediments in the Lindi area, Tanganyika, East Africa (Figs. 6, 7) and examines the biostratigraphy, proposed by Bolli (1957b, c), based on planktonic foraminifera from the uppermost Navet formation (Hospital Hill marl), San Fernando 'formation' (=Mount Moriah formation of authors) and the lowermost part of the Cipero formation (all in southern Trinidad) in the light of the evidence seen in the Lindi area. It is this part of the Tertiary succession in which the greatest discrepancies are seen between the Trinidad and East African sequences, and it is deduced here that a large and important unconformity is present in southern Trinidad. It is also considered that the San Fernando 'formation' (and the 'Globorotalia cocoaensis Zone' of Bolli, 1957c) consists of two distinct parts, the upper comprising

61

a reworked series of beds which, although initially deposited in Upper Eocene times, have been greatly disturbed, fragmented and redeposited as the basal conglomerate of the Cipero formation. This conglomerate, referred to here as the 'San Fernando conglomerate', is, together with the overlying 'Mount Moriah silt', considered to be related to the Aquitanian (Lower Miocene) transgression. This is discussed in more detail below, but reference should be made to the detailed re-appraisal of Eames, Banner, Blow and Clarke (in Part 1 of this work) concerning the age of the San Fernando formation (Mount Moriah) and the overlying lower Cipero formation; this study serves to emphasize the essential correctness of their conclusions and independent deductions.

II. Acknowledgments

The material from coastal Tanganyika described in this study results from the work of the BP–Shell Development Co. of Tanganyika Ltd, and we wish to thank this Company for enabling us to use these samples which were collected by F.C.R. Martin, geologist of the British Petroleum Co., Ltd. Because of the broad scope of this work it has been necessary to compare our material and results with those of other workers far afield. Accordingly, we wish to thank Dr D. G. Jenkins (New Zealand), Dr F. W. Anderson (H.M. Geological Survey of Great Britain), Dr T. Barnard (University College, London), Dr J. P. Beckmann (Pan-American Oil Co., Tehran), Miss Ruth Todd (U.S. Geological Survey, Washington), Professor H. V. Howe (Louisiana State University), Dr O. L. Bandy (University of Southern California), Dr E. Gasche (Naturhistorisches Museum, Basel, Switzerland), Dr F. L. Parker (Scripps Institution, La Jolla, California) and Drs C. G. Adams and R. Hedley (British Museum, Natural History, London) for the presentation or loan of type and comparative material. We are also grateful to many workers with whom we have discussed the problems involved and these particularly include Dr H. M. Bolli (Caracas), G. E. Higgins (Trinidad), W. E. Crews (Caracas), J. B. Saunders (Trinidad), Dr H. G. Kugler (lately of Trinidad, now Switzerland), Dr R. M. Stainforth (Venezuela) and Dr V. Pokorný (Prague).

We also wish to thank the Chief Geologist (N. L. Falcon, F.R.S.) and our colleagues of the British Petroleum Co., Ltd (Exploration Division) for their encouragement and assistance; F. R. Gnauck has helped with the photographic illustrations and D. J. Scandrett has drafted the maps and charts.

III. The Tanganyikan Middle Tertiary (Upper Eocene to Lower Miocene)

A. GENERAL GEOLOGY

The earliest work in the Lindi area was carried out by W. Bornhardt during the years 1896–97 whilst Tanganyika was under German administration. Other German workers included Fraas, Fahrion, Meyer and Hennig. In more recent years geologists of the British Colonial Surveys worked in the area but were mainly concerned with problems associated with water supply; these geologists included G. M. Stockley (1926[1]), F. R. Wade (1934[1]), and A. Cawley (1937[1]). Cawley (1937) was amongst the first to discuss some of the main structural elements of the region, but his Kitulo syncline in the immediate Lindi Town area is now shown by BP–Shell geologists to be non-existent. The most intensive work in the area of coastal Tanganyika has been by oil geologists employed by the BP–Shell Development Co. of Tanganyika Ltd, during the years 1951–57. These BP–Shell geologists, including, amongst others, Dr P. E. Kent and F. C. R. Martin, have written detailed private reports on the Lindi area. However, it is from the private report to BP–Shell by F. C. R. Martin (1957) that most of the geological information incorporated in this paper has been taken. Grateful acknowledgment is made here to BP–Shell for permission to use this private report. Palaeontological work for the Lindi area has been carried out on a provisional and reconnaissance basis by Dr E. Schijfsma, F. Dilley, Miss B. Graham and Miss S. de Renzy; the more detailed investigations of molluscan and larger foraminiferal faunas were made by Drs F. E. Eames and W. J. Clarke in London, whilst the detailed study of the smaller foraminiferal faunas has been the responsibility of the present authors.

[1] Unpublished, but available, reports to H.M. Colonial Geological Survey.

In the Lindi embayment (see Figs. 6 and 7) of East Africa, Cretaceous to Pliocene marine sediments are known, overlying the basement complex and upper Jurassic sediments to the west (which may not all be entirely marine). Non-marine sediments may also occur in the Lower Cretaceous. The marine Cretaceous sediments range in age from Aptian to Maestrichtian with possibly a thin and impersistent development of Danian. These Cretaceous sediments are overlain by Palaeocene to Burdigalian sediments with isolated outliers of Pliocene and subrecent deposits. The distribution of Palaeocene to Pliocene sediments in the Lindi area is shown on Fig. 7. Excellent planktonic foraminiferal faunas are present in the Cretaceous and Tertiary succession but only the Upper Eocene to lowermost Miocene is considered here. It is hoped to describe the Cretaceous to Middle Eocene faunas and the resulting biostratigraphy at a later date.

Before discussing the conditions of sedimentation and the broad outlines of the regional structure of the immediate area of Lindi, a review of the lithology for the upper part of the Middle Eocene, Upper Eocene, Oligocene and Lower Miocene is given below; this has been summarized from F. C. R. Martin's private report to BP–Shell (see Fig. 8).

(a) Middle Eocene
The Middle Eocene comprises a series of 'slabby' foraminiferal limestones, partly hard, partly marly, with buff silty marls and buff siltstones. The limestones are usually laterally impersistent and contain great quantities of *Nummulites* spp. and *Alveolina* spp.

(b) Upper Eocene
The Upper Eocene succession is little different from that of the Middle Eocene. The lowest member is a sandy reefal limestone with interbedded layers of soft marly clays. Overlying the sandy reefal limestones is a series of interbedded buff-grey clays, buff siltstones and thin (1–2 ft), hard, sometimes siliceous, foraminiferal limestones. This succession is well exposed along the Madingura River and the Kitunda Cliff section (see Fig. 7). The thickness of the Upper Eocene is estimated by F. C. R. Martin at about 220 ft.

(c) Oligocene
The Oligocene is represented by a series of impersistent, rubbly and silty limestones, soft marly limestones, buff-grey marly clays and buff silts. In general there is very little facies change as compared with the underlying Upper Eocene. The thickness of the Oligocene is estimated by F. C. R. Martin at about 290 ft.

(d) Lower Miocene (Aquitanian-Burdigalian)
The lowest Miocene beds consist of about 15 ft of soft buff silty sandstones which become increasingly calcareous upwards and contain some lenses of gypsiferous clays. In the area immediately to the east of Lindi Town (Kitunda Cliff) the silty sandstones pass upwards into massive reefal limestones but, north of Lindi Bay, clays and silts appear to replace part of the limestones.

The whole succession discussed above has a regional dip to the E.N.E. not exceeding 10°. According to Martin, sedimentation appears to have been continuous from Upper Eocene times into the overlying Oligocene and only a slight change of facies is discernible in the field. Palaeontological evidence suggests that this was probably so in some areas, but in other areas a small break is suggested with probably a part of the Lattorfian missing. Again, according to Martin, the Lower Miocene is distinctly transgressive in the area to the south of Lindi Bay (in the region of Kitunda), but field evidence suggests it is apparently conformable upon the Oligocene north of Lindi Bay. However, palaeontological evidence shows that nowhere is the Chattian present and the Aquitanian overlies the well-developed Rupelian (see below, p. 67).

The conditions under which the Upper Eocene to Lower Miocene sediments were laid down seem to have been in environments of low tectonic intensity. In the Upper Eocene and Oligocene the sea appears to have been reasonably clear and not very deep but with intermittent periods of more turbid conditions. Laterally discontinuous reef-like conditions were only established for short periods of time. In the Lower Miocene, reefal conditions became more general, but, nevertheless, conditions were such that interbedded silts, sands and clays still occurred. The region seems to have undergone gentle over-all subsidence with no violent movements, and the whole sequence of Cretaceous-Tertiary sedimentation seems to have been associated with the various cycles of broad gentle warping (as noted from geomorphological features developed on the basement complexes of the interior plateaux seen in Kenya, Western Tanganyika, and other areas of Central Africa and as discussed by Dixey (1956, and unpublished lecture to Geological Society, London, 1959)). To summarize, the Eocene to Miocene sediments were

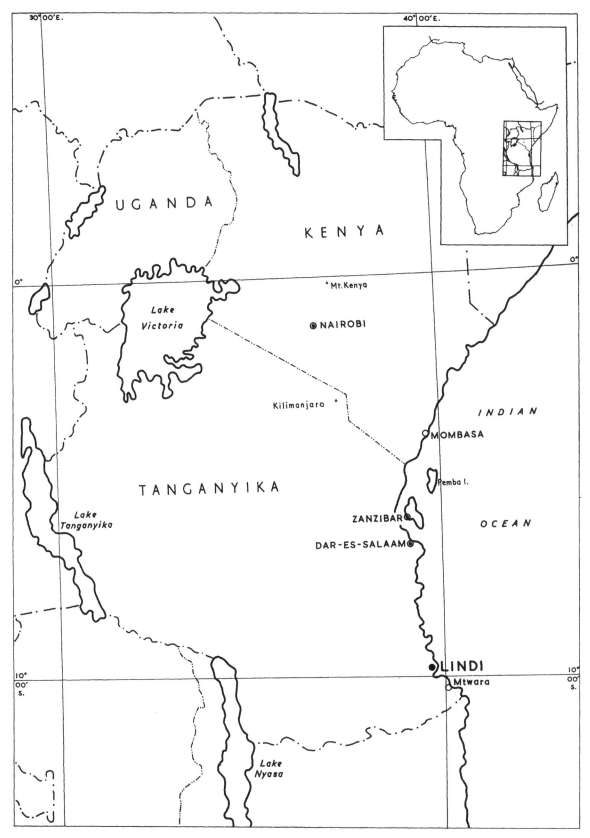

Fig. 6. Map of East Africa, showing the position of Lindi.

64

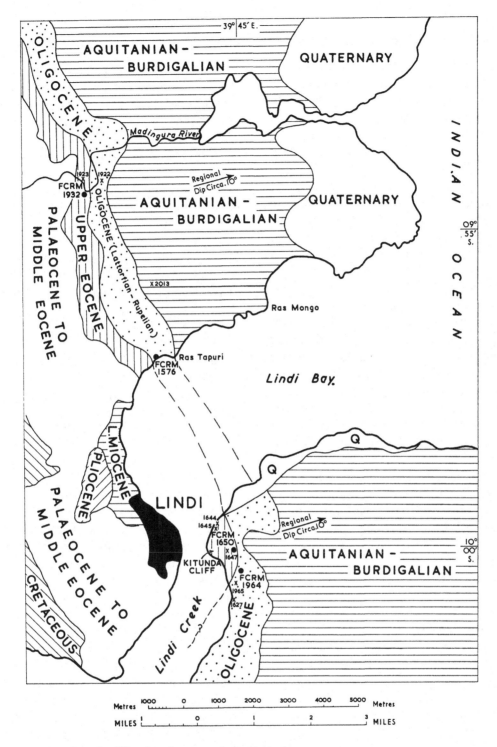

Fig. 7. Geological map of the Lindi area, Tanganyika, showing the type localities of the new biozones, and the sample-localities for those foraminifera which are now illustrated (hypotypes).

deposited on a slowly subsiding continental shelf which was stable and receiving little sediment. Sedimentation was nearly continuous but fairly slow so that no great thicknesses of sediment were built up for the interval.

B. GENERAL PALAEONTOLOGY

The different lithofacies in the Middle Eocene, Upper Eocene, Oligocene and Lower Miocene are reflected in the varying biofacies for these horizons. The presence of impersistent reef-like limestones, silts and foraminiferal clays makes it possible to correlate the various biofacies, containing different suites of organisms, one with the other. It is obvious that this type of sedimentation makes it a comparatively easy matter to check the occurrence of one fossil group against another; it is particularly instructive and valuable to do this for the larger foraminifera and planktonic foraminifera, both groups having been used for biostratigraphical studies in many different areas of the world. In this section of this Part the broad outlines of the occurrences of the stratigraphically more important larger foraminifera are noted. A few of the more important larger foraminifera seen in the Upper Eocene, Oligocene and Lower Miocene are illustrated on Plate VIII. The planktonic foraminifera are dealt with in a succeeding section below. It is emphasized that a large number of samples from the Lindi area have been investigated (over 300 samples were collected in the area of Fig. 7 alone).

(a) Middle Eocene (Lutetian)

The important larger foraminifera occurring in beds which are dated as Middle Eocene include the following:

Alveolina oblonga d'Orbigny, 1826
Amphistegina rotula (Kaufmann), 1867
Lockhartia tipperi (Davies), 1926
Nummulites beaumonti d'Archaic & Haime, 1853
N. discorbinus (Schlotheim), 1820
N. laevigatus (Bruguière), 1792
N. obtusus uranensis (de la Harpe), 1883
N. subatacicus H. Douvillé, 1919
Orbitolites complanatus Lamarck, 1801
Somalina stefaninii Silvestri, 1939

together with species of *Discocyclina* and other *Nummulites*. These larger foraminifera are asso-

ciated with planktonic foraminifera such as *Globorotalia spinulosa*, *Porticulasphaera mexicana* and *Truncorotaloides rohri*, as well as other forms described by Bolli (1957c) from the middle part of the Navet formation in Trinidad. Although the full details of the occurrences of the planktonic foraminifera have not yet been evaluated for the Middle Eocene in East Africa, preliminary results indicate that Bolli's (1957c) middle Navet zones may be recognized in beds containing the above larger foraminiferal assemblages. Possibly the Upper Eocene/Middle Eocene boundary falls within the *Truncorotaloides rohri* Zone (Bolli, 1957c), but, for convenience, it is taken at the top of this zone for the time being.

(b) Upper Eocene (Auversian–Bartonian)

The larger foraminifera noted in beds dated here as Upper Eocene include:

Chapmanina gassinensis (Silvestri), 1905
Nummulites fabianii (Prever), 1905
N. hormoensis Nuttall & Brighton, 1931
Pellatispira madaraszi (von Hantken), 1875

together with species of *Aktinocyclina*, *Discocyclina*, *Baculogypsinoides*, *Halkyardia* and *Eorupertia*. Plate VIII illustrates part of the larger foraminiferal fauna of one particular sample (FCRM 1650) which shows, by its content of Upper Eocene larger foraminifera, that the new planktonic foraminiferal fauna also present (of the, here proposed, *Globigerina turritilina turritilina* Zone) is still Upper Eocene.

(c) Oligocene (Lattorfian–Rupelian)

No palaeontological evidence has been seen in East Africa, either on foraminiferal and molluscan evidence, or from any other fossil group, which could suggest the presence of Chattian. Oligocene beds, so far seen in East Africa, seem to be referable to the Lower and Middle Oligocene (Lattorfian to Rupelian). Taken as a whole, the following larger foraminifera have been recorded from the combined Lattorfian and Rupelian:

Lepidocyclina (Eulepidina) dilatata (Michelotti), 1861
L. (Eulepidina) ephippioides (Jones & Chapman), 1900
L. (Nephrolepidina) marginata (Michelotti), 1841
L. (Nephrolepidina) tournoueri (Lem. & Douv.), 1904
Nummulites intermedius (d'Archaic), 1846

N. vascus (Joly & Leymerie), 1848
Operculina complanata (Defrance), 1842
Palaeonummulites incrassatus (de la Harpe), 1883
P. tournoueri (de la Harpe), 1879

The Lattorfian is distinguished from the Rupelian by the absence of *Lepidocyclina (Eulepidina)*, but it is pointed out here that both the Lattorfian and Rupelian are characterized by the occurrence of true *Nummulites*, especially the reticulate *N. intermedius–fichteli*. Plate VIII illustrates the occurrence of *N. intermedius* and *Lepidocyclina (Eulepidina)* in one sample (FCRM 1576), which also has the new planktonic foraminiferal fauna (*Globigerina oligocaenica* Zone, here proposed) described in detail below. As implied above, both the Lattorfian and the Rupelian stages have been recognized in the Lindi area, but so far it has not yet been possible to trace a continuous boundary between the two stages on the map. Most of the Oligocene recognized in East Africa seems to belong to the Rupelian. However, planktonic foraminiferal faunas are associated with both the Lattorfian and Rupelian samples.

(d) Lower Miocene (Aquitanian)

The Lower Miocene larger foraminifera from nearby Pemba Island have been described by A. M. Davies (1927); his assemblages are similar to, but probably younger than, those recorded here; their Miocene age was confirmed by the associated Mollusca (Cox, 1927) and Echinoidea, which included many species of the genus *Clypeaster* (Stockley, 1927).

The following larger foraminifera have been observed in the Aquitanian of the Lindi area:

Borelis pygmaea (Hanzawa), 1930
Lepidocyclina (Eulepidina) ephippioides (Jones & Chapman), 1900
    L. (Eulepidina) dilatata (Michelotti), 1861
    L. (Nephrolepidina) marginata (Michelotti), 1841
    L. (Nephrolepidina) tournoueri (Lem. & Douv.), 1904
Miogypsina globulina (Michelotti), 1841
Miogypsinella complanata (Schlumberger), 1900
Miogypsinoides dehaarti (van der Vlerk), 1924
Spiroclypeus margaritatus (Schlumberger), 1902

together with other species of *Miogypsina (s.s.)*, *Operculina* and *Heterostegina*. It should be noted that *Miogypsina (s.s.)* occurs at the base of the limestones ascribed to the Aquitanian, and Plate IV A–C

illustrates the microfauna of significant larger foraminifera occurring in one sample (FCRM 2013) taken from near the base of the limestones. Planktonic foraminifera also occur in the Aquitanian, but these are few in number and poor in preservation below the level of about the *Globigerina ouachitaensis ciperoensis* Zone. Thereafter the planktonic foraminiferal faunas become good and the evolution of the *Globigerinoides quadrilobatus* group (see p. 136) has been noted in about the middle part of the Aquitanian succession. This evolution also occurs in the *Globorotalia kugleri* Zone of the Cipero formation, Trinidad (Bolli, 1957*b*; Blow, 1959).

C. PLANKTONIC FORAMINIFERAL BIOZONES

The planktonic foraminiferal faunas are described in detail in the section dealing with 'Systematic Palaeontology', later in this Part, whilst the correlations with southern Trinidad and the evolutionary relationships of some of the planktonic foraminifera are also described below under the appropriate headings. However, some new biozones for the Upper Eocene and Oligocene are proposed here and discussed in some detail with the main diagnostic Globigerinaceae noted. The complete faunal content of each of the planktonic biozones is shown on the Range Chart (Fig. 17).

(a) Truncorotaloides rohri Zone (Bolli, 1957c) (= uppermost Middle Eocene)

This zone was proposed by Bolli for faunas seen in samples from part of the Navet formation. Similar assemblages occur in the Lindi area but are not discussed here.

(b) Globigerapsis semi-involuta Zone (Bolli, 1957c) (= lowest Upper Eocene)

Bolli (1957*c*) erected this zone for faunas seen in samples from the 'Hospital Hill marl' member of the Navet formation, southern Trinidad, and virtually identical faunas are seen in the lower part of the Upper Eocene beds of the Lindi area. These beds contain the larger foraminiferal faunas recorded above. The planktonic fauna is characterized by *Globigerapsis semi-involuta* (Keijzer) which is strictly limited to this interval, whilst *Globigerinatheka lindiensis* Blow & Banner sp.nov. is also common and limited to the interval. *Globorotalia cerro-azulensis* Cole (=*G. cocoaensis* Cushman) first occurs at the base of the zone. *Pseudohastigerina micra* (Cole) and *Hantkenina alabamensis* Cushman are both common. Morphologically primitive forms of *Cribrohantkenina danvillensis* (Howe & Wallace) first appear near the middle part of the interval. *Globigerina turritilina praeturritilina* Blow & Banner sp.nov. occurs in this zone whilst the evolution of *G. tripartita tripartita* (Koch) from *G. yeguaensis yeguaensis* Weinzierl & Applin occurs at the top of the interval. At the top of the zone some forms appearing to be transitional between *Globorotalia centralis* and *Globigerina pseudoampliapertura* Blow & Banner sp.nov. make their first appearance. No forms referable to the genus *Truncorotaloides* or any strongly hispid species of *Globorotalia* are present in the interval; indeed, all the species of *Globorotalia* present are relatively smooth forms of the subgenus *Turborotalia*. *Globigerinita africana* Blow & Banner sp.nov. and *G. globiformis* sp.nov. occur in this (and the lower part of the overlying) zone.

(c) *Cribrohantkenina danvillensis* Zone (here proposed) (=middle part Upper Eocene)

This zone is characterized by the presence of *Hantkenina* and *Cribrohantkenina*, and the complete absence of any species referable to the genera *Globigerinatheka* and *Globigerapsis*. These latter forms become extinct at the top of the underlying interval. *Globorotalia cerro-azulensis* is particularly common, and large forms with perforate pseudocarinae are well developed. *Globigerina linaperta linaperta* Finlay and *G. linaperta pseudoeocaena* (Subbotina) are common. *G. pseudoampliapertura* Blow & Banner sp.nov. is abundant, while typical *Globorotalia centralis* does not seem to range quite to the top of the zone. The first typical specimens of *G. increbescens* (Bandy) are noted near the base of the interval. No forms referable to *Globigerina ampliapertura ampliapertura* Bolli occur in this zone.

Type locality FCRM 1932, south of the Madingura River, Lindi area, East Africa (see Fig. 7).

The validity of Bolli's '*Globorotalia cocoaensis*' Zone, described and proposed from both the 'San Fernando formation' and the 'Mount Moriah silts' of southern Trinidad, is discussed below.

(d) *Globigerina turritilina turritilina* Zone (here proposed) (=uppermost Upper Eocene)

The occurrence of *Discocyclina* and *Nummulites hormoensis* (see Plate VIII) indicates that this zone is still within the Upper Eocene. The planktonic fauna is very much reduced in the number of species, but it is very rich in numbers of specimens. No forms referable to *Globorotalia centralis*, *G. cerro-azulensis*, *Hantkenina* or *Cribrohantkenina* are seen in the faunas. *Globigerina tripartita tripartita* Koch (=*G. rohri* Bolli) is common, and, together with *G. pseudoampliapertura* Blow & Banner sp.nov., makes up a considerable proportion of the total planktonic foraminiferal fauna of the interval. Typical *G. angustiumbilicata* Bolli first appears in the zone, but no forms referable to *G. ouachitaensis ciperoensis* (Bolli) occur. *G. linaperta linaperta* also occurs in the zone and does not become extinct at the top of the *Globigerapsis semi-involuta* Zone as shown by Bolli (1957c, p. 159). *Globigerina yeguaensis yeguaensis* and *G. yeguaensis pseudovenezuelana* Blow & Banner subsp.nov. are both common. *Pseudohastigerina micra* (Cole) occurs, but no forms referable to the genus *Cassigerinella* Pokorný are found. *Globigerina ouachitaensis gnaucki* Blow & Banner subsp.nov. and *G. praebulloides leroyi* Blow & Banner subsp.nov. first occur at the base of the interval. *G. turritilina turritilina* develops from *G. turritilina praeturritilina* at the base of the zone and remains fairly common within it. *Globorotalia increbescens* (Bandy) is fairly common and *Globigerina ampliapertura ampliapertura* Bolli first appears in the interval with many transitional forms present between Bandy's and Bolli's species.

Type locality FCRM 1650, Kitunda Cliff, Lindi area, East Africa (see Fig. 7).

(e) *Globigerina oligocaenica* Zone (here proposed) (=Lower to Middle Oligocene)

This zone is characterized by the occurrence of *Globigerina oligocaenica* Blow & Banner sp.nov. which is limited to its zone. *Cassigerinella chipo-*

lensis (Cushman & Ponton) is abundant and appears for the first time. *Pseudohastigerina micra* occurs and the overlap in the range of *Cassigerinella chipolensis* and *Pseudohastigerina micra* seems to define the zone. *Globorotalia increbescens* is present, as well as *Globigerina ampliapertura ampliapertura*, and transitional forms between these two species occur. *G. turritilina turritilina* is common, whilst *Globorotalia permicra* also occurs in this interval. *Globigerina ouachitaensis gnaucki* continues from the underlying zone, but *G. ouachitaensis ciperoensis* (Bolli) seems first to occur above the base of the zone. Indeed, it may be possible to erect two sub-zones for the Lattorfian–Rupelian *G. oligocaenica* Zone based on the occurrence of *G. ouachitaensis ciperoensis*. However, more work requires to be done before this can be formally proposed.

Globigerina yeguaensis yeguaensis is common and continues from the underlying Upper Eocene, whilst a rare form referred to here as *G.* aff. *yeguaensis* appears to be restricted to this interval.

Globigerinita dissimilis dissimilis (Cushman & Bermúdez) continues from the underlying Upper Eocene, but no forms referable to the new subspecies *G. dissimilis ciperoensis* Blow & Banner occur. *G. pera* (Todd) and *G. martini scandretti* Blow & Banner sp.nov., subsp.nov. are both common, the latter being restricted to this zone. *Globorotaloides suteri* Bolli, although not restricted to this interval, is particularly common.

Type locality FCRM 1964, Kitunda Cliff. Co-type locality FCRM 1576, near Ras Tapuri, Lindi area, East Africa (see Fig. 7).

(*f*) *Globigerina ampliapertura* Zone (Bolli, 1957*b*) (here considered as the basal zone of the Aquitanian)

In order to make this work as objective as possible the planktonic foraminiferal fauna of Bolli's type locality of this zone, near San Fernando Railway Station (Trinidad), was examined in great detail by the present authors. The results of this study show that, with the exception of some very small species of *Globigerina* and *Globorotalia* (*Turborotalia*), little can be added to Bolli's published description of the fauna (Bolli, 1957*b*). However, a restriction of *Globigerinita dissimilis*

(Cushman & Bermúdez) allows the new subspecies *G. dissimilis ciperoensis* Blow & Banner to be recognized, and this is described from the *Globigerina ampliapertura* Zone type locality sample. Good *G. ampliapertura* Zone to *G. ouachitaensis ciperoensis* Zone faunas have not yet been found in East Africa since the lowest Miocene is developed in a mainly reefal limestone facies; however, occasional specimens of *G. ouachitaensis ciperoensis* have been observed associated with *Miogypsina* (*s.s.*) in East Africa, whilst in Sicily (Ragusa limestone—see Part 1, p. 28) *Globigerina angulisuturalis* and *G. ouachitaensis ciperoensis* (see Fig. 9 and Plate IX) are associated with *Miogypsina* (*s.s.*) (see also Blow, 1957, and p. 84 of this Part). Furthermore, the writers' colleague, D. G. Jenkins, has shown the authors faunas from Victoria, Australia, obtained from the lowest Aquitanian. These faunas agree well with planktonic foraminiferal assemblages seen in the upper part of the *Globigerina ampliapertura* Zone in southern Trinidad, where, after the extinction of *G. ampliapertura ampliapertura* but before the first appearance of *Globorotalia opima opima* and *Globigerina angulisuturalis*, a monotonous fauna occurs. This fauna consists mainly of a form similar to *G. ampliapertura euapertura* (Jenkins) (in the press); Bolli (1957*b*, p. 100) has referred to this form as '*Globigerina* cf. *venezuelana* Hedberg' and he has noted a similar associated and monotonous fauna. In the northern part of the Lindi area and in the region of Dar-es-Salaam, Aquitanian and Burdigalian planktonic foraminiferal faunas occur and are very similar to those described by Bolli (1957*b*) and Blow (1959) for zones above the *G. ouachitaensis ciperoensis* Zone.

IV. Comparison of the Globigerinaceae of the Oligocene and Aquitanian in the Tethyan, European and Caribbean Regions

In view of the fact that the planktonic foraminiferal content of the Eocene biozones up to the *Globigerapsis semi-involuta* Zone, and the Miocene biozones above the *Globigerina ouachitaensis ciperoensis* Zone, both in Sicily and East Africa (indeed, in Australia and other areas as well), compare excellently with the faunas seen in Trinidad and

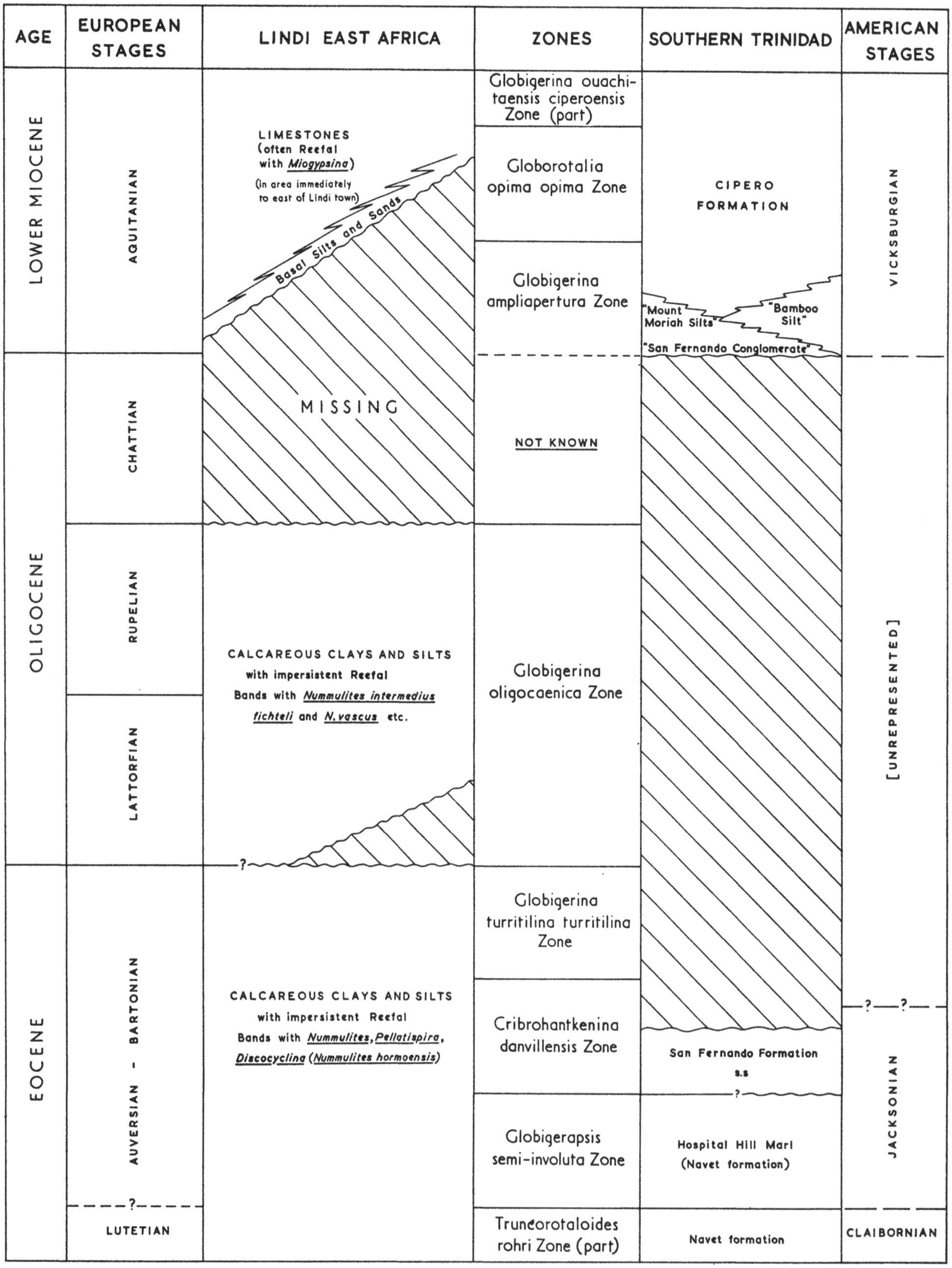

AGE	EUROPEAN STAGES	LINDI EAST AFRICA	ZONES	SOUTHERN TRINIDAD	AMERICAN STAGES
LOWER MIOCENE	AQUITANIAN	LIMESTONES (often Reefal with *Miogypsina*) (in area immediately to east of Lindi town) — Basal Silts and Sands	Globigerina ouachitaensis ciperoensis Zone (part)	CIPERO FORMATION	VICKSBURGIAN
			Globorotalia opima opima Zone		
			Globigerina ampliapertura Zone	"Mount Moriah Silts" "Bamboo Silt" "San Fernando Conglomerate"	
OLIGOCENE	CHATTIAN	MISSING	NOT KNOWN		[UNREPRESENTED]
	RUPELIAN	CALCAREOUS CLAYS AND SILTS with impersistent Reefal Bands with *Nummulites intermedius fichteli* and *N. vascus* etc.	Globigerina oligocaenica Zone		
	LATTORFIAN	?			
EOCENE	AUVERSIAN – BARTONIAN	CALCAREOUS CLAYS AND SILTS with impersistent Reefal Bands with *Nummulites, Pellatispira, Discocyclina (Nummulites hormoensis)*	Globigerina turritilina turritilina Zone		? ?
			Cribrohantkenina danvillensis Zone	San Fernando Formation s.s. ?	JACKSONIAN
			Globigerapsis semi-involuta Zone	Hospital Hill Marl (Navet formation)	
	LUTETIAN	?	Truncorotaloides rohri Zone (part)	Navet formation	CLAIBORNIAN

Fig. 8. Correlation of the biozones with the European stages and the geological successions in East Africa (Lindi area) and Trinidad, B.W.I.

eastern Falcón, Venezuela, it is difficult to postulate a 'facies' reason for the absence of the distinctive uppermost Eocene and Oligocene assemblages described here. The reason for this absence of the distinctive assemblages in Trinidad is considered to be due to stratigraphical reasons, and this is discussed below. Before this is done, however, it is useful to review the differences in planktonic foraminiferal content of the *G. oligocaenica* Zone of the Lindi area, East Africa, and the *G. ampliapertura* Zone of southern Trinidad. Furthermore, in §VII of this Part, the evolutionary relationships of the Upper Eocene, Oligocene and lowermost Miocene Globigerinaceae are discussed, and it will be seen that the description of the new faunas described in this work most satisfactorily clarifies and co-ordinates the previously published information recorded in Bolli's two outstanding contributions (1957 *b, c*).

The most easily recognized distinctions between the *Globigerina oligocaenica* Zone and the *G. ampliapertura* Zone (see also Range Chart, Fig. 17) are:

(i) The occurrence of *Globigerina oligocaenica*, *G. ouachitaensis gnaucki*, *G. turritilina turritilina* and *Globigerinita martini scandretti* in the *Globigerina oligocaenica* Zone but NOT in the *G. ampliapertura* Zone.

(ii) The occurrence of *Globigerina yeguaensis yeguaensis* in the Eocene of both Trinidad and East Africa, also in the *G. oligocaenica* Zone of East Africa, but NOT in the *G. ampliapertura* Zone of southern Trinidad (this point tells most strongly against regional 'facies' influence).

(iii) The occurrence of *Globigerina ouachitaensis ciperoensis* in the Oligocene of the Lindi area and in the Miocene of both regions (as well as in Sicily and Aquitaine), which tells most strongly against any theory involving limited access between the two areas.

(iv) The overlap in the range of *Pseudohastigerina micra* (which is very common in the Eocene of the two areas) and *Cassigerinella chipolensis* (which is common in the Miocene of Trinidad, North Africa and East Africa); this overlap in range occurs only in the Oligocene.

(v) The evidence seen in East Africa which indicates the separate evolutions of *Globorotalia centralis* → *Globigerina pseudoampliapertura*, and *Globorotalia increbescens* → *Globigerina ampliapertura ampliapertura*; this is different from the views of Bolli (1957 *b*) who considered that *G. ampliapertura ampliapertura* was directly descended from *Globorotalia centralis* (see p. 133).

(vi) The presence of *Globigerinita dissimilis ciperoensis* in the *Globigerina ampliapertura* Zone of southern Trinidad and in the Aquitanian of Sicily and East Africa but NOT in the rich planktonic foraminiferal assemblages of the *G. oligocaenica* Zone, East Africa.

Dr C. G. Adams (British Museum, Natural History) has kindly shown us a fauna from the mouth of the Benaleh River, Baram Headwaters, Fourth Division, Sarawak (Borneo), where *Nummulites fichteli* and *N. pengaronensis* are associated with *Globigerina turritilina turritilina*, *G. yeguaensis pseudovenezuelana*, *G. ampliapertura ampliapertura*, *Globorotalia increbescens* and *Globorotaloides suteri* as well as a form transitional between *Globigerina tripartita tapuriensis* and *G. oligocaenica*. This fauna compares very well with the fauna seen near the bottom of the Oligocene section of East Africa. It should also be noted that the fauna from Sarawak contains reticulate *Nummulites* of restricted Oligocene age.

Taking into account not only the similar Oligocene fauna from Sarawak, but also the other points noted above, strong evidence exists that during the Upper Eocene to Lower Miocene there existed no suite of Globigerinaceae restricted to the western Indian Ocean and unrepresented in tropical and subtropical America for any reason of 'provincialization of fauna' or any 'facies' cause. Drs H. M. Bolli and P. Bermúdez have seen our specimens of *Globigerina oligocaenica* from East Africa; they have reported to us (privately) the presence of this species in Mexico, the Dominican Republic and Cuba. This would demonstrate that the major elements of the post-Jackson–pre-Vicksburg Oligocene planktonic faunas are present in some parts of the Central American region, and that the absence of this Oligocene fauna in the American Gulf States, Trinidad and the Eastern Venezuelan Basin is due entirely to stratigraphical reasons. We can confirm the presence of this Oligocene species in the lower Alazan formation of the Tampico Embayment of Mexico, as it is present in material from these beds sent to us by Bolli and Bermúdez. The fact that this distinctive species has not been recorded (or, so far as we know, observed) by these workers or others of the same repute from the better controlled (stratigraphically) sections of the Gulf Coast, Trinidad or northern South America, etc., is most noteworthy (see p. 146).

For comparison with more temperate latitudes, Batjes (1958) has described some planktonic foraminifera from the Oligocene of Belgium and

Germany. He described and figured a form (1958, pl. 11, fig. 1) referred to *Globigerina bulloides* d'Orb., but this form is *G. praebulloides praebulloides*. The forms referred by Batjes to *G. globularis* Roemer (1958, pl. 11, figs. 3–5) are of three different specimens; however, Batjes's figs. 3a–c and 5a–c probably represent *Globorotaloides suteri* Bolli, as figs. 3a and 5a show the umbilical bulla broken off, leaving a 'ridge' encompassing the umbilicus where it was attached to the ultimate whorl. Batjes's figures should be compared to those of *Globorotaloides suteri* given in this work (Plate XIII N–P, and Fig. 11 (v)). The characteristic coarse punctae and wall structure of *G. suteri* Bolli, a feature which is most useful in the determination of the species, is also visible in Batjes's figs. 3a–c and 5a–c. The specimen illustrated by Batjes (1958) on his pl. 11, figs. 2a, b, is very closely comparable to *Globigerina officinalis* Subbotina, originally described from the Oligocene of Russia and now found to be common in the Oligocene of the Lindi area. Batjes's illustration of another specimen referred to *G. globularis* Roemer (1958, pl. 11, figs. 4a–c) is very close to *G. praebulloides leroyi* Blow & Banner subsp. nov., now described from the Oligocene of the Lindi area. The small '*Globigerina* sp.' illustrated by Batjes (1958, pl. 11, figs. 7a–c) is considered here to belong to *Globorotalia* (*Turborotalia*) *opima nana* Bolli, and it seems to show the same degree of evolutionary advancement as those specimens of this subspecies seen in the Oligocene of East Africa. Again, Batjes's figs. 8a–c (*loc. cit.*) of another small *Turborotalia*-like form (also given as '*Globigerina* sp.' by Batjes) is considered here to represent the early immature globorotaliid stage of *Globorotaloides suteri* Bolli (cf. Fig. 11 (ix) of this Part), the wall structure being most distinctive.

Most importantly, Batjes (1958, pl. 11, figs. 6a–b) illustrates a specimen referred to as '*Globigerinella micra* (Cole)' from the Septarian Clay of Pietzpuhl. Although Batjes (1958, p. 162) considered this form as being most likely reworked, this opinion was based on the views of Grimsdale (1951) as to the stratigraphical occurrence of this form. Grimsdale (1951, p. 468) pointed out that rare specimens have been seen in the Oligocene, although he considered that 'contamination' was the likely explanation for

these occurrences. It seems that Grimsdale's views were strongly influenced by the Caribbean range of this form. Whilst Batjes's specimen may not be strictly referable to the species *micra*, it is certainly referable to the genus *Pseudohastigerina* Banner & Blow, 1959, and it is probably the same form as that described here (p. 130) as *P.* aff. *micra*. Batjes's opinion as to the possible reworking of his specimens, or Grimsdale's views on contamination, are not subscribed to here as they are opposed by the evidence seen in the Lindi area, East Africa, and that reported by Russian workers. Both Myatliuk (1950) and Subbotina (1953) recorded *Globigerinella naguewichiensis* Myatliuk (= *Pseudohastigerina naguewichiensis*) from the Oligocene of Russia.

Pokorný (1955; 1958, p. 346) described *Cassigerinella boudecensis* (see p. 83 for a discussion of the synonymy of the species of the genus *Cassigerinella*) from the 'Oligocene' of Czechoslovakia, and recorded the range of the genus as 'Middle Oligocene to Burdigalian'. Contemporary work by Subbotina, Glushko & Pishvanova (1955) recorded the *nomen nudum Globalternina globoloculata* Ivanova from the Russian Oligocene; this form appears to be the same as that described as new by Bykova *et al.* (1958) as *Cassigerinella globolocula* 'Ivanova' (see p. 83). Dr Pokorný has informed us (private letter dated 27 April 1960) that his original dating of the lower range of *Cassigerinella* was incorrect, and that the genus does not occur in Europe below the Aquitanian; he goes on to state that 'the beds with *Cassigerinella* from the Soviet Carpathians, designated by Russian workers as of Upper Oligocene–Lower Miocene age, are contemporaneous with our Aquitanian beds of our Carpathian flysch'. For this reason Dr Pokorný would doubt the Oligocene age of the *Globigerina ampliapertura* Zone (*Revue de l'Inst. Français du Petrole*, in the press), as *Cassigerinella* is well known to occur abundantly in that Zone (Bolli, 1957), and there is very little doubt that the American form (*C. chipolensis*) is very closely related to, if not identical with, the European species (see also p. 83). However, we have observed *C. chipolensis* in undoubted Oligocene sediments (associated with *Nummulites fichteli*) in East Africa (see p. 81); consequently, if the lower range of *Cassigerinella* be taken as a basis for correlating the Oligo-Miocene boundary between Europe and the Caribbean, the *Globigerina ampliapertura* Zone should be regarded (as it is by Pokorný) as being of Aquitanian age; however, if the East African planktonic assemblages described in this work be considered, then, although *Cassigerinella* may be regarded as being of Oligocene to Burdigalian age, the *Globigerina ampliapertura* Zone fauna should be dated as Aquitanian. The former course, which utilizes a single index fossil, is liable to the same sources of error which have been pointed out above (p. 29, Part 1) in the case of the late appearance of *Orbulina* in the Tortonian of the Vienna Basin. It is clearly preferable to use the stratigraphical overlap of species, together with distinctive assemblages of distantly related species, to define zones; in addition, if these stratigraphical units can be associated with recognizable evolutionary sequences (as we have endeavoured to demonstrate below, §VII), then the sources of error in correlation are minimized. Therefore, we can recognize the stratigraphical overlap of the genera *Pseudohastigerina* (known from the Eocene and Oligocene of East Africa and Europe) and *Cassigerinella* (known from the Oligocene of East Africa and the Lower Miocene of Europe, North Africa, East Africa and the Caribbean) in an interval which can only be correlated with the Oligocene. Recent Russian records (Subbotina, Pishvanova & Ivanova, 1960) of *Pseudohastigerina* are, in part, clearly placed too high stratigraphically, for the succeeding beds, which contain the evolution from *Globigerinoides bisphericus* to *Orbulina*, are referred (evidently following Drooger) to the Middle Miocene.

Subbotina (1953) described *Globigerina officinalis* (= *G. parva* Bolli, *partim*) from the Oligocene of Russia. Hofker (1956) has recorded *G. ouachitaensis* from the Oligocene of Holland, and it is noteworthy that he does not record any species of the genus *Globigerinoides* Cushman, 1927, from the Dutch Oligocene. We have examined planktonic foraminiferal faunas from the German Oligocene near Offenbach (Mainz Basin), and have seen *Pseudohastigerina micra*, *Globorotalia (Turborotalia) opima nana*, and small (dwarfed?) specimens of *Globigerina praebulloides occlusa* and *G. prae-*

bulloides leroyi in these assemblages. Faunas from the south of France (Stampian) which we have examined contained only rare, small (dwarfed?) planktonic foraminifera, but specimens probably referable to *G. praebulloides praebulloides*, *G. praebulloides occlusa* and *G. officinalis* occur, as in the Oligocene of the Lindi area (see p. 146).

One of us collected a sample from the Upper Rupelian of Sportsplatz, Elmsheim, near Stadecken (14 km. south-west of Mainz), Germany. Unfortunately this sample, although containing a very rich planktonic fauna, shows the presence of much reworked Eocene and Cretaceous material. However, the planktonic fauna includes *Pseudohastigerina micra*, *Globigerina praebulloides praebulloides*, *G. praebulloides occlusa*, *G. praebulloides leroyi*, *G. officinalis*, *G. ouachitaensis ouachitaensis*, *G. angustiumbilicata*, *Globorotalia (Turborotalia) permicra*, *G. (T.) postcretacea* and the immature stages of *Globorotaloides suteri*. This assemblage could be considered representative of the Rupelian in its cold water facies and compares with the fauna described by Batjes (1958) from the Oligocene of Belgium (see also Drooger & Batjes, 1959). It is important to note the absence of any species belonging to the genus *Globigerinoides* Cushman, 1927.

Rey (1954) records '*Globigerinoides sacculifera immatura* LeRoy', amongst other forms, from a horizon in North Morocco considered by him to be of Oligocene age. However, Rey based his age-determination of this fauna on Senn's identification of *Nummulites* cf. *fichteli*, but Senn's locality is farther south than Rey's horizon, and it is not proved that the two levels have been correctly correlated. It is likely that Rey's beds at Lalla Mimouna are of Aquitanian age and overlie Senn's *N. fichteli* horizon to the south.

We know of no reliable records of the *Globigerinoides quadrilobatus* group (= *G. triloba* group, of Bolli, 1957b) below the Aquitanian (see also p. 136). This and other related points are discussed at length in Part I where a review of recent work by Drooger & Magné (1959) and Drooger & Socin (1959), relevant to this topic, is also made.

From the above notes it will be seen that the Oligocene planktonic foraminiferal faunas from

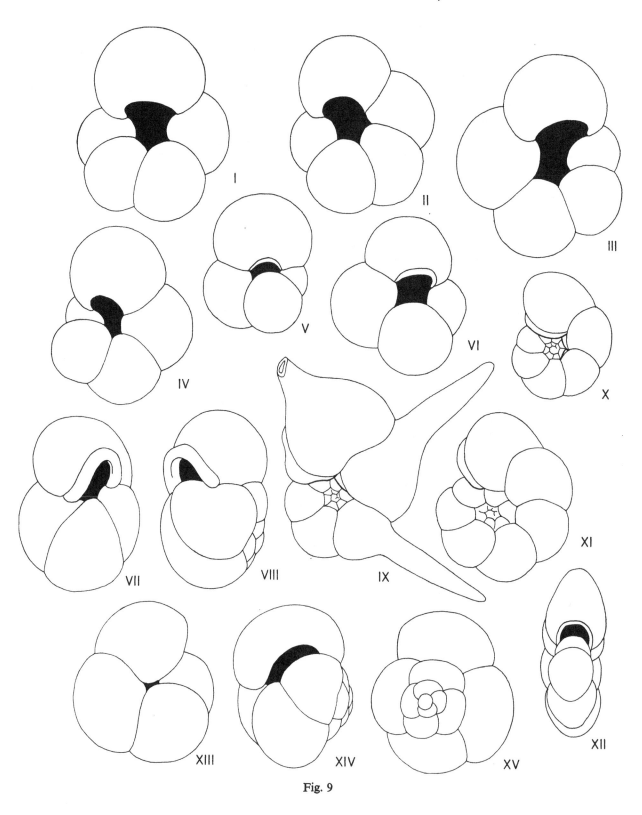

Fig. 9

widely separated areas are characterized by the continuing occurrence of species of the genus *Pseudohastigerina*, and by its co-occurrence with the genus *Cassigerinella*. It is emphasized that this forms a most useful distinction between the Oligocene and lowest Miocene foraminiferal assemblages. The indigenous co-occurrence of these two genera clearly delimits at least part of the Oligocene, the Chattian not being known in a planktonic foraminiferal facies. If this is taken in conjunction with the new species described in this work, especially the easily recognized *Globigerina oligocaenica* and *G. turritilina turritilina*, it will be seen that the lowest Aquitanian *G. ampliapertura* Zone and the Lattorfian to Rupelian *G. oligocaenica* Zone are easily separated and distinguished one from the other.

We have also examined a number of samples from the Aquitanian of south-west France, collected by Dr F. W. Anderson (H.M. Geological Survey of Great Britain) during a field excursion led by Professor J. Cuvillier and Dr Neumann during 1957. Of the samples collected, two contained planktonic foraminifera. One sample (FWA 66) from the lower Aquitanian at Escornebéou (near Dax, see also Rey, 1958, *Rev. Micropal.* **1**, 62), from the Marnière de Saubusse, contained *Globigerina angustiumbilicata*, *G. ouachitaensis ouachitaensis* and *G. ouachitaensis ciperoensis* (see Fig. 9, specimens ii, iv and vi), and *G. officinalis* and *G. praebulloides occlusa* with *Miogypsinella complanata* (see Part 1, Plate II E). Since *Globigerina ouachitaensis ouachitaensis* ranges from the Upper

Eocene to about the lower part of the *Globorotalia opima opima* Zone in southern Trinidad as well as occurring in the lower Aquitanian of south-western France, the occurrence of this form strongly suggests that the *Globigerina ampliapertura* Zone is equivalent to part of the lower Aquitanian.

The other sample containing planktonic foraminifera, collected by Dr Anderson from the middle Aquitanian of Moulin de l'Église, Gironde, contained numerous *Miogypsina* cf. *gunteri* with *Globigerina ouachitaensis ciperoensis* (see Fig. 9, specimen i), but without *G. ouachitaensis ouachitaensis*. This sample also contained *G. praebulloides occlusa* but no specimens of the genus *Globigerinoides* Cushman, the Lower Miocene to Recent species of which (i.e. *G. quadrilobatus–G. ruber* group) develop from *Globigerina praebulloides occlusa* in the *Globorotalia kugleri* Zone (see Bolli, 1957*b*; Blow, 1959; Banner & Blow, 1960*a*) and which are common in the highest Aquitanian and Burdigalian of Aquitaine. Hence the presence of *Globigerina ouachitaensis ciperoensis* and *G. praebulloides occlusa*, together with the absence of any subspecies of *Globigerinoides quadrilobatus* (= 'G. triloba', of Bolli, 1957*b*, and other authors, see Banner & Blow, 1960*a*), indicates that at least part of the middle Aquitanian and the lower Aquitanian in the type area must extend below the level of the *Globorotalia kugleri* Zone of southern Trinidad and eastern Falcón, Venezuela. The evolution of *Globigerinoides quadrilobatus* from *Globigerina praebulloides occlusa* is discussed in detail in §VII, p. 136.

Legend to Fig. 9

Fig. 9. Stratigraphically significant species from Europe, Africa and America; all specimens deposited in the British Museum (Natural History), registered (in the above order) as nos. P.44420–P.44430. i–iii: *Globigerina ouachitaensis ciperoensis* (Bolli), ventral views of specimens from (i) Moulin de l'Église, Aquitaine (Aquitanian), (ii) Escornebéou, Aquitaine (Aquitanian), and (iii) the Lower Ragusa Limestone formation, Monte Alia, Sicily (Aquitanian) (sample WACR 1326); all *ca.* × 175, P.44420–P.44422. iv: *G. angustiumbilicata* Bolli, from Escornebéou, Aquitaine (Aquitaniar), *ca.* × 175, P.44423. v: *G. praebulloides leroyi* nov.subsp., from Moulin de l'Église, Aquitaine (Aquitanian), *ca.* × 175, P.44424. vi: *G. ouachitaensis ouachitaensis* Howe & Wallace, from Escornebéou, Aquitaine (Aquitanian), *ca.* × 175, P.44425. vii, viii: '*Globigerina*' *nepenthes* Todd, from Rohrbach, near Nussdorf, Vienna Basin (Upper Vindobonian)—(vii) ventral view, (viii) axial peripheral view of same specimen, both *ca.* × 125, P.44426. ix: *Hantkenina primitiva* Cushman & Jarvis, immature specimen, from the *Cribrohantkenina danvillensis* Zone, Upper Eocene, Lindi area, Tanganyika (sample FCRM 1932), *ca.* × 175; showing the resemblance of the early portion of the test to that of *Pseudohastigerina micra* (Cole), P.44427. x: *Pseudohastigerina micra* (Cole), from the Rupelian of Offenbach, Mainz Basin, *ca.* × 175, P.44428. xi, xii: *Pseudohastigerina* aff. *micra* (Cole), from the *Globigerina oligocaenica* Zone, Lindi area, Tanganyika, sample FCRM 1922 (Rupelian/Lattorfian)—(xi) equatorial view, (xii) axial view of same specimen, *ca.* × 300, P.44429. xiii–xv: *Globorotalia* (*Turborotalia*) *increbescens* (Bandy), metatype (ex. Bandy Collection), from the Upper Jackson formation (Upper Eocene), Little Stave Creek, Clarke County, Alabama—(xiii) ventral view, (xiv) axial peripheral view, (xv) dorsal view, *ca.* × 125, P.44430.

Blow has discussed (Blow, 1957) the co-occurrence of *Globigerina ouachitaensis ciperoensis* with *Miogypsina s.s.* and *Miogypsinella complanata* in south-eastern Sicily (see also Part 1, p. 11). *Globigerina ouachitaensis ciperoensis* and *G. angulisuturalis* are here illustrated (Plate IX Aa–Cc for the latter, and Fig. 9, specimen iii, for the former) for the first time from the Sicilian Aquitanian.

The detailed evidence as to the age of the 'San Fernando formation' and lowermost Cipero formation (including the 'Mount Moriah silts') of southern Trinidad presented in Part 1 should be considered here, together with the evidence put forward in this Part 2 based largely on conclusions drawn from the stratigraphical distributions of the Globigerinaceae in Europe and East Africa. However, Kugler (1954, p. 411) makes the following important statement: 'The *Globigerina* cf. *concinna* Zone [i.e. of Cushman & Stainforth, 1945] includes the oldest *Miogypsina s.str.* bearing sediments in Trinidad.' It should be noted that the type locality of Cushman & Stainforth's (1945) *Globigerina* cf. *concinna* Zone is given by Stainforth (1948, p. 1303) as the same as for sample Rz. 90, and this sample is mentioned by Bolli (1957*b*, p. 100) as being 'in the same section' as the new type locality for Bolli's (1957*b*) *Globorotalia opima opima* Zone; it is implied that sample Rz. 90 contains the same planktonic foraminiferal fauna as found at the *G. opima opima* Zone type locality. Furthermore, Bolli (in Martin-Kaye, 1958) records *G. opima opima* from foraminiferal ashes taken from the top of the Lower Tuff Series of the Island of Carriacou (West Indies) (see Part 1, p. 43) which are associated with limestone lenses containing *Miogypsina staufferi* and *Miolepidocyclina panamensis*. This tends to confirm that the *Miogypsina s.s.* observed within the *Globorotalia opima opima* Zone of southern Trinidad are not greatly displaced either tectonically or by slumping and should be considered to have been deposited primarily within the sediments comprising the zone. Hence, it appears that *Miogypsina s.s.* occurs at least as low down in the Caribbean successions as the zone immediately overlying the *Globigerina ampliapertura* Zone; this broadly agrees with Blow's observations in Sicily (Blow, 1957). A point which must now be borne in mind concerns the distribution of *Miogypsina s.s.* and *Nummulites*. Nowhere have *Miogypsina s.s.* and true *Nummulites* been correctly recorded as co-existing in natural association and, therefore, the lower part of the *Miogypsina*-bearing limestones lying immediately above the nummulitic Lattorfian–Rupelian in the Lindi area, East Africa, may be correlated *approximately* with the *Globorotalia opima opima* Zone of the Cipero formation, southern Trinidad. As discussed above, the planktonic foraminiferal faunas of the Lattorfian–Rupelian in East Africa are distinctly different from those of the *Globigerina ampliapertura* Zone of the Cipero formation, southern Trinidad, and are undoubtedly older than the faunas from that zone, not only for biostratigraphical reasons, but also for reasons deduced from evolutionary principles evaluated on the evidence of a rapidly evolving group, i.e. the Globigerinaceae. Hence, the biostratigraphical evidence points to the fact that there is a hiatus between the highest Eocene in Trinidad and the overlying *Globigerina ampliapertura* Zone, Cipero formation. This hiatus is represented in the Lindi area, East Africa, by the Lattorfian–Rupelian *G. oligocaenica* Zone as well as the highest Eocene *G. turritilina turritilina* Zone.

Whilst it cannot be directly shown on the planktonic foraminiferal evidence given in this Part 2 that the Chattian is also missing in Trinidad, the evidence presented in Part 1 concerning the recognition and redescription of the genus *Pliolepidina* should be considered, and from this and other evidence put forward in Part 1, the *Globigerina ampliapertura* Zone is considered here most likely to be Aquitanian.

The stratigraphical position of the 'San Fernando formation' cannot be entirely resolved on palaeontological grounds, but the presence in East Africa of an Uppermost Eocene planktonic fauna not seen in Trinidad (the *Globigerina turritilina turritilina* Zone fauna) indicates that the position of the 'San Fernando formation' is distinctly different from that hitherto considered by Suter (1951–52), Kugler (1953) and Bolli (1957*b, c*). This is now discussed in the succeeding § V below.

76

V. The Stratigraphical Position of the 'San Fernando Formation', Southern Trinidad

Before discussing the stratigraphical position of the 'San Fernando formation' of southern Trinidad, we recapitulate on the succession seen within the Upper Eocene of the Lindi area, East Africa. In this area there is a complete and continuous sequence of sediments wherein the only subdivisions that can be made are in terms of faunal content. The biozones recognized and erected in this work are sharply defined on their faunal content; nowhere is there any known evidence to suggest that the boundaries of any of the Upper Eocene zones are markedly influenced or defined by a stratigraphical discontinuity except, perhaps, between the uppermost Eocene *Globigerina turritilina turritilina* Zone and the base of the Oligocene *G. oligocaenica* Zone. The zones of the Upper Eocene in the Lindi area are as follows:

(iii) *Globigerina turritilina turritilina* Zone,
(ii) *Cribrohantkenina danvillensis* Zone,
(i) *Globigerapsis semi-involuta* Zone.

As pointed out previously, the *Globigerapsis semi-involuta* Zone of the Lindi area correlates perfectly with the same zone of the uppermost Navet formation of Trinidad (Hospital Hill marl), and with planktonic faunas seen in the lower Jacksonian of the American Gulf States.

Bolli (1957*c*) established the *Globorotalia cocoaensis* Zone (*vel. G. cerro-azulensis* Zone) comprising the 'San Fernando formation' (= Mount Moriah). Bolli's '*Globorotalia cocoaensis*' Zone (i.e. the 'San Fernando formation') overlies the *Globigerapsis semi-involuta* Zone of the Hospital Hill marl and has an assumed unconformable contact with this latter member of the Navet formation (Kugler, 1953; Bolli, 1957*c*) as well as transgressing on to Lower and Middle Eocene rocks of the lower parts of the Navet formation as well as on to the Palaeocene Lizard Springs formation (Bolli, 1957*c*, p. 160). Furthermore, the 'San Fernando formation' contains much reworked material including boulders of Cretaceous, Palaeocene and Lower to Middle Eocene rocks (Kugler, 1953). The reworking of foraminifera ranging from Cretaceous to Eocene is also well known to be common in the 'San Fernando formation' and we, ourselves, have seen reworked planktonic foraminifera ranging in age from Campanian to Middle Eocene in one single sample from the 'San Fernando formation'.

No faunas comparable to those of the *Globigerina turritilina turritilina* Zone, of the Lindi area, have been recognized in southern Trinidad. In general the faunas of the *Cribrohantkenina danvillensis* Zone are similar to those described by Bolli (1957*c*) from his '*Globorotalia cocoaensis* Zone'. However, Bolli (1957*c*, p. 159, fig. 26) continues the range of *G. centralis* right to the top of the presumed Eocene succession of southern Trinidad. In the Lindi area, typical *G. centralis* does not continue to the top of the *Cribrohantkenina danvillensis* Zone and certainly not into the *Globigerina turritilina turritilina* Zone, but evolves into, and is replaced by *G. pseudoampliapertura*. Bolli, however, does illustrate specimens believed by him to be transitional between *Globorotalia centralis* and *Globigerina ampliapertura* (Bolli, 1957*c*, pl. 36, figs. 9, 10), but which are here considered to be transitional between *Globorotalia centralis* and *Globigerina pseudoampliapertura*. The hypotype illustrated by Bolli (1957*c*, pl. 36, figs. 8a–b) as *G. ampliapertura* from the 'San Fernando formation' appears to be different from the holotype and paratypes of *G. ampliapertura* as illustrated by Bolli (1957*b*, pl. 22, figs. 5–7, *not* figs. 4a–b) from the Cipero formation. The hypotype from the 'San Fernando formation' seems to be a small specimen of *G. pseudoampliapertura* (see p. 95). Since in East Africa there is evidence that there is a difference in time between the evolutions of *G. pseudoampliapertura* and *G. ampliapertura* from their respective ancestors (see p. 131) it now appears that neither the range of *G. ampliapertura* nor the previously presumed transitional forms (of Bolli) between *Globorotalia centralis* and *Globigerina ampliapertura* can now be used as evidence to deduce that there was no great time break between the 'San Fernando formation' and the lowermost Cipero formation. However, Bolli (1957*b*, pp. 99–100) makes the following remarks: 'The basal part of the Cipero formation (*Globigerina ampliapertura* Zone) often appears as a dark silt. Lithologically it then becomes almost indistinguishable from the similar facies of the Mount Moriah silt member of the Upper Eocene "San Fernando formation". Those beds which have a Mount Moriah silt aspect but do not contain any Eocene foraminiferal markers are here placed into the Oligocene part of the Cipero formation.' (Also, compare Kugler, 1954, p. 413.)

The Mount Moriah silts are younger than the true 'San Fernando formation' and are separated from it by the 'San Fernando conglomerate', which we believe to represent the base of the Miocene transgression. We have indicated, in this work, that the *Globigerina ampliapertura* Zone (i.e. the 'Oligocene' part of the Cipero formation of Bolli, 1957*b*) is Aquitanian and is separated from the Upper Eocene not only by the uppermost Eocene *G. turritilina turritilina* Zone but also by the Lattorfian–Rupelian *G. oligocaenica* Zone, both of which are not represented in Trinidad. Here then is an apparent paradox; (*a*) an apparently continuous sequence of deposition from the 'Mount Moriah silt member' into the dark silts of the lower Cipero formation, as compared to (*b*) strong palaeontological and biostratigraphical evidence indicating a large hiatus between the 'San Fernando formation' and the basal Cipero formation. In this case, Bolli's implied opinion that no great time break exists between the two Trinidad formations is considered here to be based (i) on species or species-groups which are long ranging, or (ii) on a mis-interpretation of the evolution of *G. ampliapertura*, which is shown below not to originate directly from *Globorotalia centralis*.

The paradox stated above must be resolved on other evidence, comprising: (*a*) the distribution of larger foraminifera within the 'San Fernando' and Cipero formations, (*b*) the geological field-relationships of those beds which have been considered members of the 'San Fernando formation' (both correctly and incorrectly) both to the underlying Hospital Hill marl (Navet formation) and to the overlying lower Cipero formation, and (*c*) the biostratigraphical sequence described here from the Lindi area, East Africa. The last point (*c*) has already been discussed, and the first point (*a*) has been dealt with in Part 1, but some points are again briefly mentioned below. However, it is point (*b*)

which must now be discussed in detail; all three aspects of the problem must be taken into account and a reasonable agreement between them deduced.

Kugler (1953, p. 45) has pointed out that there is a change of facies from the 'open sea facies' of the Hospital Hill marl to the 'neritic near-shore facies' of the 'San Fernando formation', and that the 'San Fernando formation' is probably unconformable on the Hospital Hill marl. This must be considered together with the fact that a part of the 'San Fernando formation' (*s.l.*) is mainly a 'coarse block conglomerate' which is followed by impure sands and some orbitoidal 'bioherms'; this part of the succession is included in the 'Mount Moriah silt' member. The lithologies are distinctly different for the Navet formation (Hospital Hill marl) and the 'San Fernando formation'; the former is a white or greyish marl, whilst the latter is often highly glauconitic and, on alteration and decomposition, gives a highly iron-stained, limonitic and brownish coloration to the sediments; this is also reflected in the preservation of the microfauna. Dr H. G. Kugler has kindly informed us (private letter to Dr F. E. Eames, dated 29 May 1959) of the relationships seen in a recently opened and new exposure near San Fernando, southern Trinidad. Dr Kugler remarks as follows (the italics are the present authors'):

Large-scale excavations for a building project have lately exposed the contact of the *ampliapertura* beds with the Mt. Moriah Silts. Large slipmasses of *penecontemporaneous* Upper Eocene rocks such as glauconitic sands, silts and sandstones, associated with older rocks, were again found in the Mt. Moriah silt with its typical assemblage of foraminifera of the *cerro-azulensis* zone. The well bedded dark silty clays of the *ampliapertura* beds rest with sharp contact on this Mt. Moriah wildflysch. Not more than 30 feet above the contact one observes large blocks of Naparima Hill Formation and other Cretaceous and Palaeocene *slipmasses in the lower Cipero silts* of the *ampliapertura* and *opima* zones. On top of the Mt. Moriah Hill one observes the contact of the San Fernando Formation with the Navet Formation. The well bedded highly glauconitic orbitoidal sandstones of the San Fernando Formation rest with sharp contact on the white chalk of the Navet Formation. About 80 feet above this contact, and thus above the glauconitic sandstones, follows a 10 foot boulder bed mainly of argillites of the Naparima Hill Formation. This boulder bed is *considered to be a part of the San Fernando Formation* but at this spot could already belong to the *ampliapertura* zone with its own slip and block masses,....

An important point seen from Dr Kugler's letter is that blocks and slipmasses occur in the *Globigerina ampliapertura* Zone, and, indeed, up to the overlying *Globorotalia opima opima* Zone, both of which zones have been shown here to be lower Aquitanian. It is recalled here that on p. 76 it was shown that *Miogypsina s.s.* occurs in the *Globorotalia opima opima* Zone in both Trinidad and Carriacou. It also seems from Dr Kugler's remarks that these 'blocks' within the two lower Miocene zones of the Cipero formation are genetically related to the 'blocks' seen in the supposedly Upper Eocene 'San Fernando formation'. Another point is that Dr Kugler refers (in his letter quoted above) to 'slipmasses of *penecontemporaneous* Upper Eocene rocks'; the important word here seems to be '*penecontemporaneous*' for it is agreed that the rocks represented by the blocks were fragmented after diagenesis and consolidation of the sediments involved. It seems necessary now to consider a mechanism which would have created the conditions which caused the deposition of the 'San Fernando formation' conglomerates and the blocks of older rocks which are emplaced in the Aquitanian lowermost Cipero formation. A reasonable geological explanation which would fit the observed conditions in Trinidad and the succession seen in East Africa seems likely to be along the lines suggested below. It must be remembered first, however, that the fauna described by Bolli (1957c) for his *Globorotalia cocoaensis* Zone is not that which occurs in the highest Eocene of the Lindi area.

We suggest that at the end, or towards the close, of Eocene times there was a period of emergence (evidence for this is certainly seen in the neritic character of material *within* the blocks of autochthonous Upper Eocene) followed by warping and strong folding which gradually became more intense during the Oligocene. During the period of emergence and folding, diversification of the Oligocene land surface occurred with a period of subaerial denudation. It is likely, from the distribution of the 'San Fernando formation' (see Kugler, 1953), that the Oligocene folding followed older structural lines initiated and active in Cretaceous and Palaeocene/Lower Eocene times. It is also likely that during the Oligocene diversification of the

land-surface, resistant masses of Cretaceous/Palaeocene/Lower Eocene rocks would be exposed and that around the peripheral margins of these resistant Cretaceous/Lower Palaeogene masses (which may have formed peaks and/or promontories from the sea or land-surface) the Upper Eocene beds would be protected. Furthermore, the Upper Eocene beds as originally deposited would be of a more resistant and different lithology over, and in the vicinity of, these resistant areas, which probably began to rise before the close of Eocene times. In this case, the Upper Eocene beds away from the structurally high and resistant areas would probably be of a more silty or clay-like nature than their contemporary sediments being deposited over them and in their vicinity. In this case the softer basinal sediments of Upper Eocene times would be more easily swept away during the uppermost Eocene–Oligocene emergence or simply reworked and redeposited as the basal silt of the advancing Aquitanian transgression. In the case of the areas around the structural highs and resistant masses, which would form islands or promontories within the advancing and deepening Aquitanian seas, the protective nature of such islands or promontories would serve to preserve the skirting remnants of the now fragmented and eroded Upper Eocene beds. As the Aquitanian seas progressed over the Oligocene land-surface together with differential deepening of the basinal areas, blocks from these resistant masses would be emplaced by slumping into the deeper water sediments of the lower Aquitanian (compare the blocks present in the *Globorotalia opima opima* Zone, as discussed by Dr Kugler in his letter quoted above). This explanation would also account for the gradation of the conglomerates both upwards and laterally into sands and silts, together with the shallow-water 'biohermal' limestones. The absence of Aquitanian planktonic foraminifera between the redeposited 'Mount Moriah–San Fernando conglomerate' sediments is readily explained in terms of the turbid environment which would be associated with the advancing seas, as well as the probable restricted environment with only limited access to the open ocean. It is not supposed in this explanation that the sole cause of the fragmentation of the Upper

Eocene beds was entirely due to the Aquitanian transgression but was also due to tectonic activity, the whole period of the Oligocene being, no doubt, extremely active tectonically. This explanation does not depend on the actual nature of the structurally high and resistant masses and the explanation would equally apply to 'resistant masses' which are themselves 'root-less exotic blocks' ('morros', of Kugler, 1953). Even if the 'morros' were not emplaced until Aquitanian times the explanation would still hold, since the 'morros' would most likely have originated from a structurally high area and carry with them the skirting remnants of Upper Eocene (cf. also Bailey & Weir, 1933).

This interpretation of events is put forward as a likely solution, but we appreciate that it must be examined in the light of all the field evidence. However, this interpretation of events would account for the distribution of the conglomerates which, according to G. E. Higgins (private discussions), seem to be absent in the basinal areas but well developed over areas near structural highs (Soldado and Plaisance conglomerates) and around 'massifs' composed of older resistant rocks (Naparima Hill). If this interpretation of events is followed, the 'wildflysch' sedimentation of the 'San Fernando conglomerate' is seen to belong, with the Cipero and Nariva formations, to essentially the same major cycle of post-*danvillensis* Zone tectonic activity and sedimentation. Kugler (1953) has pointed out the high and intense tectonic environment which prevailed during Cipero (i.e. Lower Miocene) times (compare the wildflysch of the Nariva formation), and all that has been proposed here is to unite the tectonic conditions indicated by the wildflysch of the 'San Fernando conglomerate' with that indicated by the essentially similar middle Aquitanian Nariva formation to form one major cycle of tectonic activity, this major period of tectonic activity having secondary cycles of tectonic activity superimposed upon it.

If this interpretation of events is not followed, then all the (San Fernando) events noted above must have taken place within the Jacksonian (Upper Eocene), and they must have occurred in a very short part of Jacksonian times. It must be remembered that the Hospital Hill marl is Jacksonian, and the 'San Fernando conglomerate' would also be Jacksonian if our interpretation is not followed and if the dominant part of its fauna were considered to be *in situ*. From the evidence of the Lindi area not all the Upper Eocene is present in Trinidad. Hence, if the San Fernando wildflysch deposits are to be considered as Eocene, they must all have been deposited in the very little Upper Eocene time which would have been left available after the supposed intra-Jacksonian (San Fernando/Navet) unconformity and the interval of the uppermost Eocene *Globigerina turritilina turritilina* Zone and the Oligocene *G. oligocaenica* Zone. Further, the present writers cannot think of an alternative explanation of the stratigraphical position of the Lindi succession, and this applies even if all the correlations with the European stages involved are incorrect. This last point is most important since the correlations are independent of correlation with Europe, it being the correlations and relative dating between East Africa and Trinidad which are important. However, we are confident that the faunas seen in East Africa are more than adequate to confirm our views as to the dating of the beds relative to the classical European stages. We also feel that an intra-Jacksonian time break is not long enough to account for all the observed geological phenomena associated with the problem; however, the whole of the Oligocene would suffice. We prefer to consider the 'San Fernando conglomerate' as the basal conglomerate of the Miocene Cipero formation, containing remanents and remanent beds of the Upper Eocene which have been fragmented during the Oligocene emergence. The evidence afforded by palaeontology does not contradict the interpretation of events put forward above, since the faunas of the 'San Fernando conglomerate' would, in fact, be those of the Upper Eocene, not only from the blocks but also from the associated silts which would have a largely derived fauna. This is especially so for the planktonic foraminifera, and it should be remembered that the Miocene forms would be rare, since planktonic foraminifera are scarce in neritic deposits for many reasons. Further, any Miocene forms actually occurring would be heavily swamped by the re-worked Eocene forms. With regard to the larger

80

foraminifera, many Eocene forms would also be derived into the associated silts; indeed, owing to their shape and size, they may even be concentrated into bands giving the appearance of *in situ* 'bioherms' or biostromes. However, in Part 1, *Pliolepidina tobleri* is shown to be an Aquitanian form, and the genus *Pliolepidina* is considered to range from Aquitanian to Burdigalian only. Therefore there is no palaeontological contradiction for regarding the stratigraphical position of the 'San Fernando conglomerate' as stated above.

Because it is considered here that the 'San Fernando conglomerate' is a redeposited series of beds genetically and environmentally associated with the overlying Cipero formation (Miocene), it seems desirable to discontinue the use of the term 'Zone' to define or describe the beds, especially since they are at present defined in terms of a derived Eocene foraminiferal species. This is undesirable, and leads to confusion as to the true nature and stratigraphical position of the beds in question. It is advisable to restrict the term 'San Fernando formation', in view of the fact that the upper conglomerate, originally referred to this formation, is most likely redeposited, and to use the term 'San Fernando formation' purely as here restricted to the *in situ* Upper Eocene in its type section (also see Kugler, International Stratigraphical Lexicon, Trinidad), and to refer the 'Mount Moriah silt' to the lower Cipero formation. In view of our remarks above, we have erected a biozone defined in terms of the clear biostratigraphical succession seen in the Lindi area, East Africa, and this is termed the '*Cribrohantkenina danvillensis* Zone, Upper Eocene', the type locality of which is shown on Fig. 7. Only the lower part of the Zone is present *in situ* in Trinidad, although much material which may be referable to higher parts of this Zone exists as derived fragments; both the *in situ* beds and these remanents were referred to by Bolli (1957c) as his '*Globorotalia cocoaensis* Zone, San Fernando formation'.

J. B. Saunders (private letter) has informed us that *Pliolepidina tobleri* (which we consider to be a Miocene form) occurs at the type locality of Bolli's '*Globorotalia cocoaensis*' Zone at San Fernando Railway Station. However, it should be remembered that the type locality of the '*G. cocoaensis*' Zone is not the same as the type locality of the San Fernando formation at Mount Moriah Hill and which (*teste* Kugler, *verb.*) is considered to be *in situ* Upper Eocene.

Although *Pliolepidina tobleri* has not been found at the type locality of the Mount Moriah Silt exposed along the Alexander Road, Kugler (*verb.*) believes the Point Bontour beds, from which the type specimens of *P. tobleri* were collected and described, to be approximately of the same stratigraphic level as the type Mount Moriah Silts. Our conclusions as to the Lower Miocene age of the Mount Moriah Silts are not invalidated by the fact that the Point Bontour beds are only a slip mass in the Cipero formation.

VI. Systematic Palaeontology

The classification used here follows that discussed by Banner & Blow (1959, 1960b). However, the subfamily Catapsydracinae Bolli, Loeblich & Tappan, 1957, is now shown to be an unnecessary grouping, and those genera originally placed within it are now assigned to either the Globigerininae Carpenter, Globorotaliinae Cushman or the Orbulininae Schultze. This is discussed at length under the genera *Globigerinita* and *Turborotalita* below. A glossary of descriptive terms employed has been published by us previously (1959); reference should also be made to Fig. 10.

All figured specimens (including holotypes, paratypes and hypotypes, etc.) are deposited in the British Museum (Natural History), London, and the registered numbers of these specimens can be referred to on the Plate explanations.

Order: **FORAMINIFERA** d'Orbigny, 1826

Superfamily: ***GLOBIGERINACEAE*** Carpenter, 1862, *nom. trans.*

Family: GLOBIGERINIDAE Carpenter, 1862

Subfamily: Cassigerinellinae Bolli, Loeblich & Tappan, 1957

Genus: *Cassigerinella* Pokorný, 1955

Type species: *Cassigerinella boudecensis* Pokorný, 1955 = *C. chipolensis* (Cushman & Ponton), 1932

Cassigerinella chipolensis (Cushman & Ponton) (Plate XV m, n)

Cassidulina chipolensis Cushman & Ponton, 1932, *Florida Geol. Surv. Bull.* **9**, 98, pl. 15, figs. 2a–c.

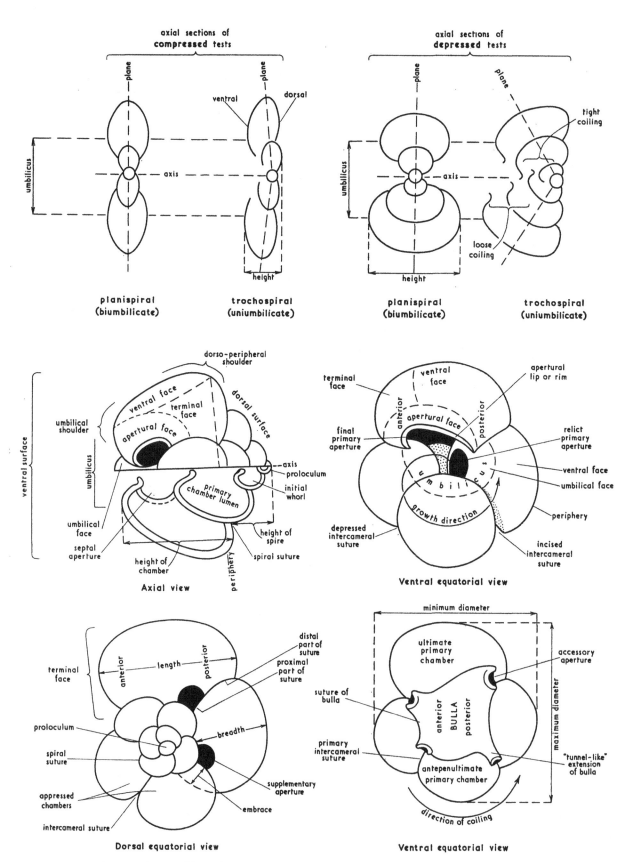

Fig. 10. Diagrams illustrating the descriptive morphological terminology of the Globigerinidae.

Cassigerinella boudecensis Pokorný, 1955, *Věstník, Ústred. Úst. Geol.* 30, 136–40, text-figs. 1–3.

C. globolocula Ivanova, 1958, in Bykova, *Mikrofauna SSSR*, sb. 9, vip. 115; *Trudi, Vses. Neft. Nauchno-Issled. Geol. Inst. (VNIGRI)*, p. 57, pl. 11, figs. 1–3.

C. chipolensis (Cushman & Ponton), Blow, 1959, *Bull. Amer. Paleont.* 39, no. 178, p. 169, pl. 7, figs. 30a–c.

REMARKS

We have compared our specimens from the Oligocene of East Africa with those from the Cipero formation of southern Trinidad and with metatypes of *C. boudecensis* kindly sent to us by Dr V. Pokorný. We have also compared our specimens with near topotype material from the American Gulf Coast kindly supplied by Dr T. Barnard. No differences can be seen between populations from East Africa and those observed in the Caribbean and American Gulf States. Specimens of *C. chipolensis* from East Africa and Trinidad were sent to Dr Pokorný, who writes as follows (letter dated 24 May 1960): 'after seeing your specimens I think that *boudecensis* is probably conspecific with *chipolensis*. As to *C. globolocula* Ivanova, 1958, it is certainly conspecific with *boudecensis*, belonging to the same palaeogeographic area and to the corresponding stratigraphical level in which the Moravian form is found.' We agree with Dr Pokorný and note that the Moravian specimens are less well preserved and that the walls appear superficially less smooth than in the American and East African specimens. The differences in the wall surface and texture are, at the most, no more than subspecific, but may be entirely due to the state of preservation.

Ivanova (1958) described *Cassigerinella globolocula* from the 'Upper Oligocene' beds of the Russian Carpathians, but we are informed by Dr Pokorný (see p. 72) that these beds are probably of Aquitanian age.

STRATIGRAPHICAL RANGE

In East Africa, this species ranges from the base of the Oligocene to high in the Burdigalian. In North Africa, Sicily and Malta, it occurs in both Aquitanian and Burdigalian sediments. In the Cipero formation of southern Trinidad, it ranges from the *Globigerina ampliapertura* Zone to near the top of the '*Globorotalia fohsi robusta* Zone' which is considered here to be uppermost Burdigalian.

Subfamily: GLOBIGERININAE Carpenter, 1862, *nom. trans.*

Genus: *Globigerina* d'Orbigny, 1826

Type species: *Globigerina bulloides* d'Orbigny, 1826 (lectotype selected by Banner & Blow, 1960*a*)

Globigerina ampliapertura Bolli

Globigerina ampliapertura ampliapertura Bolli
(Plates XI A–D, XVII c; Fig. 12*b*)

Globigerina ampliapertura Bolli, 1957 (part), *Bull. U.S. Nat. Mus.* 215, 108, pl. 22, figs. 5a–7b (*not* figs. 4a–b).

DESCRIPTION

The fairly small test consists of about three whorls of slowly enlarging, inflated, moderately embracing chambers. The adult typically possesses four chambers in the last whorl, but four to five chambers may be present in the earlier whorls. The chambers are coiled in a low trochospire; they are slightly convex dorsally and are more strongly convex ventrally, being ovoid in axial view. The equatorial profile of the test is subcircular and the equatorial periphery is weakly lobulate. The axial profile is ovoid, tumid to globose, and the axial periphery is broadly and smoothly rounded. The dorsal surface of the test is distinctly convex but less so than the ventral surface. The sutures are distinctly but not strongly depressed. The ventral intercameral sutures are nearly radial. The dorsal intercameral sutures are initially curved, meeting the spiral suture nearly at right angles, becoming subradial during ontogeny. The spiral suture is moderately lobulate. In dorsal aspect the chambers are initially semicircular in outline, but they become distinctly longer circumferentially than broad radially during ontogeny and become increasingly depressed. The ventral umbilicus is shallow, and not sharply delimited; the last-formed chamber covers the umbilicus formed by the preceding four chambers, and this 'penultimate umbilicus' may be seen within the last-formed aperture (compare *Globorotalia (Turborotalia) increbescens*). The 'exterior umbilicus' is consequently composed of parts of the ventral walls (but not the apertural faces) of the earlier chambers of the last whorl. The aperture is typically fairly high, an intraumbilical, asymmetrical arch, broadest and most steeply arched at its anterior end; extreme forms of this species may possess an aperture which is almost asymmetrically arched. The aperture is furnished with a distinct, thickened rim-like lip. The apertural face is narrow, convex and often relatively smooth. The wall is thick relative to the size of the test, and it is uniformly and fairly coarsely perforate. The surface of the test has a rough, 'granular' and hispid appearance.

REMARKS

We have compared our specimens from East Africa with topotype material from the *Globigerina ampliapertura* Zone, Cipero formation, southern Trinidad. This form should be carefully compared to, and contrasted with, *G. pseudoampliapertura* sp. nov. (see p. 95).

STRATIGRAPHICAL RANGE

This form first appears in the *Globigerina turritilina turritilina* Zone, uppermost Eocene, and ranges throughout the *G. oligocaenica* Zone, Oligocene, in the Lindi area, East Africa. It does not quite range to the top of its zone (lowermost Aquitanian) in the Cipero formation (see Bolli, 1957 *b*, p. 100) where it is replaced by *G. ampliapertura euapertura* (Jenkins).

Globigerina ampliapertura euapertura (Jenkins)
(Plate XI E–G)

Globigerina ampliapertura Bolli, 1957 (part), *Bull. U.S. Nat. Mus.* **215**, 108, pl. 22, figs. 4 a, b (*not* figs. 5 a–7 b).

G. ampliapertura Bolli; Drooger & Batjes, 1959, *Proc. K. Akad. Wet. Amst.*, Ser. B, **62**, no. 3, 174, pl. 1, fig. 1.

G. euapertura Jenkins (in the press).

GENERAL REMARKS

Jenkins is already describing this form from the lowest Aquitanian of Victoria, southern Australia, and we have compared our specimens with his types. We believe Bolli's reference to *Globigerina* cf. *venezuelana* (Bolli, 1957 *b*, p. 100) in the upper part of the *G. ampliapertura* Zone of the Cipero formation, southern Trinidad, to be the same form. This form appears to be subspecifically related to *G. ampliapertura ampliapertura*, but differs mainly in the more depressed later chambers and in the lower, more slit-like aperture as well as in having a slightly more open umbilicus in which the relict aperture of the penultimate chamber may be observed. This form lacks any trace of an umbilical tooth but may have a reduced apertural rim.

STRATIGRAPHICAL RANGE

In the Lindi area, East Africa, this subspecies ranges throughout the Oligocene, whilst in southern Trinidad it occurs throughout the whole of the *G. ampliapertura* Zone, the *Globorotalia opima opima* Zone, and probably also in the basal part of the *Globigerina ouachitaensis ciperoensis* Zone, Cipero formation.

Globigerina angulisuturalis Bolli
(Plate IX Aa–Cc)

Globigerina ciperoensis angulisuturalis Bolli, 1957, *Bull. U.S. Nat. Mus.* **215**, 109, pl. 22, figs. 11 a–c.

REMARKS

The morphology of this form was described by Bolli (1957 *b*). However, it is considered here to be specifically distinct from *Globigerina ouachitaensis ciperoensis* as it possesses not only characteristically 'U'-shaped, excavated intercameral sutures but also a much smaller umbilicus into which a smaller and more restricted aperture, lacking a rim, directly opens; it also possesses a rougher (more hispid) wall. The specimens from Sicily have been directly compared with topotypic specimens from Trinidad.

STRATIGRAPHICAL RANGE

G. angulisuturalis ranges from the base of the *Globorotalia opima opima* Zone to the top of the *Globigerina ouachitaensis ciperoensis* Zone in the Cipero formation, southern Trinidad, and in their equivalents in Venezuela and Barbados. Our figured specimen is from sample WACR 1326, collected from a measured section in the Lower Ragusa limestone formation at Monte Alia ($1\frac{1}{2}$ km north of Monte Rosso) south-east Sicily. The Lower Ragusa limestone formation also contains *Lepidocyclina* (*Eulepidina*), *Miogypsina* and *Spiroclypeus* and is therefore believed to be of Aquitanian age. This species has not been observed either in the rich Oligocene planktonic foraminiferal faunas of East Africa (where its ancestral form *Globigerina ouachitaensis ciperoensis* does occur) or yet in the Oligocene faunas examined from Germany and France.

Globigerina angustiumbilicata Bolli
(Plate IXx–z; Figs. 9 (iv), 16 (vi, vii))

Globigerina ciperoensis angustiumbilicata Bolli, 1957, *Bull. U.S. Nat. Mus.* **215**, p. 109, pl. 22, figs. 12a–13c; ?p. 164, pl. 36, figs. 6a–b.

G. angustiumbilicata Bolli, Blow, 1959, *Bull. Amer. Paleont.* **39**, no. 178, 172, pl. 7, figs. 33a–c.

?*G. ciperoensis* Bolli; Drooger & Batjes, 1959 (*not* Bolli, 1954), *Proc. K. Akad. Wet. Amst.* Ser. B, **62**, no. 3, 179, pl. 1, fig. 10.

REMARKS

This form differs from *Globigerina ouachitaensis ciperoensis* in possessing a smaller umbilicus and a smaller and less highly arched aperture which is typically distinctly asymmetrical and is furnished with a well marked lip. The shape of the aperture in all the subspecies of *G. ouachitaensis* shows a characteristic arching at the posterior end whereas in *G. angustiumbilicata* (as in *G. officinalis*) the aperture is strongly arched at the anterior end only. In the subspecies of *G. ouachitaensis* the last-formed chambers, especially, become subovoid in axial view with a flattening of the apertural face; in *G. angustiumbilicata* (as in *G. officinalis*) the chambers remain subspherical and the apertural face remains smoothly convex. The chambers of *G. angustiumbilicata* are more closely appressed and more tightly coiled, giving a narrower and shallower umbilicus than in *G. ouachitaensis ciperoensis*. The surface texture of *G. angustiumbilicata* is closely similar to that of *G. officinalis* but is distinctly less coarsely perforate and punctate than in *G. ouachitaensis ciperoensis*.

Transitional forms from *Globerigina officinalis* to *G. angustiumbilicata* occur in the Upper Eocene and Oligocene of the Lindi area (Fig. 16). These forms demonstrate that *G. angustiumbilicata* differs from *G. officinalis* by an increased growth rate (i.e. five chambers per whorl instead of four), concomitant with a slight increase in size of the umbilicus.

STRATIGRAPHICAL RANGE

Globigerina angustiumbilicata was described from the *G. ouachitaensis ciperoensis* Zone, Cipero formation; we have carefully compared our material with topotypic specimens. Forms referable to this species first occur in the topmost part of the *Cribrohantkenina danvillensis* Zone, Upper Eocene, and, as previously pointed out by Blow (1959, p. 172), it ranges to the *Globigerina bulloides* Zone, Pozón formation, Venezuela, which is of probable Tortonian age. It occurs in the upper Vindobonian of Nussdorf, Austria, and it is possible that this species ranges still higher. It is probable that many of the forms recorded by non-American writers as '*Globigerina ciperoensis*' may in fact be referable to *G. angustiumbilicata*, especially since this species is particularly common in the Burdigalian of south-west France (Pont Pourquey, Aquitaine).

Globigerina linaperta Finlay
Globigerina linaperta linaperta Finlay
(Plate XIH)

Globigerina velascoensis Cushman var. *compressa* White, 1928 (*not G. compressa* Plummer, 1926), *J. Paleont.* **2**, 196 (?*not* pl. 28, figs. 3a–b).

G. linaperta Finlay, 1939, *Trans. Proc. Roy. Soc., N.Z.*, **69**, 125, pl. 13, figs. 54–57.

G. eocaenica Terquem var. *eocaenica* Terquem, Subbotina, 1953, *Trudi, VNIGRI*, N.S., no. 76, p. 80, pl. 11, figs. 8, 10, 11 (?*not* Terquem, 1882).

G. linaperta Finlay; Said & Kenawy, 1956, *Micropaleontology*, **2**, 157, figs. 27a–b.

Globorotalia tortiva Bolli, 1957 (*nom. nov.*), *Bull. U.S. Nat. Mus.* **215**, 78 (*not* pl. 19, figs. 19–21).

Globigerina linaperta Finlay, Bolli, 1957, *Bull. U.S. Nat. Mus.* **215**, 163, pl. 36, figs. 5a–b.

G. inaequispira Subbotina; Loeblich and Tappan, 1957 (*not* Subbotina, 1953), *Bull. U.S. Nat. Mus.* **215**, 181, pl. 49, figs. 2a–b; pl. 62, figs. 2a–c.

G. triloculinoides Plummer; Loeblich & Tappan, 1957 (part, *not* Plummer, 1926), *Bull. U.S. Nat. Mus.* **215**, 183–184, pl. 62, figs. 3a–c (lectotype of *Globorotalia tortiva* Bolli, 1957), figs. 4a–c; pl. 52, figs. 3, 4, 5; pl. 56, figs. 8a–c.

G. linaperta Finlay, Hornibrook, 1958, *Micropaleontology*, **4**, 33, pl. 1, figs. 19–21 (holotype redrawn).

DESCRIPTION

The fairly small test consists of about three whorls of inflated and moderately embracing chambers. In the early whorls about four chambers are present; these are slowly enlarging, but in the later growth stages the chambers enlarge more rapidly so that only three are typically clearly visible on the ventral side of the adult. The chambers are arranged in a low trochospire; they are slightly convex dorsally and rather more strongly convex ventrally, being broadly ovoid in axial view. The equatorial profile of the test is subcircular to subquadrate and the periphery of the test is broadly and weakly lobulate. In axial profile the test is broadly ovoid to oval and the axial periphery is broadly and smoothly rounded. The dorsal surface of the test is distinctly less convex than the ventral surface. The dorsal sutures are initially somewhat obscure but later become distinctly though not strongly depressed. The dorsal intercameral sutures are initially curved, meeting the moderately lobulate spiral suture at broad angles, but in the last whorl they become subradial. The ventral sutures are subradial and distinctly depressed. In dorsal view the early chambers are semicircular to reniform, but in the last whorl they become characteristically depressed and longer than broad. In ventral view the chambers are subcircular to oval, being similarly depressed terminally. The umbilicus is very small, not sharply delimited, and is usually almost closed by the umbilical overlap of the last chamber. The aperture is a long, very low arch, usually symmetrically disposed about the umbilicus and extending to its broadest limits. The aperture is furnished throughout its length with a distinctive broad lip; the lip is often of uniform breadth throughout, but in some specimens it may broaden slightly at its mid-point. The wall of the test is thick relative to its size, and it is uniformly and coarsely perforate. The coarse perforations give the surface of the test a 'granular' appearance recalling that of *Globigerina ampliapertura ampliapertura*, but distinct from the reticulate, strongly cancellate texture characteristic of the wall of *Globigerinoides quadrilobatus*. The surface of the test is smooth, except for the umbilical margins and the ventral surface where it may be weakly hispid.

REMARKS

Hornibrook (1958, p. 34) has commented on the 'coarse reticulate pattern' of the surface of *Globigerina linaperta*. However, we consider that a distinction should be made between wall structures of this type where the coarse pores equal in diameter the superficial punctae (which are separated by pustules representing the rounded exterior ends of crystal bundles, giving the test surface a 'granular' appearance) and the wall structure of *Globigerinoides quadrilobatus* (s.l.), where a reticulate (cancellate) superficial pattern is produced by the ridge-like margins of juxtaposed coarse punctae of much greater diameter than the pores which open into them, giving a favose texture to the test surface.

Le Calvez (1949, p. 9) has correctly stated that the type specimens of *Globigerina eocaenica* Terquem, 1882, are lost and destroyed. With the assistance of MM. Pierre Marie and M. Lys we endeavoured to obtain topotypes, but the type locality of Terquem's species is no longer accessible. Consequently, we consider that Terquem's taxon *G. eocaenica* should be considered *nomen dubium* and not used.

Loeblich & Tappan (1957) selected and illustrated the lectotype of *Globigerina velascoensis* var. *compressa* White, 1928. They considered it to be synonymous with *G. triloculinoides* Plummer, 1926, but we believe that the latter species should be considered distinct from both *G. linaperta* and White's specimen; it probably belongs to a group of forms which possesses a slower rate of chamber enlargement, a more open umbilicus, a higher aperture (often with a more medially pointed apertural lip) and less depressed chambers. Bolli (1957a, p. 78) noted that the taxon *G. velascoensis* var. *compressa* White, 1928, was a homonym of *G. compressa* Plummer, 1926, and he therefore proposed the new name *Globorotalia tortiva* for White's species. However, as noted by Loeblich & Tappan (1957b, p. 184), the specimen illustrated by Bolli as *G. tortiva* was both specifically and generically distinct from the lectotype of White's species (and therefore of *G. tortiva*) which they had subsequently selected (Loeblich & Tappan, 1957b, p. 184). As an appreciation of Bolli's original work on the Globigerinaceae, we propose the new name *G. (Turborotalia) hansbollii* for Bolli's illustrated form. The holotype of *G. (T.) hansbollii* is that specimen illustrated by Bolli (1957a, pl. 19, figs. 19–21); it was obtained from the *G. pseudomenardii* Zone, Lower Lizard Springs formation, southern Trinidad, and is deposited in the U.S. National

Museum, registered number P. 5066. The diagnosis of the characters of *G. (T.) hansbollii* has been published by Bolli (1957a, p. 78) as for the taxon *G. tortiva*.

Globigerina protoreticulata Hofker (1956b) may represent a particularly coarsely punctate variant of *G. linaperta linaperta* Finlay.

Through the courtesy of Miss R. Todd (U.S. Geological Survey), we have been allowed to examine a paratype (Cushman Collection number 64484a) of *Globigerina patagonica* Todd & Kniker, 1952 (from the Eocene of Chile); this form differs from *G. linaperta linaperta* in possessing less depressed chambers, a weak apertural lip, a much more finely punctate wall, and a higher aperture.

STRATIGRAPHICAL RANGE

In East Africa, this form is known to range from beds of upper Palaeocene age (with a planktonic foraminiferal fauna similar to that of the *Globorotalia pseudomenardii* Zone, Lizard Springs formation, southern Trinidad) to the top of the *Globigerina turritilina turritilina* Zone, uppermost Eocene. It does not occur in the Oligocene of East Africa. In Trinidad, it also ranges from the Palaeocene, but extends only to the top of the Navet formation. It may be significant, in the light of the comments in this work concerning the stratigraphical position of the 'San Fernando formation', that Bolli (1957c) did not record this form from this latter horizon.

Globigerina linaperta pseudoeocaena (Subbotina)
(Plate XIM)

Globigerina pseudoeocaena var. *pseudoeocaena* Subbotina, 1953. *Trudi, VNGRI*, N.S., no. 76, pp. 67–68, pl. 4, figs. 9a–c (?not pl. 5, figs. 1–2).

REMARKS

This subspecies differs from *Globigerina linaperta linaperta* Finlay in possessing a slower rate of chamber enlargement (with four chambers clearly visible ventrally), a more open umbilicus, a more laterally restricted and slightly higher aperture, and less depressed, more globose chambers in the final whorl. All these characters appear to follow from the essential fact that the juvenile characters of *G. linaperta linaperta* persist into the adult of *G.*

linaperta pseudoeocaena, notwithstanding that the maximum size attained by *G. linaperta linaperta* is often less than that attained by *G. linaperta pseudoeocaena*. This may prove to be one of the rare examples of neotonous development in the foraminifera.

The holotypic or lectotypic specimen of *Globigerina triloculinoides* Plummer, 1926, has never been selected unambiguously, in so far that, although Plummer referred to her fig. 10a (plate 8) 'as the type', she also figured another specimen and stated that the 'cotypes' were deposited in the Walker Museum, Chicago. Only the dorsal view of the specimen represented by her fig. 10a was illustrated, and the types of the species have never been redescribed. As the taxonomy of the Globigerinidae has become considerably more refined since Plummer's publication, we have no means by which we may confidently assume that both Plummer's illustrated specimens were even conspecific. However, two superficially similar but actually distinct species are known to exist in the Palaeocene and Eocene, and these distinct species-groups have been recognized by most authors (e.g. Bolli 1957a, p. 70). The stratigraphically younger form is *G. linaperta linaperta* (known from upper Palaeocene to uppermost Eocene), whereas the stratigraphically older form has been recorded from the Danian to the Palaeocene by many authors as *G. triloculinoides* Plummer. In consequence, although we cannot be certain of the identity of the older forms with *G. triloculinoides*, we are referring to them by that name until further evidence becomes available from a re-examination of the types. For the time being we are basing our concept of the species *G. triloculinoides* Plummer upon the topotype figured by Loeblich & Tappan (1957b, pl. 43, figs. 9a–b); *G. triloculinoides* may then be distinguished from *G. linaperta pseudoeocaena* in possessing a more rapid rate of chamber enlargement, more deeply depressed sutures (especially between the last two or three chambers), and a higher, more laterally restricted aperture with a broader, more flaring lip.

Globigerina triloculinoides may prove to be the ancestor of *G. linaperta* (s.l.) for it appears to possess almost all the separate morphological characters seen in the grade from *G. linaperta linaperta* to *G. linaperta pseudoeocaena* but in a different and distinct combination. We suspect that the forms described by Subbotina (1953, pl. 11, figs. 12–14) as *G. eocaena* var. *irregularis* may be morphological intermediates between *G. linaperta linaperta* and *G. linaperta pseudoeocaena*.

STRATIGRAPHICAL RANGE

Subbotina (1953) has recorded this form from the Middle and Upper Eocene of the Caucasus. In East Africa it also ranges from the Middle Eocene but does appear to occur above the *Cribrohantkenina danvillensis* Zone. A similar stratigraphical range is also believed to occur in Trinidad.

Globigerina officinalis Subbotina
(Plate IX A–C; Fig. 16)

Globigerina officinalis Subbotina, 1953 (part), *Trudi, VNIGRI*, N.S. no. 76, p. 78, pl. 11, figs. 1 a–c (holotype), 2 a–c, 6 a–7 c; ?figs. 5 a–c (*not* figs. 3 a–4 c).

G. paratriloculinoides Hofker, 1956, *J. Paleont.* **30**, no. 4, 956, text-fig. 99.

G. parva Bolli, 1957, *Bull. U.S. Nat. Mus.* **215**, 108, pl. 22, figs. 14 a–c (holotype) (*not* hypotype pl. 36, fig. 7).

G. bulloides d'Orbigny, Batjes, 1958 (part) (*not* d'Orbigny, 1826), *Mém. Inst. Sci. nat. Belg.* no. 143, pp. 161–162, pl. XI, figs. 2 a–c.

G. parva Bolli; Drooger & Batjes, 1959, *Proc. K. Akad. Wet. Amst.* ser. B, **62**, no. 3, 175, pl. 1, fig. 5.

DESCRIPTION

The small trochospiral test consists of about three whorls of subglobular to ovoid, inflated chambers arranged four to a whorl. The equatorial profile is subquadrate and the equatorial periphery is lobulate. The axial profile is suboval and the axial periphery is broadly rounded. The dorsal surfaces of the chambers are less inflated than the ventral surfaces. The chambers are constant in shape and are regularly enlarging, with the occasional exception of the last-formed chamber which may be no larger than the penultimate chamber. The sutures are depressed; the spiral suture is lobulate, the dorsal intercameral sutures are slightly curved to nearly radial and the ventral sutures are subradial. The umbilicus is small (almost closed), subrectangular in outline and shallow. Relict apertures are usually not visible within the umbilicus. The aperture of the last chamber is a low to very low asymmetrical arch, highest in its anterior part which is outside the deepest part of the umbilicus. A narrow lip or apertural rim is typically present. The wall is uniformly and finely perforate and hispid, being most hispid in the umbilical region; the apertural face is smoothly convex and as hispid as the remainder of the test.

REMARKS

This form differs from *Globigerina ouachitaensis ouachitaensis* principally in lacking a broad open umbilicus and in its tighter coiling. This form was recorded as *G. parva* sp.nov. by Bolli (1957b) from the *G. ampliapertura* Zone, Cipero formation, southern Trinidad.

Globigerina pseudoedita Subbotina, 1956, may prove to be a transitional form between *G. officinalis* Subbotina and *G. angustiumbilicata* Bolli; this is probably the same form as that illustrated by Drooger & Batjes (1960, p. 1, fig. 5) as *G. angustiumbilicata*.

STRATIGRAPHICAL RANGE

Subbotina (1953) recorded *Globigerina officinalis* from the Upper Eocene and Lower Oligocene of the Caucasus. In the Lindi area, East Africa, it ranges from at least the *Truncorotaloides rohri* Zone to the top of the *Globigerina oligocaenica* Zone. It occurs in the Stampian of south-west France (Cambes, Gironde) and in the Oligocene (Rupelian) of Offenbach and Elmsheim, Germany. Closely similar forms occur in the lower Aquitanian of Escornebéou, south-west France and in the *Globigerina ampliapertura* Zone, Cipero formation, southern Trinidad.

Globigerina oligocaenica Blow & Banner sp.nov.
(Plate X G, L–N)

DIAGNOSIS

The large test consists of about three whorls of rapidly enlarging, moderately inflated, partially embracing chambers coiled in a low trochospire. Four chambers are typically present in the early whorls, characteristically reducing to three in the last whorl. The equatorial profile of the test is subcircular to subquadrate and the equatorial periphery is weakly and broadly lobulate. In axial profile the test is subconical; the dorsal surface is only slightly convex, whereas the ventral side is strongly vaulted. The axial periphery is broadly and smoothly rounded, but there is a distinct, although broad, dorso-peripheral shoulder. Dorsally, the sutures are initially obscure but they become more distinctly depressed during ontogeny; the dorsal intercameral sutures are initially curved, meeting the moderately lobulate spiral suture at broad angles, but between the last two chambers the suture may become subradial. In ventral view only three chambers are visible; the last chamber is of distinctly reniform shape, whilst the earlier two chambers are subreniform and the suture between them is subradial and deeply depressed. In dorsal view the early chambers are semicircular but they become reniform, depressed, and longer than broad during ontogeny. The dorsal surfaces of the chambers are of uniform and gentle convexity throughout ontogeny. The umbilicus is small and is often almost completely covered by the last formed chamber. The umbilicus is typically of triangular shape, but this may vary; it is open and deep but is not usually sharply delimited. The apertural face is a fairly broad, flattened, reniform re-entrant in the ventral face of the last chamber; it is often distinctly less hispid than the remainder of the chamber wall. The aperture extends the width of the umbilicus, along the length of the re-entrant in the ventral face of the last chamber; it is often less hispid than the remainder of the chamber wall. The aperture may vary from a very low to a moderately high arch, but it is always clearly restricted laterally to the re-entrant of the apertural face, which is frequently bounded by lateral lobe-like expansions of the ventral surface of the last chamber. The aperture is furnished with a lip which sometimes is merely rim-like, but which may broaden medially to form a weak 'umbilical tooth'. The wall of the test is fairly thick, moderately coarsely perforate and uniformly and strongly hispid. Maximum diameter of holotype: 0·65 mm; from sample FCRM 1964, Lindi area.

REMARKS

In any observed population of *Globigerina oligocaenica* there exists slight but noteworthy variation in the rate of enlargement of the later chambers; this affects the over-all shape of the test in so far that some specimens may be more globose whilst others may be more quadrate than the holotype. In consequence, there is a comparable degree of variation in the breadth of the later chambers as seen in dorsal view and a similar variation in the degree of overlap of the ventral side of the last chamber over the umbilicus and early chambers of the last whorl. The hispidity and pore size are reduced on the apertural face of some specimens, and this reduction also occurs on the septa.

Globigerina oligocaenica differs from *G. tripartita tripartita* in having more rapidly enlarging chambers which are much more strongly hispid, and in possessing an aperture which is distinctly limited in lateral extent to a well marked re-entrant in the apertural face.

Globigerina oligocaenica differs from *G. tripartita tapuriensis* in possessing more strongly depressed chambers, a less lobulate periphery, a more flattened dorsal surface, a smaller umbilicus, a more restricted intraumbilical aperture set at the base of a distinctly flattened apertural face, and in possessing a much more strongly hispid test (see p. 146).

STRATIGRAPHICAL RANGE

This species has only been observed in beds containing *Nummulites intermedius–fichteli* and *N. vascus* in the Lindi area, which are considered to be Lattorfian–Rupelian (Lower to Middle Oligocene). It has not been observed in numerous Upper Eocene samples from either the East African or Caribbean regions. It does not occur in the *Globigerina ampliapertura* Zone, Cipero formation (lowest Aquitanian) and has not been seen in any lowest Aquitanian sample from the Mediterranean or in any sample from the Vicksburgian and Jacksonian of the American Gulf States. A transitional form between *G. tripartita tapuriensis* and *G. oligocaenica* has been observed in a sample from the Lower Oligocene of Sarawak (see p. 71). The occurrence of this species is reported to us by Bolli and Bermúdez (privately, see p. 71) from the Dominican Republic and Cuba, in beds which should be considered to be of pre-*ampliapertura* Zone age.

We have seen and confirmed their specimens of this species from the lower Alazan formation of Mexico, which indicates that the lower part of this formation is of Oligocene age.

Globigerina ouachitaensis Howe & Wallace

Globigerina ouachitaensis ouachitaensis Howe & Wallace
(Plate IXD, H–K; Fig. 9(vi))

Globigerina ouachitaensis Howe & Wallace, 1932, *Bull. La Conserv. Geol.* no. 2, p. 74, pl. 10, figs. 7a–b.

G. ouachitaensis Howe & Wallace, Hofker, 1956, *Spolia zool. Mus. Hauniensis*, **15**, 220, pl. 33, figs. 9–14.

G. parva Bolli, 1957 (part), *Bull. U.S. Nat. Mus.* **215**, 108 and 164, pl. 36, figs. 7a–c (hypotype) (*not* pl. 22, figs. 14a–c (holotype)).

G. bulloides d'Orbigny; Drooger & Batjes, 1959 (*not* d'Orbigny, 1826), *Proc. K. Akad. Wet. Amst.*, ser. B, **62**, no. 3, 175, pl. 1, fig. 3.

DESCRIPTION

The small trochoid test consists of about four whorls of subglobular inflated chambers arranged four to a whorl throughout growth. The equatorial profile is subquadrate to subtrapezoidal and the equatorial periphery is strongly lobulate. In axial profile, the test is high spired with a strongly depressed spiral suture; the axial periphery is broadly rounded, the dorsal surfaces of the chambers being smoothly convex. Except for the last chamber which may sometimes be abortively reduced (see Plate IXD) all the chambers are constant in shape, rapidly and regularly enlarging. The intercameral sutures are distinctly depressed and nearly radial in direction both ventrally and dorsally. The spiral suture is strongly lobulate. The umbilicus is broad, deep, open and characteristically quadrate in outline. The aperture of the last chamber is a fairly low arch, symmetrical in shape and laterally restricted to the breadth of the umbilicus which it directly faces. The relict apertures of the preceding two or three chambers are visible within the umbilicus. Both these relict apertures and the final aperture are each bordered by a narrow thickened rim. The wall is fairly thin, uniformly and finely perforate; the surface is hispid and the hispidity is strongest on the earlier chambers and particularly on their umbilical margins. The apertural face of the last chamber is smooth.

REMARKS

We are grateful to Professor H. V. Howe for sending us topotypic material from the 'Upper Horizon, Danville Landing, Ouachita River, Catahoula River, Louisiana', with which our figured specimens have been compared. The holotype illustrated by Howe & Wallace (1932) appears to have a slightly reduced final chamber. This is common in our material, and Plate IXD illustrates a form with a reduced final chamber which compares very well with Howe and Wallace's holotype.

We believe that *Globigerina parva* is a junior synonym of *G. officinalis* Subbotina, 1953 (see p. 88); however, Bolli's figured hypotype of *G. parva* from the *Globigerapsis semi-involuta* Zone, Hospital Hill marl, southern Trinidad, is considered to be conspecific with *Globigerina ouachitaensis ouachitaensis*.

STRATIGRAPHICAL RANGE

This form is common in the Upper Eocene and Oligocene of East Africa and has also been observed in the Rupelian of Offenbach and Elmsheim, Germany. It also occurs in the *Globigerina ampliapertura* Zone of southern Trinidad. It also occurs in the lower Aquitanian of Escornebéou, southwest France, but not in the middle Aquitanian of Moulin de l'Église. Published records of this species include Hofker's reference (1956a) from the Dutch Oligocene.

Globigerina ouachitaensis ciperoensis (Bolli)
(Plate IXE–G; Fig. 9 (i–iii))

Globigerina cf. *concinna* Reuss; Cushman & Stainforth, 1945, *Spec. Publ. Cushman Found. Foram. Res.* no. 14, p. 67, pl. 13, figs. 1a–b.

G. ciperoensis Bolli, 1954, *Contr. Cushman Found. Foram. Res.* **5**, pt. 1, p. 1, text-figs. 3–6.

G. ciperoensis ciperoensis Bolli, 1957, *Bull. U.S. Nat. Mus.* **215**, 109, pl. 22, figs. 10a–b.

REMARKS

This form differs typically from *Globigerina ouachitaensis ouachitaensis* in the possession of five chambers in each whorl throughout ontogeny, although some stratigraphically early forms show a transition from the four-chambered *G. ouachitaensis*

ouachitaensis. This closely parallels the relationship seen between *G. officinalis* and *G. angustiumbilicata* (see p. 139) as well as the relationship between *G. bulloides bulloides* and *G. bulloides concinna* in the Tortonian. Concomitant with the increase in the number of chambers per whorl, the umbilicus loses its regularly quadrate outline and approaches an asymmetric pentagon, but it remains broad and deep. The aperture retains its height and intra-umbilical restriction. Dorsally, the spire often appears lower than in typical *G. ouachitaensis ouachitaensis*, but this seems to be related to the increase in number of chambers per whorl. The apertural rim is less well marked than in *G. ouachitaensis ouachitaensis* and may be absent, especially in stratigraphically younger forms.

We have compared both the East African and Sicilian specimens with topotypic material from southern Trinidad.

STRATIGRAPHICAL RANGE

This subspecies has not been recorded or observed in Upper Eocene beds either in East Africa or Trinidad. However, it occurs in samples containing *Nummulites intermedius–fichteli* and *Lepidocyclina* (*Eulepidina*) in the Lindi area, and also in samples with *Miogypsina*, *Lepidocyclina* (*Eulepidina*) and *Spiroclypeus* in a measured section at Monte Alia, south-east Sicily. Occasional specimens of *G. ouachitaensis ciperoensis* occur in beds with *Miogypsina* in East Africa. It also occurs in the Aquitanian of Moulin de l'Église and Escornebéou, Aquitaine, south-west France. It does not seem to occur in the very lowest part of the Oligocene of the Lindi area (? = Lattorfian); accordingly, the range of this form is now considered to be at least Rupelian to lower Aquitanian.

Globigerina ouachitaensis ciperoensis occurs abundantly, with a fauna similar to that of the *G. ouachitaensis ciperoensis* Zone of southern Trinidad and Sicily, in samples sent to us by Professor M. N. Bramlette (Scripps Institution, University of California) from the Relizian of California. These samples included one from the lower Relizian (Adelaide Quadrangle, Salinas Valley) and one from the *Siphogenerina hughesi* Zone (Bradley Quadrangle, section 22, T. 9 S–R 24. W, Salinas Valley).

Renz (1948, pp. 78, 108) equated the lower Relizian of California to the European upper Aquitanian and the upper Relizian to the Burdigalian. Thus, following the correlations of Renz, the *Globigerina ouachitaensis ciperoensis* Zone is admittedly of Lower Miocene age. We believe that it is in fact a zone belonging to the lower part of the Aquitanian (Lower Miocene) (see also p. 35 in Part 1 of this work).

Globigerina ouachitaensis gnaucki Blow & Banner
subsp.nov.
(Plate IX L–N)

DIAGNOSIS

The fairly small test consists of about three whorls of subglobular to slightly ovoid, inflated chambers arranged four to four and a half in each whorl in a moderately high trochospire. The rate of enlargement of the chambers increases during ontogeny; in dorsal aspect the early chambers are of semi-circular outline, being distinctly longer circumferentially than they are broad radially, but in the last whorl the chambers broaden more rapidly than they lengthen and become slightly ovoid in axial profile. The sutures are distinctly and broadly depressed; the dorsal intercameral sutures are initially curved but become nearly radial. The spiral suture and the equatorial periphery are strongly lobulate. The axial periphery is broadly rounded. The umbilicus is fairly broad, open and deep and even in four-chambered forms loses the square outline typical of *Globigerina ouachitaensis ouachitaensis*. The apertural face is less flattened than in *G. ouachitaensis ouachitaensis* and consequently the umbilicus is less clearly delimited from the rest of the ventral surface. The intraumbilical aperture is a fairly low arch, highest at its anterior end, bordered by a distinct rim. Relict apertures of the last two or three chambers are visible within the umbilicus. The wall is fairly thin, uniformly and finely perforate; its surface is cancellate and hispid. The hispidity is strongest on the earlier parts of the test, being reduced on the last chamber where the apertural face is smooth. Maximum diameter of holotype: 0·36 mm; from sample FCRM 1965, Lindi area.

REMARKS

This form differs from *Globigerina ouachitaensis ouachitaensis* in its loss of a quadrate umbilicus and symmetrical aperture, and in its acquisition of a changed growth rate and a distinctly more hispid test. It differs from *G. ouachitaensis ciperoensis* in having a higher-spired test with fewer chambers per whorl, a more strongly developed apertural rim, more rapidly enlarging chambers and a more hispid test. It differs from *G. praebulloides praebulloides* in having more, less rapidly enlarging chambers per whorl, a smaller aperture, and a more hispid test.

This subspecies is named for F. R. Gnauck, Palynologist, The British Petroleum Co. Ltd, in recognition of his help with the photographs (Plates XVI and XVII).

STRATIGRAPHICAL RANGE

Globigerina ouachitaensis gnaucki ranges from the *G. turritilina turritilina* Zone, Upper Eocene, to the *G. oligocaenica* Zone, Oligocene, in the Lindi area. It has not been seen in any Aquitanian sediments either in Europe or the Caribbean region.

Globigerina praebulloides Blow
emended

Globigerina praebulloides praebulloides Blow
(Plate IX o–q)

Globigerina praebulloides Blow, 1959, *Bull. Amer. Paleont.* **39**, no. 178, p. 180, pl. 8, figs. 47a–c; pl. 9, fig. 48.

G. cf. *trilocularis* d'Orbigny, Bolli, 1957 (part) (*not G. trilocularis* Deshayes, 1832), *Bull. U.S. Nat. Mus.* **215**, 110, pl. 22, figs. 8a–c (*not* figs. 9a–c).

DIAGNOSIS

The fairly small test consists of two to three whorls of subglobular to ovoid, inflated but little embracing chambers arranged four to a whorl in a low trochospire. The equatorial profile (in ventral view) is subtrapezoid and the equatorial periphery is strongly lobulate. The axial profile approaches an oval shape. The axial periphery is broadly rounded; the chambers are slightly elongate ventrally, the ventral surfaces being strongly convex and the dorsal surfaces slightly flattened. The chambers are regularly and rapidly enlarging; in dorsal view the

early chambers are of reniform outline, but the later chambers become semicircular and are at least as broad radially as they are long circumferentially. The sutures are distinctly, broadly and deeply depressed. The dorsal intercameral sutures are initially curved but increasingly approach a radial direction during ontogeny. The spiral suture is lobulate, and the spire projects above the plane of the dorsal surface. The umbilicus is broad, deep and open; as the apertural faces of the chambers are but slightly flattened, the margins of the umbilicus are not sharply delimited and the umbilicus itself does not possess a clear quadrangular outline. The intraumbilical aperture of the last chamber is a high assymmetrical arch, highest and most steeply arched at its anterior end; it is bordered by a thin rim-like lip broadest at its mid-point. The relict aperture of the penultimate chamber can be seen within the umbilicus. The wall is fairly thick, uniformly and finely perforate and often distinctly but fairly finely hispid.

REMARKS

This form is the ancestor of *Globigerina bulloides* d'Orbigny (see Banner & Blow, 1960a) and the differences between them have already been discussed by Blow (1959, pp. 180, 181). *G. praebulloides praebulloides* differs from *G. ouachitaensis ouachitaensis* in possessing more rapidly enlarging chambers so that in equatorial view the test appears distinctly longer than broad (i.e. not equidimensional with regard to the diameters of the test) and in possessing a convex hispid apertural face to each chamber so that the umbilicus lacks a distinct quadrate outline. The aperture of *G. praebulloides praebulloides* is asymmetrically arched whilst that of *G. ouachitaensis ouachitaensis* is symmetrical.

STRATIGRAPHICAL RANGE

This form ranges from the base of the *Globigerapsis semi-involuta* Zone, Upper Eocene, to the middle part of the *Globorotalia cultrata/Globigerina nepenthes* Zone (Helvetian). Above this level populations of *G. praebulloides praebulloides* are replaced by *G. bulloides* but juveniles of the latter are difficult to separate from *G. praebulloides praebulloides*. This subspecies has been observed in the Rupelian of Elmsheim and Offenbach (Germany), the

Stampian of Cambes (Gironde) and the Aquitanian of Moulin de l'Église and Escornebéou (south-west France), and references by many authors to the occurrence of *G. bulloides* in these beds should correctly be ascribed to *G. praebulloides*.

Globigerina praebulloides leroyi Blow & Banner
subsp.nov.
(Plate IX R–T; Fig. 9 (v))

Globigerina officinalis Subbotina, 1953 (part, *not* holotype), *Trudi*, *VNIGRI*, N.S., **76**, 78, pl. 11, figs. 4a–c (*not* figs. 1a–3c, 5a–7c).

G. globularis Roemer, Batjes, 1958 (part) (*not* Roemer, 1838) *Mém. Inst. Sci. nat. Belg.* no. 143, pp. 161–162, pl. XI, figs. 4a–c (*not* figs. 3 and 5).

DIAGNOSIS

The fairly small test consists of about two to three whorls of subglobular to very slightly ovoid, inflated, partially embracing chambers arranged four to a whorl in a low trochospire. The equatorial profile (in ventral view) is subquadrate to subtrapezoid and the equatorial periphery is lobulate. The axial profile is suboval; the axial periphery is broadly rounded, the ventral sides of the chambers being only slightly more convex than the dorsal sides. The chambers are regularly and fairly rapidly enlarging; in dorsal view the early chambers are broadly reniform, becoming hemispherical during ontogeny. The sutures are distinctly and broadly depressed. The dorsal intercameral sutures are initially curved but become increasingly radial during ontogeny. The spiral suture is moderately lobulate, and the spire projects slightly above the plane of the dorsal surface. The umbilicus is small, but deep and open; the apertural faces of the chambers are but slightly flattened and the umbilicus is not sharply delimited and does not possess a clear quadrangular outline. The intraumbilical aperture of the last chamber is a low symmetrical arch highest at its mid-point; it is bordered by a distinct rim-like lip. The relict aperture of the penultimate chamber can be seen within the umbilicus. The wall is fairly coarsely perforate relative to the size of the test and it is hispid, often most strongly so on the dorsal side and on the margins of the umbilicus. Maximum diameter of holotype: 0·265 mm; from sample FCRM 1965, Lindi area.

REMARKS

This subspecies is distinguished from *Globigerina praebulloides praebulloides* principally by its smaller umbilicus, lower and more symmetrical aperture, slower rate of chamber enlargement and greater degree of embrace between its chambers.

It is distinguished from *Globigerina ouachitaensis ouachitaensis* by its much smaller umbilicus which lacks a distinct quadrate shape, by its smaller aperture and by its relatively unflattened and distinctly hispid apertural face. It is distinguished from *G. officinalis* by its deeper umbilicus, symmetrical aperture, its more coarsely perforate and hispid test.

This subspecies is named for Professor L. W. LeRoy (Colorado School of Mines) in recognition of his work on foraminiferal biostratigraphy.

STRATIGRAPHICAL RANGE

This subspecies is widely distributed in both the Caribbean and East African regions. In the Lindi area it ranges from the *Globigerina turritilina turritilina* Zone, Upper Eocene, to the top of the Oligocene. Elsewhere in East Africa it occurs in beds of Aquitanian age. In Trinidad it ranges from the *G. ampliapertura* Zone to, at least, the upper Aquitanian *Globigerinatella insueta* Zone. It has been observed in the Aquitanian of Moulin de l'Église and Escornebéou, south-west France, and also in the Oligocene of Elmsheim and Offenbach, Germany. Batjes (1958) illustrated a form as *Globigerina globularis* Roemer from the Oligocene of Belgium; this form seems to be the same as *G. praebulloides leroyi*. We consider *G. globularis* Roemer to be *nomen dubium*. Subbotina's specimen, referred by her to *G. officinalis*, came from the uppermost Eocene (*Bolivina* Zone) of the Caucasus and the *Bolivina* Zone may prove to be equivalent to the *G. turritilina turritilina* Zone of East Africa.

Globigerina praebulloides occlusa Blow & Banner
subsp.nov.
(Plate IX U–W; Fig. 14 (i–ii))

Globigerina trilocularis d'Orbigny, Roemer, 1838 (?*not* d'Orbigny, 1826, *nom. nud.*; not *Globigerina trilocularis* Deshayes, 1832), *Neues Jb. Min. Geol. Paläont*, p. 390, pl. 3, fig. 41a.

G. cf. *trilocularis* d'Orbigny, Bolli, 1957 (part) (*not G. trilocularis* Deshayes, 1832), *Bull. U.S. Nat. Mus.* **215**, 110, 163, pl. 22, figs. 9a–c (*not* figs. 8a–c); pl. 36, figs. 3a–b.

G. globularis Roemer; Drooger & Batjes, 1959 (*not* Roemer, 1838), *Proc. K. Akad. Wet. Amst.*, ser. B, **62**, no. 3, p. 174, pl. 1, fig. 2.

DIAGNOSIS

The fairly small test consists of about two to three whorls of subglobular to ovoid, fairly inflated chambers arranged in a low trochospire. The equatorial profile in ventral view is subquadrate to subtrapezoid and the equatorial periphery is strongly lobulate. The axial periphery is broadly rounded, the ventral sides of the chambers being more strongly convex than the dorsal sides. The chambers are regularly and fairly rapidly enlarging; in ventral view they are distinctly ovoid, whilst in dorsal view they are reniform, becoming hemispherical. The sutures are distinctly, broadly and deeply depressed. The dorsal intercameral sutures are initially curved but become radial during ontogeny. The spiral suture is lobulate and the spire projects above the plane of the dorsal surface. The umbilicus is open but narrow and shallow; as the apertural faces of the chambers are slightly flattened, the umbilicus is fairly sharply delimited, although it does not possess a clear quadrangular outline. The intraumbilical aperture of the last chamber is a very low, asymmetrical arch, more steeply arched at its anterior end; it lacks a lip or rim. Typically, relict apertures of the earlier chambers cannot be seen within the umbilicus. The wall is fairly thick, uniformly perforate and distinctly hispid; advanced forms become distinctly punctate. The hispidity is strongest over the peripheral parts of the chambers, on the dorsal surface and within the umbilicus. Maximum diameter of holotype: 0·37 mm; from sample FCRM 1922, Lindi area.

REMARKS

Globigerina praebulloides occlusa differs from *G. praebulloides praebulloides* in possessing a smaller, shallower umbilicus, a smaller, lower aperture which lacks a lip or rim and in possessing a slightly thicker, rougher and more coarsely perforate wall.

G. praebulloides occlusa is distinguished from *G. praebulloides leroyi* in possessing a shallower umbilicus and an asymmetrical aperture which lacks a lip. The chambers of *G. praebulloides occlusa* are more embracing than those of *G. praebulloides praebulloides*, but are less tightly embracing than those of *G. praebulloides leroyi*.

It is probable that the form illustrated by Roemer (1838) as '*G. trilocularis* d'Orbigny' from marine sands of North Germany, near Osnabrück (probably Oligocene), is synonymous with *G. praebulloides occlusa* subsp.nov. However, Roemer's specimen seems distinct from that drawn by d'Orbigny (published by Fornasini, 1897), and in any case *G. trilocularis* d'Orbigny, 1826, was *nomen nudum*. With the assistance of MM. Marie and Lys, we searched the d'Orbigny collections in Paris, but were forced to conclude that all d'Orbigny's specimens which he referred to *G. trilocularis* have been lost. As the name was first validly described by Deshayes in 1832, the taxon should be credited to him and not to d'Orbigny, and only Deshayes' specimens can now be considered to rank as available syntypes. Unfortunately, Deshayes did not illustrate his form and we have not been able to locate his collections. Consequently it is not possible to use the name *G. trilocularis* Deshayes until his material is redescribed. The published illustrations of Fornasini (1898) and Roemer (1838) are, together with d'Orbigny's 'planches inédites', published by Fornasini, without status.

STRATIGRAPHICAL RANGE

This subspecies is known to occur in the Middle Eocene and ranges throughout the Upper Eocene and Oligocene to about the middle part of the Aquitanian (*Globorotalia kugleri* Zone) (see also §VII, p. 136).

It has been observed in the Oligocene of Offenbach and Elmsheim, the Stampian of south-west France (Cambes, Gironde) as well as in the Aquitanian of Moulin de l'Église and Escornebéou, Aquitaine. It is widely distributed in both the Caribbean and East African regions.

Globigerina pseudoampliapertura Blow & Banner sp.nov.
(Plates XII A–C, XVII A, E; Fig. 12 *c*)

?*Acarinina centralis* (Cushman & Bermúdez), Subbotina, 1953 (part), pp. 237–239, pl. 25, figs. 10a–11c only.

'Specimens transitional between *Globorotalia centralis* Cushman & Bermúdez and *Globigerina ampliapertura* Bolli', Bolli, 1957, *Bull. U.S. Nat. Mus.* **215**, 164, pl. 36, figs. 9, 10.

DIAGNOSIS

The fairly large test consists of about three whorls of slowly enlarging, moderately inflated, partially embracing chambers coiled in a low trochospire. The adult test possesses about four chambers in the last whorl, the number of chambers per whorl often reducing slightly during ontogeny. The chambers are only slightly convex dorsally but are relatively very convex ventrally, being strongly ovoid to rounded-triangular in axial view. The equatorial profile is subcircular and the equatorial periphery is weakly lobulate. In axial profile the test shows a marked dorsal flattening and a strongly convex ventral side; the axial periphery is broadly and smoothly rounded, but a distinct dorso-peripheral shoulder is present. The ventral intercameral sutures are distinctly but not strongly depressed, and are subradial. Dorsally, the sutures are initially indistinct but become increasingly depressed during ontogeny. The dorsal intercameral sutures are initially curved, meeting the spiral suture at acute angles, but the later dorsal intercameral sutures become straighter although remaining oblique. In dorsal aspect, the chambers are initially semicircular in outline, but during ontogeny they become increasingly longer than broad; often the anterior part of the chamber is narrower than the posterior part. The dorsal sides of the chambers tend to become more inflated and less depressed during ontogeny. The ventral umbilicus is small but deep, and is not covered by the umbilical margin of the last chamber. The aperture is interiomarginal, intraumbilical, a fairly high, somewhat asymmetrical arch sometimes bordered by a weak and narrow lip or rim. The apertural face is narrow, somewhat flattened and often hispid. The relict aperture of the penultimate chamber may typically be observed within the umbilicus. The wall is fairly thin relative to the size of the test and is uniformly and finely perforate.

The surface of the test is smooth except for weakly hispid areas around the umbilicus and aperture and sometimes on the earlier whorls seen dorsally. Maximum diameter of holotype: 0·60 mm; from sample FCRM 1923, Upper Eocene, Lindi area.

REMARKS

Globigerina pseudoampliapertura sp.nov. is distinguished from *G. ampliapertura* by the presence of the smooth 'non-granular', finely perforate wall (see Plate XVII A–E) in the former species. The chamber shape in *G. pseudoampliapertura* is typically asymmetrical anteriorly-posteriorly, and resembles that of *Globorotalia centralis*; this contrasts with *Globigerina ampliapertura* which has a chamber symmetry resembling that of *Globorotalia increbescens*. The umbilicus of *Globigerina pseudoampliapertura* is more open than in *G. ampliapertura* and the coiling looser. Populations studied suggest that the average size attained by adults of *G. pseudoampliapertura* is distinctly greater than that attained by adults of *G. ampliapertura*. Not only does *G. pseudoampliapertura* differ from *Globorotalia centralis* in possessing an intraumbilical aperture but also by developing a more open umbilicus (in which relict apertures can be clearly seen) and more inflated dorsal sides of the later chambers.

STRATIGRAPHICAL RANGE

In the Lindi area, the species evolves from and replaces *Globorotalia centralis* at about the middle part of the *Cribrohantkenina danvillensis* Zone. It ranges to the top of the *Globigerina turritilina turritilina* Zone, Upper Eocene, but does not occur in Oligocene or younger beds. Specimens apparently referable to this species occur in the *Bolivina* Zone, Upper Eocene, of the Caucasus (Subbotina, 1953). Forms recorded by Bolli (1957 *c*) from the 'San Fernando formation' may be from derived blocks of Upper Eocene age.

Globigerina senilis Bandy
(Plate XI R–U)

Globigerina ouachitaensis Howe & Wallace var. *senilis* Bandy, 1949, *Bull. Amer. Paleont.* **32**, no. 131, p. 121, pl. 22, figs. 5a–c.

REMARKS

The holotype of *Globigerina senilis* Bandy can be seen, from the original illustration, to possess a slightly reduced final chamber which possesses a very low aperture. Such specimens form a high proportion of our own material and one such form is figured here for comparison (Plate XI U). However, regularly developed specimens of the species (Plate XI R–T) possess a broader umbilicus and a more highly arched, asymmetrical aperture, bordered by a narrow rim-like lip, than in Bandy's figured holotype. We consider this species to be distinct from *G. ouachitaensis ouachitaensis* because of its thicker, more coarsely perforate wall which possesses a 'granular' surface texture. It also differs from *G. ouachitaensis ouachitaensis* by possessing more depressed chambers, a less distinctly quadrate umbilicus, and an asymmetrical final aperture situated at the base of a distinctly convex apertural face.

STRATIGRAPHICAL RANGE

This species was originally described from the Upper Jacksonian of Alabama. In East Africa it ranges from the base of the *Cribrohantkenina danvillensis* Zone, Upper Eocene, to the top of the *Globigerina oligocaenica* Zone, Oligocene. We consider it likely that Bandy's records of this species from the Red Bluff and Mint Springs (Vicksburgian) refer to specimens reworked from the underlying Eocene.

Globigerina tripartita Koch *emended*

Globigerina tripartita tripartita Koch
(Plate X A–F; Fig. 18)

Globigerina bulloides d'Orbigny var. *tripartita* Koch, 1926, *Ecl. geol. Helv.* **19** (1925–26), no. 3, p. 746, text-figs. 21 a–b (p. 737).

Globigerina rohri Bolli, 1957, *Bull. U.S. Nat. Mus.* **215**, 109, pl. 23, figs. 1 a–4 b.

DESCRIPTION

The moderately large test consists of about three whorls of rapidly enlarging, moderately inflated, partially embracing chambers coiled in a low trochospire. Four chambers are typically present in the early whorls, often reducing to three in the last whorl. The equatorial profile of the test is subcircular and the equatorial periphery is weakly and broadly lobulate. In axial profile the test is subconical; the dorsal surface is only slightly convex, whereas the ventral side is highly vaulted. The axial periphery is smoothly and broadly curved, there being no sharply angular peripheral shoulder between the weakly convex dorsal chamber sides and the strongly vaulted ventral ones. Dorsally, the sutures are initially obscure, but they become distinctly depressed during ontogeny; the dorsal intercameral sutures are initially curved, meeting the moderately lobulate spiral suture at broad angles, but they later become subradial. The ventral intercameral sutures are moderately deeply depressed. The chambers of the last whorl are strongly depressed. In dorsal view the early chambers are semicircular to reniform but, during ontogeny, they become very much longer than broad; they often possess a narrow, steeply sloping, 'bevelled' area adjacent to their sutures. In ventral view the chambers are subreniform and often only three are visible, even when it can be seen dorsally that four chambers are present in the last convolution. The umbilicus is deep and small but open; it is sharply delimited by the steeply sloping umbilical margins of the chambers of the last whorl. The apertural face of the last chamber is narrow, distinctly flattened, and is often quite sharply delimited from the rest of the chamber surface. The aperture is a very low arch, entirely intraumbilical in position; it is almost symmetrical and is bordered by a strong flap-like lip which often becomes sharply pointed medially and projects into the umbilicus. The wall of the test is fairly thick, moderately coarsely perforate and hispid. The hispidity is usually strongest on the ventral sides of the earlier chambers of the last whorl, especially around the umbilicus; the hispidity is often greatly reduced on the surface of the last chamber which is often smooth away from its sutures. Maximum diameter of Koch's holotype 0·59 mm.

REMARKS

Populations of this subspecies, which possess many individuals of apparent maximum size and which are often characterized by an abundance of specimens with only three chambers in the last whorl, seem to contain especially large numbers of specimens with aborted end-chambers. These aborted final chambers usually retain their normal position in the progression of the spire but are reduced in volume, are variable in shape (although retaining the fundamental form of the normal chamber), and often partially or wholly cover the umbilicus. Such a misshapen final chamber is present in the holotype of *Globigerina bulloides* var. *tripartita* Koch (Plate XA–C) which we have been allowed to examine and illustrate through the kindness of Dr E. Gasche, Naturhistorisches Museum, Basel. An even more abortive final chamber is present on the holotype of the synonymous species *G. rohri* Bolli. We have been able to study topotypic populations of *G. rohri* from the *Globorotalia opima opima* Zone, Cipero formation, southern Trinidad, and have concluded that there is no reason to separate Bolli's species from *Globigerina tripartita tripartita* Koch. The Cipero specimens usually possess a somewhat abraded surface on which the hispidity is only present within the shelter of the umbilicus, as is the case in Koch's holotype.

Later evolutionary forms of *Globigerina tripartita tripartita* show the reduction from four to three chambers per whorl earlier in ontogeny; apparently associated with this is the frequent development of aborted final chambers.

Globigerina tripartita tripartita has evolved from *G. yeguaensis yeguaensis* (see p. 141); Fig. 15a illustrates intermediate types between the two forms.

Globigerina tripartita tripartita differs from *G. yeguaensis pseudovenezuelana* in possessing fewer chambers per whorl, which are also more rapidly enlarging and more depressed; the umbilicus of *G. tripartita tripartita* is smaller and more restricted. Similar characters serve to distinguish this form from *G. venezuelana* Hedberg. *G. exima* Todd may be closely related to *G. tripartita tripartita*.

STRATIGRAPHICAL RANGE

Koch (1926) first described this form from the lower beds of the 'Globigerinenmergel' of East Borneo; this horizon was probably Upper Eocene or Lower Oligocene. Bolli (1957b) has recorded this form (as *Globigerina rohri*) from the *G. ampli-* *apertura* Zone to the '*Catapsydrax*' *dissimilis* Zone, Cipero formation of southern Trinidad, but his stratigraphically higher records (i.e. those above the *Globorotalia kugleri* Zone) are probably, like those of Blow (1959), referable to *Globoquadrina dehiscens praedehiscens* subsp.nov. (p. 116). In East Africa, *Globigerina tripartita tripartita* occurs in the topmost *Truncorotaloides rohri* Zone (possibly top Middle Eocene), and ranges throughout the Upper Eocene and Oligocene. Elsewhere in East Africa it has been observed rarely in the lower part of the Aquitanian. It has not been observed in any of the continental European Oligocene or Aquitanian which has been studied, and it may prove to be a tropical species.

Globigerina tripartita tapuriensis Blow & Banner subsp.nov.
(Plate XH–K)

DIAGNOSIS

The moderately large test consists of about three whorls of rapidly enlarging, inflated, little embracing chambers coiled in a low trochospire. Four chambers are present in the early whorls, typically reducing to three in the last whorl. The equatorial profile of the test is subcircular to subquadrate and the equatorial periphery is broadly and smoothly lobulate. In axial profile, the test is subovoid; the dorsal surface is moderately convex whereas the ventral side is strongly convex. The axial periphery is smoothly and broadly rounded with no distinct dorso-peripheral shoulder. Dorsally, the sutures are initially obscure, but they become distinctly depressed during ontogeny; the dorsal intercameral sutures are initially curved, meeting the lobulate spiral suture at broad angles, but they later become subradial. The ventral intercameral sutures are moderately deeply depressed. In dorsal view the chambers are semicircular to reniform throughout ontogeny, but become rather longer than broad. In ventral view the chambers are subcircular to subreniform. The umbilicus is indistinctly delimited but appears to be moderately broad and of triangular shape; it is open and deep. The apertural face of the last chamber is convex and not clearly delimited from the rest of the chamber surface; however, in umbilical view the apertural face may be seen to occupy an ill-defined, roughly reniform re-entrant. The aperture is a fairly low but long arch extending symmetrically across the full width of the umbilicus, and along the full length of the apertural face. The aperture is furnished throughout its length with a distinct narrow rim-like lip broadest at its mid-point. The relict aperture of the penultimate chamber can be seen within the umbilicus. The wall of the test is fairly thick, moderately coarsely perforate and uniformly hispid. The hispidity is usually strongest around the umbilical margins. Maximum diameter of holotype 0·73 mm; from sample FCRM 1964, Lindi area.

REMARKS

This subspecies is distinguished from *Globigerina tripartita tripartita* by its much less depressed chambers, more convex dorsal surface, much more open umbilicus, higher, more convex apertural face, and larger, more open aperture with a reduced apertural lip.

STRATIGRAPHICAL RANGE

This subspecies occurs only in the *Globigerina oligocaenica* Zone, Oligocene, Lindi area. Like *G. tripartita tripartita*, it may be restricted to tropical areas. It has not been recorded from either the 'San Fernando formation' or the Cipero formation of southern Trinidad; we have examined much material from all the zones of the Cipero formation, and have failed to find it. An intermediate form between this subspecies and *G. oligocaenica* has been observed in a Lower Oligocene sample from Sarawak (see p. 89).

Globigerina turritilina Blow & Banner sp.nov.

Globigerina turritilina turritilina Blow & Banner subsp.nov.
(Plate XIII D–G)

DIAGNOSIS

The large or fairly large test consists of about three whorls of rapidly enlarging, inflated, weakly embracing chambers, arranged in a high trochospire with four chambers in each whorl. The equatorial profile of the test is subquadrate and the equatorial periphery is strongly lobulate. In axial profile the test is ovoid, the dorsal surface being strongly convex (high-spired) and the ventral surface being inflated but relatively weakly convex. The axial profile of the test is strongly lobulate, and the axial profile of each chamber is broadly and smoothly curved, no dorso-peripheral shoulder being apparent. The sutures are broadly and deeply depressed. The spiral suture is strongly lobulate. The dorsal intercameral sutures are weakly curved to subradial throughout, owing to the height of the spire and the strong inflation of the chambers. The ventral intercameral sutures are subradial. In dorsal view, the chambers are semicircular to reniform and in ventral and axial views they are subovoid. The umbilicus is moderately broad, deep, open and of quadrate outline. It is not sharply delimited, as the apertural faces of the chambers are only weakly flattened. The aperture of the last chamber is a fairly low arch, practically symmetrical in shape, and it is entirely restricted to a position within the umbilicus which it directly faces. The relict aperture of the penultimate chamber may be visible within the umbilicus. The final aperture is bordered by a distinct lip throughout its length, often broadest at its mid-point. The wall is fairly thick, uniformly and fairly coarsely perforate. The surface is weakly cancellate and finely hispid. The hispidity is strongest on the ventral sides of the chambers and their umbilical margins. Maximum diameter of holotype 0·60 mm; from sample FCRM 1964, Lindi area.

REMARKS

The only known members of the Globigerinidae which grossly resemble *Globigerina turritilina turritilina* are *G. turritilina praeturritilina* subsp. nov. (see p. 99), *G. helicina* d'Orbigny, 1826 (see Banner & Blow, 1960a), and *Globigerinoides mitrus* Todd (1957; from the Tagpochau limestone, Burdigalian of Saipan), which differs generically and is probably quite unrelated. *Globigerina helicina* d'Orbigny has less globular chambers, a less lobulate periphery, a lower dorsal spire and a less hispid test than *G. turritilina turritilina*.

The larger specimens of *Globigerina turritilina turritilina* from both the uppermost Eocene and Oligocene populations frequently possess aborted final chambers which are reduced in size and are of variable shape and position within the limits of the ventral side. The holotype may be seen (Plate XIII E–G) to have lost such an aborted final chamber, although the paratype (Plate XIII D) still possesses one. (See also p. 146.)

STRATIGRAPHICAL RANGE

Globigerina turritilina turritilina has been found to occur, in East Africa, only in the uppermost Bartonian and in the Lattorfian-Rupelian (cf. *G. turritilina praeturritilina*).

Dr C. G. Adams has kindly shown us a sample containing this form from the mouth of the Benaleh River, Baram Headwaters, Sarawak, where it is associated with the Oligocene *Nummulites fichteli* (see also p. 71).

This species has not been recorded or observed by us in the *Globigerina ampliapertura* Zone, Cipero formation, southern Trinidad, or in equivalent horizons in the Caribbean–Central American regions, including the Vicksburgian of the American Gulf States. Similarly, it has not been recorded from, or observed in, the Upper Eocene of these areas, so that, as the form appears to be a tropical or subtropical species, this is confirmatory evidence for the absence of both the uppermost Eocene and Oligocene in the Central American Region.

Globigerina turritilina praeturritilina Blow & Banner
subsp.nov.
(Plate XIII A–C)

DIAGNOSIS

This subspecies is similar to *Globigerina turritilina turritilina* but differs in possessing more depressed (slightly less inflated) chambers, a much broader umbilicus, a higher final aperture and a slightly less coarsely perforate wall. The height of the aperture is not always much greater in relation to the shape of the chamber but always clearly opens more to the exterior, less directly into the umbilicus, than in *G. turritilina turritilina*. The apertural face of *G. turritilina praeturritilina* is narrowly but distinctly flattened. Maximum diameter of holotype 0·48 mm; from sample FCRM 1645, Lindi area.

REMARKS

In *Globigerina turritilina praeturritilina* the umbilicus may be so open and deep that the relict apertures of almost all the earlier chambers of the last and preceding whorls may be observed within it. Frequently, the primary aperture is a high, semicircular arch, much higher than in the holotype; this seems to occur where the umbilicus is less open, and the combination of these two characters serves as a ready distinction from *G. turritilina turritilina*. Aborted end-chambers often occur in the larger individuals; they are reduced in size and are of irregular shape, but usually retain their position within the normal progression of the spire and do not usually tend to cover the umbilicus as in *G. turritilina turritilina*. The maximum size attained by individuals of *G. turritilina praeturritilina* seems to be less than that attained in populations of *G. turritilina turritilina*.

STRATIGRAPHICAL RANGE

In East Africa, this subspecies ranges from the *Globigerapsis semi-involuta* Zone to the lower part of the *Globigerina turritilina turritilina* Zone, Upper Eocene, where it evolves into, and is replaced by, *G. turritilina turritilina*. Although it has not been recorded from other localities, similar forms have been observed by us in the Upper Eocene of southern Trinidad.

Globigerina yeguaensis Weinzierl & Applin *emended*

Globigerina yeguaensis yeguaensis Weinzierl & Applin
(Plate XIII H–M; Fig. 18)

Globigerina yeguaensis Weinzierl & Applin, 1929, *J. Paleont.* **3**, 408, pl. 43, figs. 1 a–b.

cf. *Globigerina yeguaensis* Weinzierl & Applin, Bolli, 1957 (part), *Bull. U.S. Nat. Mus.* **215**, 163, pl. 35, figs. 15 a–c (*not* figs. 14 a–c).

DESCRIPTION

The moderately large test consists of about two to three whorls of fairly rapidly enlarging, inflated, little embracing chambers which are coiled in a fairly low trochospire with four chambers in each whorl. In ventral view the chambers are subcircular to oval; in axial view they are ovoid, being slightly depressed, and in dorsal view they are semicircular to broadly reniform. The equatorial periphery of the test is strongly lobulate; the axial periphery is broadly rounded. In axial profile the test is subovoid; the chambers are moderately convex dorsally and strongly convex ventrally (with no distinct dorso-peripheral shoulder) and the spire projects slightly above the dorsal surface of the test. All the sutures are deeply depressed. The dorsal intercameral sutures meet the strongly lobulate spiral suture almost at right angles; initially they are curved but become subradial during ontogeny. The ventral umbilicus is moderately broad, open and deep; it is well defined by the steeply sloping apertural faces of the last and penultimate chambers and the umbilical margins of the earlier chambers of the last whorl. The aperture is a low, almost symmetrical, intraumbilical arch, bordered by a very strong, flap-like lip which often projects over almost half the width of the umbilicus. Similar 'umbilical teeth' may be observed covering relict apertures within the umbilicus. The apertural face of the last chamber is only weakly flattened and is not clearly delimited from the rest of the chamber surface. The wall of the test is fairly thick, moderately coarsely perforate and is uniformly and fairly coarsely hispid. Often the hispidity is strongest around the umbilical margins.

REMARKS

The specimen illustrated here (see Plate XIII K–M) is virtually identical to others which we sent to Miss R. Todd (U.S. Geological Survey). In her reply (dated 7 May 1959), Miss Todd wrote as follows: 'I have compared the two specimens you sent with the holotype of *Globigerina yeguaensis* Weinzierl & Applin and would judge them to be identical. They seem equal in every respect (except color—the holotype is black) and, in fact, are closer to the holotype in shape and in wall surface than the two hypotypes (P.5507 and 8) figured by Bolli which I have also looked at.'

Bolli (1957c, pl. 35, figs. 14a–c, 15a–c) illustrated the specimens mentioned above by Miss Todd. We regard the specimen illustrated in Bolli's figs. 15a–c from the *Globigerapsis semi-involuta* Zone, Hospital Hill marl (Navet formation), southern Trinidad, to be an intermediate form in the evolution towards *Globigerina tripartita tripartita* (see p. 141), whilst the specimen illustrated in Bolli's figs. 14a–c from the Middle Eocene, *Porticulasphaera mexicana* Zone, Navet formation, is very close to *Globigerina*

yeguaensis pseudovenezuelana subsp.nov. (see below).

Notwithstanding the well developed umbilical teeth present in this species, we do not consider it to be referable to the genus *Globoquadrina* Finlay, since it is considered that the diagnostic character of that genus is the change in apertural position during ontogeny (see Banner & Blow, 1959, p. 5).

STRATIGRAPHICAL RANGE

This species was originally described from the upper Claiborne, Middle Eocene, of Texas. In East Africa it ranges from the Middle Eocene to the top of the Oligocene, but in the Caribbean it does not occur above the San Fernando beds. It has not been observed by us in any Lower Miocene samples either in the Old or New World.

Globigerina yeguaensis pseudovenezuelana Blow & Banner subsp.nov.
(Plate XI J–L, N, O)

Globigerina venezuelana Hedberg, Bolli, 1957 (*not* Hedberg, 1937), *Bull. U.S. Nat. Mus.* 215, 164, pl. 35, figs. 16a–17.

DIAGNOSIS

The fairly large test consists of about three whorls of rapidly enlarging, moderately inflated, partially embracing chambers which are coiled in a low trochospire with about four chambers in each whorl. The equatorial profile of the test is subcircular to subquadrate and the equatorial periphery is weakly lobulate. The axial profile of the test is broadly ovoid and the axial periphery is broadly rounded. The chambers are ovate in both ventral and axial views; in dorsal view, the chambers are initially semicircular, but become reniform and distinctly longer than broad during ontogeny. The test, like the later chambers, is depressed and is much more strongly convex ventrally than dorsally. All the sutures are distinctly depressed. The dorsal intercameral sutures meet the strongly lobulate spiral suture almost at right angles; initially they are curved, but become subradial during ontogeny. The ventral umbilicus is moderately broad, open and deep; it is clearly delimited by the steeply sloping apertural faces of the last and penultimate chambers and the umbilical margins of the earlier chambers. The apertural face is narrow, distinctly flattened and is parallel to the axis of coiling, directly facing the umbilicus. The aperture is a very low arch, entirely within the umbilicus. The aperture is furnished with a very strongly developed flap-like lip which often projects over almost half the width of the umbilicus. The lip may be broad throughout its length, but it is always distinctly broadest at its mid-point; sometimes the lip is of distinct triangular shape ('umbilical tooth'). Relict apertures, with their umbilical teeth, may be seen within the umbilicus. The wall of the test is fairly thick, moderately coarsely perforate and is uniformly and fairly coarsely hispid. The hispidity is usually strongest around the umbilical margins. Maximum diameter of holotype 0·51 mm; from sample FCRM 1923, Lindi area.

REMARKS

This subspecies is distinguished from *Globigerina yeguaensis yeguaensis* by its slower rate of chamber enlargement, more depressed and appressed chambers, flatter apertural face, deeper umbilicus and less lobulate periphery.

Globigerina venezuelana Hedberg very commonly has aborted final chambers which obscure essential specific characters. Such a form was illustrated as the type of *G. venezuelana* by Hedberg (1937). We have searched through near-topotypic material and have observed specimens with a normal growth rate; such a specimen is here figured for comparison (Fig. 11, xv). It can be seen that *G. venezuelana* Hedberg differs from *G. yeguaensis pseudovenezuelana* in possessing a smoother test, more depressed (less rounded) chambers in equatorial profile, less strongly depressed sutures, practically no hispidity, even on the umbilical margins and apertural face, less deeply depressed sutures and narrower, more proximally restricted umbilical teeth. '*G. venezuelana*-like' forms are therefore present in the Middle Eocene to Oligocene (*G. yeguaensis pseudovenezuelana*), are absent in the lowest Aquitanian (*G. ampliapertura* Zone, where only the form *G. ampliapertura euapertura* is present which could be confused), are present from the upper part of the *Globorotalia opima opima* Zone probably to at least the Tortonian (i.e. *Globigerina venezuelana* itself) and from the Plio-

cene to Recent (*G. conglomerata* Schwager, see Banner & Blow, 1960a). All these forms seem to be stable end-forms of independent phylogenetic lineages and closely homeomorph one another. *G. conglomerata* Schwager (see Banner & Blow, 1960a, p. 7, pl. 2, fig. 3) is close to *G. venezuelana* Hedberg, differing principally from it in its smaller and more restricted umbilicus; as we have previously stated (*loc. cit.*) these two forms may be only subspecifically distinct.

Through the kindness of Dr E. Gasche (Naturhistorisches Museum, Basel), we have been able to examine the holotype of *Globigerina bulloides* var. *quadripartita* Koch, 1926. This form was originally described from the middle Tertiary of Borneo. Unfortunately, the holotype is very badly damaged, the ventral side being entirely destroyed; however, as the dorsal side resembles *G. yeguaensis pseudovenezuelana*, it is figured here (Plate IX Gg), although we consider that the name *G. bulloides quadripartita* Koch should be considered *nomen dubium*.

Dr Orville L. Bandy has kindly sent us metatypic specimens of his form *Globigerina rotundata* var. *jacksonensis* from the upper part of the Jackson formation, Little Stave Creek, Alabama. These metatypes, like his illustrated holotype, possess peculiarly deformed final chambers which obscure the characters we consider to be of specific importance. Although it is impossible to disprove that *G. yeguaensis pseudovenezuelana* is conspecific with *G. jacksonensis* Bandy, we consider that his species is incapable of adequate determination and consequently cannot be used in detailed studies.

STRATIGRAPHICAL RANGE

In the Lindi area, *G. yeguaensis pseudovenezuelana* ranges from the Middle Eocene to the top of the Oligocene. In Trinidad, it has been recorded by Bolli (1957c, pp. 159 and 164) as '*G. venezuelana* Hedberg' (part) from the *Porticulasphaera mexicana* Zone, Navet formation (Middle Eocene), to the top of the Eocene.

Globigerina aff. yeguaensis Weinzierl & Applin
(Plate XI P, Q)

aff. *Globigerina yeguaensis* Weinzierl & Applin, 1929, *J. Paleont.* **3**, 408, pl. 43, figs. 1 a–b.

REMARKS

Rather rare forms, so far only observed in the Oligocene, possess a morphology very close to that of *Globigerina yeguaensis yeguaensis*. However, they possess a wider umbilicus, looser coiling and rather more globose chambers with a higher spire.

All the specimens observed had a broken last chamber and it is not yet possible fully to describe or define the form. It is recorded purely because of its possible stratigraphical value.

STRATIGRAPHICAL RANGE

This form is, as yet, known only from the *Globigerina oligocaenica* Zone, Oligocene, Lindi area, East Africa.

Genus: *Globigerinita* Brönnimann, 1959
emended

Type species: *Globigerinita naparimaensis* Brönnimann, 1951

Synonyms: *Tinophodella* Loeblich & Tappan, 1957a; *Catapsydrax* Bolli, Loeblich & Tappan, 1957.

EMENDATION AND DISCUSSION

The genus *Globigerinita* Brönnimann, 1951, was based upon the type species *G. naparimaensis* Brönnimann, 1951; this form is essentially a *Globigerina* in which the umbilicus and primary aperture of the final chamber are covered by a bulla, the only openings to the exterior being small accessory apertures situated in the suture of the bulla with the primary chambers.

In April 1957, Loeblich & Tappan emended and restricted Brönnimann's species to that form represented by the holotype only of *G. naparimaensis*. All the paratypes of *G. naparimaensis* were then referred to *Tinophodella ambitacrena* Loeblich & Tappan (gen.nov., sp.nov.), the type species of their new genus *Tinophodella*. *Tinophodella* was distinguished from *Globigerinita* by possessing 'a distinct and separate supplementary plate over the umbilicus, extending somewhat along the sutures, with numerous accessory apertures along all the margins'. The holotype of *G. naparimaensis* was believed by them to possess a 'modified final chamber which extends across the umbilical region', containing at its margins 'one or more

arched supplementary (*vel* accessory) apertures'. It appears that Loeblich & Tappan considered the 'supplementary plate' (i.e. the bulla) of *Tinophodella* to be distinct from the 'modified final chamber' of *Globigerinita*. However, as will be seen from Bolli, Loeblich & Tappan's (1957, pl. 8) re-illustration of the holotype of *G. naparimaensis*, compared to Loeblich & Tappan's illustration of the holotype of *Tinophodella ambitacrena* (1957a), the dorsoventral extent of the last-formed structure is essentially the same in each, *T. ambitacrena* differing merely in possessing far more accessory apertures along the margins of the bulla. These accessory apertures are sometimes placed at the ends of 'tunnel-like' prolongations of the bulla, which is less inflated in *T. ambitacrena* than is the equivalent structure in *Globigerinita naparimaensis*. We are indebted to Dr F. L. Parker for supplying us with recent populations of *Tinophodella ambitacrena*; in these assemblages every morphological grade exists between the structures described by Loeblich & Tappan for the genera *Globigerinita* and *Tinophodella*.

The term 'bulla' was originally proposed by Bolli *et al.* (1957, p. 13) and redefined by us (1959) as 'a perforate, inflated, plate-like structure which covers the umbilicus'; we see no reason to regard the 'modified final chamber' of *Globigerinita* as anything other than a bulla, or to consider it to be other than homologous to the 'supplementary plate' of *Tinophodella*. *Globigerinita incrusta* Akers, 1955, is clearly no more than subspecifically distinct from *Tinophodella ambitacrena* (cf. Blow, 1959, pp. 206–207, text-fig. 4), yet it possesses a much more simplified bulla with fewer accessory apertures than does the more complex holotype of *T. ambitacrena*; we have seen an intergrade between them at many horizons within the Miocene and Recent. The position and extent of the primary aperture, in both *Globigerinita naparimaensis* and *Tinophodella ambitacrena*, are given by their authors as 'interiomarginal and umbilical', which confirms our own observations. Consequently, we consider that both *Globigerinita* and *Tinophodella* are essentially merely bulla-bearing *Globigerina*, and are synonymous.

The genus *Catapsydrax* was erected by Bolli *et al.*

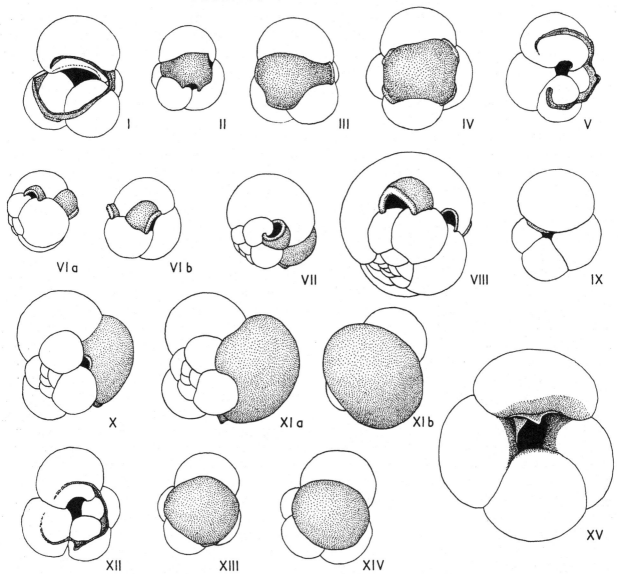

Fig. 11. i–xiv: Variation in bullate Globigerinidae; the bullae are stippled. i–iv: *Globigerinita africana* sp.nov.; specimens from sample FCRM 1645, *Globigerapsis semi-involuta* Zone, Upper Eocene, Lindi area, Tanganyika; all *ca.* × 90 —(i) adult specimen with bulla dissected, showing form of the primary aperture; (ii) entire but immature specimen; (iii)–(iv) adult specimens, showing the extremes of observed variation in the form of the bulla (from three to four accessory apertures). v, ix: *Globorotaloides suteri* Bolli; umbilical views of two specimens which lack bullae, showing the extremes of observed variation in the shape and extent of the primary aperture; both specimens from sample FCRM 1922, *Globigerina oligocaenica* Zone (Rupelian/Lattorfian), Lindi area, Tanganyika, *ca.* × 90. vi–viii: *Globigerinatheka lindiensis* sp.nov.; specimens from sample FCRM 1645, *Globigerapsis semi-involuta* Zone, Upper Eocene, Lindi area, Tanganyika, all *ca.* × 90—(via) and (vib) axial and ventral views of an immature specimen; (vii) axial view of a later ontogenetic stage; (viii) axial view of an adult specimen, which lacks bullae over all but one of the final apertures. These specimens demonstrate the constancy of chamber form through-out ontogeny, together with the gradually increasing tightness of coiling of the later primary chambers. x–xiv: *Globigerinita howei* sp.nov.; specimens from sample FCRM 1645, *Globigerapsis semi-involuta* Zone, Upper Eocene, Lindi area, Tanganyika, all *ca.* × 90—(xii) specimen with bulla dissected, showing form of the primary aperture; (x) and (xia) two specimens in oblique axial view, showing gross morphological affinities to *Globigerinita naparimaensis* (xia) and to '*Catapsydrax*' *dissimilis* s.l. (x); (xib), ventral view of adjacent specimen (xia), which, with (xiii) and (xiv), shows the range of observed variation in the form and size of the bulla. xv: *Globigerina venezuelana* Hedberg, near-topotype, from the *Globorotalia lobata lobata* Zone (Burdigalian), Pozón Formation, Falcón, Venezuela (sample RM 19444); ventral view, with lips and umbilical margins of the chambers stippled, *ca.* × 150. This specimen has been carefully compared to less well preserved conspecific topotypic material from the middle Carapita formation, Venezuela, and shows the development of *Globoquadrina*—like 'umbilical teeth'. All specimens deposited in the British Museum (Natural History), registered (in the above order) as numbers P.44478–P.44492.

late in 1957, subsequent to the proposal of the taxon *Tinophodella*. *Catapsydrax* was based upon the type species *Globigerina dissimilis* Cushman & Bermúdez, 1937, and it was distinguished from *Globigerina* by 'the presence of the umbilical bulla covering the primary aperture, and in having the accessory infralaminal apertures'. This genus also is clearly a *Globigerina* which bears an umbilical bulla. Unfortunately, Bolli *et al.* (1957) did not distinguish *Catapsydrax* from *Tinophodella*, and their references to *Globigerinita* are to the broad, unrestricted concept of the genus as originally described by Brönnimann (1951). In consequence, distinction between *Catapsydrax* and *Globigerinita*, as emended by Loeblich & Tappan (1957a), cannot be seen from their work. In an attempt to retain the taxonomic *status quo*, Banner & Blow (1959, pp. 5–6) retained the subfamily Catapsydracinae and attempted to distinguish between *Catapsydrax* and *Tinophodella* using the characters of the chamber and bulla walls, as well as the restriction of the accessory apertures, as seen in the type species of the two genera. We have now been led to the conclusion that these distinctions are of no more than specific character, for all gradations between the '*Catapsydrax*'-form and the '*Tinophodella*'-form may be seen within many assemblages of a single species (e.g. *Catapsydrax stainforthi* Bolli, Loeblich & Tappan, *Globigerinita africana* and *G. howei* Blow & Banner spp.nov.). We think that the distinctions made by Bolli *et al.* (1957) (as well as by Loeblich & Tappan, 1957a) between the form-genera *Globigerinita*, *Tinophodella* and *Catapsydrax* are unreal, in so far as distinctions between them are biologically, stratigraphically and taxonomically impractical and/or insignificant. We consider that all the *Globigerina*-like forms which possess true bullae and accessory apertures should be placed within the genus *Globigerinita*, belonging to the subfamily Globigerininae.

The differences in the structure of the walls of the bullae of different species now referred to *Globigerinita* (e.g. '*Catapsydrax*' *dissimilis* and *G. naparimaensis*) are believed to be reflexions of the wall structure of the primary chambers. The more complex bullae are usually of younger stratigraphical age, and we appreciate that extreme forms are often strikingly different; however, these differences are only of degree, not of kind, and we do not divide *G. ambitacrena* from *G. dissimilis* at generic level on their gross morphology any more than we would similarly subdivide *Globigerina linaperta* from *G. turritilina turritilina*.

The only possible distinctions between the so-called 'modified final chamber' of *naparimaensis* and the bullae of *ambitacrena* and *dissimilis* are that in *naparimaensis* the final structure is almost as inflated as the last true chamber, and that it possesses no accessory apertures at its dorso-peripheral margin. If the degree of inflation be considered to be of generic value, subdivision of these forms would become chaotic. If the absence of dorso-peripheral accessory apertures be used, then '*Catapsydrax*' *unicavus* Bolli, Loeblich & Tappan, 1957, should be referred to *Globigerinita*; similarly, *Globigerina pera* Todd would be referred to *Globigerinita*, although this species also clearly belongs to the Bolli *et al.* (1957) concept of *Catapsydrax*; Fig. 11 (x–xiv) illustrates such variation within the single species *Globigerinita howei*. On the other hand, following Loeblich & Tappan's concept of *Tinophodella*, the presence of tunnel-like multiple extensions of the bulla, associated with small accessory apertures placed above the sutures of the earlier chambers, would place the species '*Catapsydrax*' *stainforthi* of Bolli *et al.* into the prior genus *Tinophodella*. If the morphology of *stainforthi* is compared to *Globigerinita incrusta* Akers, it can be seen that there is no fundamental character of kind to separate them at generic level. In the species ascribed to our new genus *Turborotalita* (see p. 122 below) we find that the bullae are similarly furnished in some cases with tunnel-like extensions (e.g. *T. cristata*) but in other cases they are absent (e.g. *T. humilis*). Since, in the group of bullate globigerinids, there appear to be no fundamental distinctions either in the structure or the morphology of the bullae, and as the morphology of the primary test is in every case generically the same, *Tinophodella* and *Catapsydrax* are to be considered as full junior synonyms of the genus *Globigerinita* Brönnimann, 1951, here emended.

We have previously pointed out (Banner & Blow, 1960a, p. 37) that the whole subfamily

'Catapsydracinae' needs detailed revision. We believe that we were correct when we described this subfamily 'as a polyphyletic group of specialized end-forms which possesses unity only in the uniformity of its specialized characters' (Banner & Blow, 1959, p. 25), and we now consider that *Globigerinita* and *Globigerinoita* may be referred to the Globigerininae, and that *Globigerinatheka* and *Globigerinatella* can be placed in the Orbulininae. All these forms are more or less specialized bulla-bearing representatives of previously well recognized fundamental morphological groups. Forms like *Turborotalia* which bear bullae, and are now referred to *Turborotalita* Blow & Banner (see p. 122), belong to the subfamily Globorotaliinae; they show, even more clearly than their analogues in the Globigerininae, that they should be referred to a supra-specific group distinct from the parent genus, for the species placed in *Turborotalita* (*Globigerinita parkerae* Loeblich & Tappan, 1957, *Globigerina cristata* Heron-Allen & Earland, 1929, and *Truncatulina humilis* Brady, 1884) could not be confused with any known species of *Globorotalia*, but display close morphological convergence in characters other than the bulla (such as their small size, many chambers, low spire, etc.).

Globigerinita africana Blow & Banner sp. nov.
(Plate XV A–C; Fig. 11 (i–iv))

DIAGNOSIS

The moderately large test consists of about three whorls of fairly rapidly and regularly enlarging, inflated, slightly embracing chambers, arranged in an increasingly high trochospire with about four chambers in each whorl. The equatorial profile of the test is subcircular to subquadrate, and the equatorial periphery is broadly and distinctly lobulate. In axial profile, the dorsal surface of the early whorls is slightly convex and the initial trochospire is fairly low; however, the later chambers are added in an increasingly high trochospire and the dorsal surface becomes increasingly more convex. In axial view, the periphery of the last whorl is broadly and smoothly rounded; the deeply depressed spiral suture between the last two whorls gives a distinctly lobulate axial profile. The primary chambers are initially subglobular, but become subovoid in the last whorl, where they are distinctly more inflated ventrally than dorsally. A broadly rounded dorso-peripheral shoulder may develop in the later chambers. In dorsal view, the chambers are semicircular to reniform; they may become slightly depressed and longer than broad in the later stages of ontogeny. The dorsal sutures are initially obscure but they later become distinctly and often sharply depressed, often narrowly so. The dorsal intercameral sutures are slightly curved or subradial, meeting the lobulate spiral suture at broad angles. The ventral sutures are equally distinctly depressed, and are subradial or slightly curved. The ventral umbilicus (beneath the bulla) is narrow and almost closed, and is not clearly delimited because of the uniform convexity of the umbilical margins of the ventral sides of the chambers. The umbilicus and the primary aperture are completely covered by a broad inflated bulla, which protrudes beyond the ventral surfaces of the primary chambers. The bulla typically extends beyond the limits of the umbilicus, especially anteriorly and posteriorly; its outline follows that of the umbilicus, but is modified by extensions along the areas marked by the intercameral sutures of the primary chambers, and by the fact that the bulla usually encroaches more upon the posterior part of the ventral face of the last chamber than upon the anterior part. At least three accessory apertures are present; two are present on the posterior side of the bulla, one over the suture between the last-formed and the penultimate chambers, and one over the suture between the penultimate and antepenultimate chambers. On the anterior side of the bulla, a single broad accessory aperture is usually present (see Fig. 11, iii), extending the full breadth of the ventral face of the first chamber of the last whorl, but in some exceptionally loosely coiled specimens, where more of the ventral surface of that chamber is exposed, this accessory aperture may be divided and replaced by an accessory aperture over each of the intercameral sutures limiting the first-formed chamber of the last whorl (see Fig. 11, iv); this produces a maximum of four accessory apertures to the bulla. The accessory apertures are typically furnished with distinct thin lips and are fairly high arches. The sutures between the primary chambers and the bulla are distinctly but not deeply depressed. The primary aperture of the final chamber is a high symmetrical arch, sometimes furnished with an indistinct rim-like thickening. The wall of the primary chambers is fairly thick and is uniformly and moderately coarsely perforate; its surface is punctate and distinctly, densely and finely hispid. The wall of the bulla is much thinner than that of the primary chambers; it is more finely perforate and is very finely hispid. Maximum diameter of holotype 0·38 mm; from sample FCRM 1645, Lindi area.

REMARKS

It appears that the increased height of the trochospire in the later ontogeny of this species is associated with relatively tighter coiling in the last whorl and a narrowing of the umbilicus and the bulla. Occasional specimens which do not show such a marked change in the tightness of coiling and height of the trochospire possess bullae which are broader than the typical; it is these forms which may possess four accessory apertures (Fig. 11, iv). This species should be compared to *Globigerinita howei* sp.nov. (p. 109).

STRATIGRAPHICAL RANGE

In East Africa, this species is known to range from the base of the *Globigerapsis semi-involuta* Zone to about the middle part of the overlying *Cribrohantkenina danvillensis* Zone, Upper Eocene. It may also occur in the upper part of the Middle Eocene, *Truncorotaloides rohri* Zone.

Globigerinita dissimilis (Cushman & Bermúdez) *emended*

Globigerinita dissimilis dissimilis (Cushman & Bermúdez)
(Plate XIV D)

Globigerina dissimilis Cushman & Bermúdez, 1937, *Contr. Cushman Lab. Foram. Res.* **13**, pt. 1, p. 25, pl. 3, figs. 4–6.

Catapsydrax dissimilis (Cushman & Bermúdez), Bolli, Loeblich & Tappan, 1957 (part), *Bull. U.S. Nat. Mus.* **215**, 36, pl. 7, figs. 6a–c (refigured holotype), 7a–b (*not* figs. 8a–c).

?*Catapsydrax* cf. *dissimilis* (Cushman & Bermúdez), Bolli, 1957, *Bull. U.S. Nat. Mus.* **215**, 166, pl. 37, figs. 6a–b.

DESCRIPTION

The test consists of about three whorls of fairly rapidly enlarging, inflated, partially embracing chambers. The first whorl consists of five chambers, but only four are present in the second and third whorls. The chambers are arranged in a low trochospire. In equatorial profile the test is subcircular to subquadrate and the equatorial periphery is broadly lobulate. In ventral view the primary chambers are subglobular to suboval. In axial view the test is oval to ovate; the ventral side is slightly more convex than the dorsal side. In axial profile the chambers are initially subglobular, but later become ovoid, being more strongly inflated ventrally than dorsally. There is no distinct dorsoperipheral shoulder. In dorsal view the sutures are distinctly depressed throughout; the dorsal intercameral sutures, beyond the first whorl, are subradial, meeting the lobulate spiral suture and the periphery almost at right-angles. The ventral intercameral sutures are deeply depressed and are subradial. In dorsal view the chambers are reniform and are very slightly depressed; they usually appear to retain their form throughout the last two whorls. The ventral umbilicus is fairly broad and deep; it is delimited by the slightly flattened umbilical shoulders of the later chambers of the last whorl. The primary aperture of the last chamber is a moderately high intraumbilical arch. The relict part of the primary aperture of the penultimate chamber is visible within the umbilicus, and is bordered by a narrow rim-like lip. The bulla extends from the apertural face of the last chamber across to the umbilical margin of the opposing chamber, partially covering the umbilicus. Two broad and high accessory apertures are present, one being situated between the bulla and the entire umbilical margin of the first chamber of the last whorl and the other being situated between the bulla and the entire umbilical margin of the penultimate chamber. In some forms (especially from stratigraphically earlier horizons) the breadth of the bulla is less than that of the umbilical limits, but in others (from stratigraphically later horizons) the bulla may extend to, and even beyond, the umbilical limits. The bulla is only slightly convex, and, when

narrow, appears to be merely arched; it is typically bordered by narrow but distinct lips over the accessory apertures. The wall of the primary chambers is fairly thick; it is uniformly and fairly finely perforate, although the surface may show fairly coarse punctae. The surface of the primary chambers is uniformly and finely hispid. The wall of the bulla is thin and very finely perforate; it is weakly hispid, especially on the lips of the accessory apertures.

REMARKS

In the Tertiary sections of Venezuela, Trinidad and East Africa, it has been observed that forms referable to *Globigerinita dissimilis dissimilis* show slight but noteworthy modifications with time. Stratigraphically younger specimens (e.g. those from the lower Cipero formation) typically possess thicker and more coarsely perforate walls in both the primary chambers and the bulla, although the bulla always remains relatively thinner and less coarsely perforate. Concomitantly, the primary chambers become more depressed and embracing, whilst the bulla broadens laterally to cover the umbilicus completely. These trends are of obvious stratigraphical interest, and lead to *G. dissimilis ciperoensis*.

The narrow bulla, characteristic of stratigraphically early forms, can be clearly seen in the otherwise imperfectly preserved Eocene holotype of *G. dissimilis dissimilis*. This feature, together with the distinct wall structure of *G. dissimilis*, indicates that its lineage is probably unrelated to that of *G. pera–G. unicava*.

Although Cushman and Bermúdez originally described '*Globigerina*' *dissimilis* from the Eocene of Cuba, Bolli (1957 c, p. 166) referred his Eocene specimens to '*Catapsydrax* cf. *dissimilis*' apparently because they differed from those of 'Oligocene–Lower Miocene' Cipero age 'in having somewhat more globular chambers. The umbilical bullae have commonly only two and more rarely only one infralaminal accessory aperture, whereas the bullae of the Oligocene–Miocene specimens often display three or four accessory apertures.' However, the Eocene forms (which include the holotype) must be considered typical of *Globigerinita dissimilis dissimilis*, and the stratigraphically later forms mentioned by Bolli display the evolution into *G. dissimilis ciperoensis* subsp.nov.

STRATIGRAPHICAL RANGE

Globigerinita dissimilis dissimilis ranges from at least the upper part of the *Truncorotaloides rohri* Zone, Navet formation, Middle Eocene, to the top of the *Globigerinita stainforthi* Zone (Aquitanian) in southern Trinidad. In Tanganyika it is known to occur also from the upper part of the *Truncorotaloides rohri* Zone, ranging up into the Aquitanian. In south-east Sicily it has also been seen in samples from Aquitanian horizons from the Ragusa limestone formation.

Globigerinita dissimilis ciperoensis Blow & Banner
subsp.nov.
(Plate XIV A–C)

Catapsydrax dissimilis (Cushman & Bermúdez), Bolli, Loeblich & Tappan, 1957 (part), *Bull. U.S. Nat. Mus.* **215**, 36, pl. 7, figs. 8 a–c (*not* figs. 6 and 7).

Catapsydrax dissimilis (Cushman & Bermúdez), Blow, 1959, *Bull. Amer. Paleont.* **39**, no. 178, p. 203, pl. 12, figs. 88–90.

DIAGNOSIS

This subspecies differs from *Globigerinita dissimilis dissimilis* (Cushman & Bermúdez, 1937) in possessing an anterio-posteriorly broad bulla which is attached to the umbilical margins of the ventral surfaces of not only the last-formed and the antepenultimate chambers, but also to the corresponding surface of at least one other chamber in the last whorl, so that at least one of the broad accessory apertures present in *G. dissimilis dissimilis* is divided to produce two arched accessory apertures situated over the intercameral sutures of the primary chambers of the last whorl. Three or four accessory apertures are thus present in the suture of the bulla of *G. dissimilis ciperoensis*.

Globigerinita dissimilis ciperoensis possesses the thicker and more coarsely perforate walls, together with the more depressed adult chambers and less lobulate periphery, typical of advanced forms of *G. dissimilis dissimilis*, from which it appears to have descended.

Maximum diameter of holotype 0·50 mm; from the type locality of the *Globigerina ampliapertura* Zone, southern Trinidad (see Bolli, 1957b, p. 100).

REMARKS

As discussed above (*Globigerinita dissimilis dissimilis*, p. 106), the Lower Miocene (Cipero) forms referred by Bolli *et al.* (1957, p. 36) and by Bolli (1957*b*, p. 116; 1957*c*, p. 166) to '*Catapsydrax dissimilis*', contain forms now referred to *G. dissimilis ciperoensis* which differ significantly from the Eocene forms of Cushman & Bermúdez (1937).

The acquisition of a greater number of accessory apertures appears to result from the anterioposterior broadening of the bulla (cf. *Globigerinita africana*, p. 105); this may be associated with a narrowing umbilicus and tighter coiling, trends which are observable to a lesser degree in the evolution of *G. dissimilis dissimilis*. We have made an arbitrary distinction at the point in the evolution of *G. dissimilis s.l.* where more than two accessory apertures are present, as this appears to be the clearest way of distinguishing the most advanced forms (characteristic of the Aquitanian) from the other, more or less primitive, forms which range from the Eocene to the Aquitanian.

STRATIGRAPHICAL RANGE

In southern Trinidad this subspecies ranges from the *Globigerina ampliapertura* Zone to the top of the '*Catapsydrax*' (*vel Globigerinita*) *stainforthi* Zone, Cipero formation, Aquitanian; it may also occur in the San Fernando beds, but this has not yet been published. It also occurs over the same interval in eastern Falcón, Venezuela. In East Africa, it does not occur in either the Eocene or the Oligocene, but it has been observed in Aquitanian sediments. It also occurs in the Aquitanian of south-east Sicily (Ragusa limestone formation).

Globigerinita globiformis Blow & Banner sp.nov.
(Plate XIV s–u)

DIAGNOSIS

The fairly small test consists of two to three whorls of weakly inflated, partially embracing, regularly enlarging chambers, coiled in a low trochospire with about four chambers in each whorl. The equatorial profile of the test is subcircular and the equatorial periphery is weakly and broadly lobulate. In axial profile the test is subglobular to sub-oval, the dorsal spire being almost as equally convex as the ventral bullate side. In axial view the periphery is broadly and smoothly rounded, the chambers being subglobular to subovoid, lacking a distinct dorso-peripheral shoulder. In dorsal view the chambers are initially semicircular to reniform but they become increasingly depressed and longer than broad during ontogeny. The dorsal sutures are initially obscure, but they later become distinctly, although still weakly and broadly, depressed. The dorsal intercameral sutures are slightly curved or subradial, meeting the weakly lobulate dorsal spiral suture almost at right angles. The ventral intercameral sutures are weakly depressed and subradial. Three and a half or four primary chambers are visible ventrally. The ventral umbilicus is poorly delimited owing to the uniform convexity of the umbilical margins of the ventral sides of the chambers; even in the deepest part (beneath the bulla) the umbilicus is shallow and nearly closed. The umbilicus and the primary aperture are completely covered by a broad inflated bulla, which protrudes beyond the ventral surfaces of the primary chambers. The bulla typically extends beyond the limits of the umbilicus, extending almost halfway across the ventral chamber surfaces, often possessing extensions along the intercameral sutures of the primary chambers. The bulla usually encroaches more upon the posterior part of the ventral face of the last chamber than upon the anterior part, but this need not necessarily occur. Three accessory apertures are typically present; two are present on the posterior side of the bulla, over the intercameral sutures between the antepenultimate, penultimate and final chambers, and a single accessory aperture is present anteriorly over the ventral surface of the first chamber of the last whorl. The accessory apertures are low to moderately high, small, semicircular arches, typically furnished with very narrow, thin lips. The sutures between the primary chambers and the bulla are weakly but distinctly depressed. The primary aperture of the final chamber is a very small arch, typically as high as broad, situated at the deepest point of the umbilicus; it may be furnished with an indistinct rim-like lip. The wall of the primary chambers is very thick relative to the size of the test,

and it is uniformly and fairly coarsely perforate; its surface is punctate, densely and coarsely hispid. The wall of the bulla is distinctly thinner than that of the primary chambers; it is much more finely perforate and is sparsely and very finely hispid. Maximum diameter of holotype 0·24 mm; from sample FCRM 1645, *Globigerapsis semi-involuta* Zone, Upper Eocene, Lindi area.

REMARKS

Globigerinita globiformis is characterized by its globose shape, which results from tight coiling, depressed and embracing chambers, weakly depressed sutures and an extensive (but not highly inflated) bulla, which tends to cover much of the ventral side of the test. It is distinguished from *G. africana* sp.nov. by its lower trochospire, less lobulate periphery, much tighter coiling and relatively coarser hispidity. The maximum size attained by individuals of *G. globiformis* is distinctly less than that attained by individuals of *G. africana*.

STRATIGRAPHICAL RANGE

In the Lindi area this species ranges from the base of the *Globigerapsis semi-involuta* Zone to about the middle of the *Cribrohantkenina danvillensis* Zone, Upper Eocene. It may occur in the uppermost part of the *Truncorotaloides rohri* Zone, Middle Eocene.

Globigerinita howei Blow & Banner sp.nov.
(Plate XIV P–R; Fig. 11 (x–xiv))

DIAGNOSIS

The moderately large test consists of about three whorls of fairly rapidly enlarging, inflated, slightly embracing chambers arranged in a fairly low trochospire, with about four chambers in each whorl. The equatorial profile of the test is subcircular, with a tendency to be subtriangular or subquadrate; the equatorial periphery is distinctly and broadly lobulate. In axial profile, the dorsal surface is moderately convex whilst the very highly inflated bulla gives a very strong convexity to the ventral surface. In axial view the primary chambers are subglobular to ovoid, the ventral surfaces of the later chambers being more strongly inflated than the dorsal surfaces; no clear dorso-peripheral

shoulder is present. In dorsal view the chambers are semicircular to reniform, being of almost constant shape throughout ontogeny. The dorsal sutures are initially obscure, but they later become distinctly and broadly depressed. The dorsal intercameral sutures are initially slightly curved, meeting the lobulate spiral suture at broad angles, but they later become subradial. The ventral intercameral sutures are subradial; they are deeply and narrowly depressed, and this depression increases towards the periphery, where the sutures are typically sharply incised, giving an almost 'bevelled' appearance to the chamber margins. The umbilicus and primary aperture are completely covered by a very broad and very strongly inflated bulla, which protrudes far beyond the ventral surfaces of the primary chambers. The bulla extends equally far beyond the limits of the umbilicus; it is attached at or near the periphery of the first chamber of the last whorl, approximately halfway across the ventral surfaces of the succeeding two chambers and at the highest point of the apertural face of the final chamber. The bulla is often of approximately the same volume as the last-formed primary chamber, but may exceed it. Two or three accessory apertures are present at the margins of the bulla; two are present at the posterior part of the bulla, one over the suture between the last-formed and the penultimate chambers, and the other over the suture between the penultimate and antepenultimate chambers. On the anterior side of the bulla, a single accessory aperture may be present, situated at the suture between the bulla and the first chamber of the last whorl (Fig. 11, x and xi a). The two accessory apertures on the posterior margin of the bulla are typically high arches, bordered by thin but distinct lips. On the anterior side of the bulla, when an accessory aperture is present, it varies from a small, unlipped opening to a high, lipped arch (similar to the other accessory apertures), extending, in its maximum development, along the full breadth of the first chamber of the last whorl. The primary aperture (Fig. 11, xii) is a very low arch, laterally restricted to, and symmetrically arched about, the fairly small, subquadrate, deep umbilicus. Removal of the bulla discloses that the relict part of the primary aperture of the penultimate primary chamber is often visible within the umbilicus. The primary aperture may, in some cases, be furnished with a thin, more or less distinct, rim-like thickening. The wall of the primary chambers is fairly thick, and it is uniformly and finely perforate; its surface is weakly cancellate and uniformly hispid. The wall of the bulla is slightly but distinctly thinner than that of the primary chambers, and it is more finely perforate and hispid. Maximum diameter of holotype: 0·39 mm; from sample FCRM 1645, *Globigerapsis semi-involuta* Zone, Upper Eocene, Lindi area.

REMARKS

As may be seen from the illustrations (Fig. 11, x–xiv), every possible morphological intermediate exists between forms which possess two accessory apertures and those with three. Some specimens possess a bulla which extends laterally to the periphery of the first primary chamber of the last whorl, and at the suture at this point there is no trace of an accessory aperture (Fig. 11, xi*a*); these specimens are morphologically indistinguishable from the genus *Globigerinita* Brönnimann as restricted by Loeblich & Tappan (1957*a*). Other specimens of comparable size, and of otherwise identical gross morphology, from the same assemblage, possess a very small accessory aperture at the suture between the bulla and the periphery of the first chamber of the last whorl (Fig. 11, x). In yet other specimens of comparable size and form from the same assemblage, this accessory aperture is equivalent in size and shape to the others in the bulla, and these forms agree with the description of the genus *Catapsydrax* given by Bolli *et al.* (1957); this agreement is, of course, most marked where the anterior accessory aperture is large, extending the full breadth of the first chamber of the last whorl, so that the bulla can no longer reach the periphery of that chamber and appears to be more 'umbilical' in position (Fig. 11, xiv). Specimens of this species are also frequent (again, even in the same assemblage) in which the bulla possesses slight but definite extensions along the intercameral sutures of the ventral surface; the accessory apertures are situated at the end of these extensions and are frequently furnished with distinct lips; these forms, although clearly conspecific with the holotype of *Globigeri-*

nita howei, would appear to be referable to the taxon *Tinophodella* Loeblich & Tappan, 1957. As we have noted above (p. 102), the multiplicity of accessory apertures in *Tinophodella ambitacrena* is exceptional, for specimens of the clearly closely related *Globigerinita incrusta* Akers possess only three or four such apertures. The range of variation within *G. howei* thus supports our belief that the taxa *Tinophodella* Loeblich & Tappan, 1957, and *Catapsydrax* Bolli, Loeblich & Tappan, 1957, possess no valid distinctions from the prior taxon *Globigerinita* Brönnimann, 1951.

Globigerinita howei possesses an exceptionally inflated, large and extensive bulla, which distinguishes it from all other known species of *Globigerinita*; however, it may, in consequence, be confused with the species described here as *Globigerapsis tropicalis* (see p. 124) which is distinguished primarily by its possession of a distinct change in coiling mode *before* the emplacement of the final adult chamber; the final chamber of *Globigerapsis* follows and intensifies the coiling mode demonstrated by the penultimate and antepenultimate chambers of the test, and there is no sudden change in wall structure or surface texture.

Globigerinita howei differs from the other species of its genus described in this work in possessing sutures which are sharply incised into the periphery of the last whorl.

This species is named for Professor H. V. Howe, in honour of his pioneering work in the study of the Globigerinaceae.

STRATIGRAPHICAL RANGE

In the Lindi area, this species ranges from the *Truncorotaloides rohri* Zone, Middle Eocene, to about the middle part of the *Cribrohantkenina danvillensis* Zone, Upper Eocene. In southern Trinidad, it has been observed by us in samples from blocks within the 'San Fernando conglomerate', which are considered here to be derived from Upper Eocene, and which form the basal beds of the Miocene (Cipero) transgression.

Globigerinita martini Blow & Banner sp.nov.

Globigerinita martini martini Blow & Banner subsp.nov.
(Plate XIV o)

DIAGNOSIS

The small test consists of about three whorls of chambers coiled in a low trochospire. The first whorl consists of five relatively slowly enlarging, moderately inflated chambers; in succeeding whorls four chambers only are present, and they become increasingly inflated and more rapidly enlarging, until (in the largest specimens) only three and a half chambers are visible ventrally. In equatorial profile the test is subcircular to subtriangular and the equatorial periphery is broadly and moderately strongly lobulate. In axial profile the test is oval to slightly ovate; the chambers are subglobular, becoming slightly ovoid and more inflated ventrally in the final part of the test. There is no distinct dorso-peripheral shoulder. In dorsal view, the earlier chambers are semicircular, but they become increasingly inflated and become as broad as long. The dorsal sutures are initially obscure but they later become increasingly, but still relatively weakly, depressed. The dorsal intercameral sutures are initially curved, meeting the lobulate spiral suture at broad angles, but they later become subradial. The ventral intercameral sutures are subradial and distinctly depressed. The ventral umbilicus is narrow and shallow; it is not clearly delimited, as the umbilical margins of the chambers are of uniform convexity. Relict parts of earlier primary apertures are not visible. A strongly inflated but fairly small bulla covers the umbilicus and the primary aperture of the final chamber. The bulla is attached to the umbilical margins of the first and second and final chambers of the last whorl. One single large accessory aperture is present; it is situated at the base of the posterior side of the bulla, above the umbilical margin of the penultimate chamber, and it frequently extends along the anterior part of the umbilical face of the antepenultimate chamber, and the most posterior part of the apertural face of the last chamber. The accessory aperture is furnished with a very narrow and thin, but usually distinct, lip along the associated margin of the bulla. The primary aperture of the final chamber may sometimes be observed through the accessory aperture, where it may be seen to be a very low intraumbilical arch, furnished with a very narrow rim-like thickening. The bulla is so strongly inflated that it very markedly protrudes above the ventral surfaces of the primary chambers; it is of subglobular form and is usually restricted to the umbilical limits, although in some specimens it extends almost to the periphery of the first chamber of the last whorl. The fairly thick wall of the primary chambers is quite coarsely perforate relative to the size of the test; the surface is densely and relatively coarsely hispid. The wall of the bulla is distinctly thinner and more finely perforate than that of the primary chambers; usually its surface shows little trace of hispidity, but some exceptionally large specimens may possess a uniformly hispid bulla. Maximum diameter of holotype 0·19 mm; from sample FCRM 1932, *Cribrohantkenina danvillensis* Zone, Upper Eocene, Lindi area.

REMARKS

This subspecies should be compared to *Globigerinita martini scandretti*, described below. The small size of the adult test, the form of the adult chambers, the structure and surface of the wall, the shape and position of the bulla, are all among the many characteristic features which serve to distinguish this species.

This species is named for F. C. R. Martin, geologist of the British Petroleum Co. Ltd, whose careful collections and field work made this study possible.

STRATIGRAPHICAL RANGE

In the Lindi area, this subspecies ranges from at least the upper part of the *Truncorotaloides rohri* Zone (Middle Eocene) to the top of the *Globigerina turritilina turritilina* Zone (Upper Eocene). It has not been found to occur in the *G. oligocaenica* Zone (Oligocene).

Globigerinita martini scandretti Blow & Banner
subsp.nov.
(Plate XIV v–x)

DIAGNOSIS

This subspecies differs from *Globigerinita martini martini* in possessing more rapidly enlarging chambers which are more strongly embracing, so that only three and a half to four chambers occur in

111

the second whorl, and that the equatorial periphery is much less strongly lobulate. This subspecies is also characterized by its possession of a much broader, but much less inflated, bulla, which frequently extends well beyond the limits of the small ventral umbilicus; the bulla typically reaches the dorso-ventral periphery of the first chamber of the last whorl, but does not protrude above the ventral surfaces of the primary chambers as strongly as does the bulla of *G. martini martini*. The accessory aperture has the same lateral extent as in *G. martini martini*, but is much lower and is often slit-like. The wall of the primary chambers of *G. martini scandretti* is thinner and more finely perforate than in *G. martini martini*. Maximum diameter of holotype: 0·15 mm; from sample FCRM 1922, *Globigerina oligocaenica* Zone (Oligocene), Lindi area.

REMARKS

This subspecies is named for D. J. Scandrett, British Petroleum Co. Ltd, in recognition of his assistance in the compilation of many of the text-figures in this work.

STRATIGRAPHICAL RANGE

In the Lindi area, this subspecies occurs only in the *Globigerina oligocaenica* Zone, Oligocene.

Globigerinita pera (Todd)
(Plate XIV E–H)

Globigerina inflata d'Orbigny, Subbotina, 1953 (part) (*not* d'Orbigny, 1839), *Trudi, VNIGRI*, N.S., **76**, 72, pl. 15, figs. 6a–c, ?figs. 4, 5 (*not* plates 7, 8).

G. pera Todd, 1957, *U.S. Geol. Surv., Prof. Paper*, **280**–H, p. 301, pl. 70, figs. 10, 11.

DESCRIPTION

The moderately large test consists of about three whorls of fairly rapidly enlarging, inflated, partially embracing chambers. About four primary chambers are present in each whorl throughout ontogeny, and they are coiled in a fairly low trochospire. In equatorial profile, the test is subcircular to subquadrate, and the equatorial periphery is moderately lobulate. In axial profile the test is tumid and broadly oval; the ventral and dorsal sides are approximately equally convex, as the height of the dorsal spire is almost as great as that of the ventral

bulla. In axial profile, the chambers are sub-globular to ovoid, their dorsal surfaces being almost as strongly inflated as the ventral surfaces, and the axial periphery is weakly depressed but smoothly and broadly rounded. The sutures are distinctly and broadly depressed. The dorsal intercameral sutures are initially curved, meeting the lobulate spiral suture almost at right angles, but they become subradial in the last whorl. The ventral intercameral sutures are weakly curved to subradial. In dorsal view, the chambers are reniform to semicircular, and are of almost uniform outline throughout ontogeny. The ventral umbilicus is deep and broad, but it is covered by a broad, inflated bulla. Deeply depressed sutures mark the attachment of the bulla to the last-formed primary chamber and to the two earliest chambers of the last whorl. A single broad accessory aperture is present between the margin of the bulla and the surface of the penultimate chamber; this accessory aperture extends for the full breadth of the penultimate chamber, and also extends along the most anterior part of the ventral surface of the antepenultimate chamber. This accessory aperture is often narrowest at its mid-point; it is unrestricted by any lip or rim-like thickening. The bulla is of approximately quadrate outline; it completely covers the maximum extent of the umbilicus and is of approximately the same outline. The concealed primary aperture is an intraumbilical arch. The wall of the primary chambers is thick, and is uniformly and coarsely perforate; its surface is coarsely punctate and approaches a reticulate (cancellate) pattern. Traces of hispidity may be present. The wall structure of the bulla is fundamentally similar to that of the primary chambers, but it is slightly thinner and very slightly less coarsely perforate and punctate.

REMARKS

Our figured hypotypes of this species display much the same range of variation as shown by Todd (1957) between her holotype and illustrated paratype. Subbotina (1953) illustrated specimens from the Upper Eocene of the Caucasus which she referred to *Globigerina inflata* d'Orbigny, but which we believe to belong to *Globigerinita pera*. Her form, like Todd's, appears to be distinct from

Globigerina corpulenta Subbotina, 1953 (also a *Globigerinita*) in possessing more depressed primary chambers, a less high spiral side and a broader bulla which completely covers the umbilicus, being attached to three chambers of the last whorl (not two only as in *Globigerina corpulenta*). We think it probable that *G. corpulenta* Subbotina is a junior synonym of *G. bulloides* var. *cryptomphala* Glaessner, 1937. *Globigerinita cryptomphala* (Glaessner) is recorded from the Upper Eocene of the Western Caucasus; it has not been observed either in East Africa or in the Central American region, and it may prove to be a more boreal species.

Globigerinita pera should be compared to *G. unicava s.l.* (pp. 113, 114). *G. pera* is distinguished from *G. dissimilis s.l.* in possessing more inflated and better separated primary chambers, a higher spire, and a bulla with a single accessory aperture.

STRATIGRAPHICAL RANGE

Todd (1957) described this species from the Upper Eocene Hagman and Densinyama formations of Saipan. Bolli (1957c, p. 166) mentioned '*Catapsydrax* cf. *dissimilis*' from the uppermost Middle Eocene and Upper Eocene parts of the Navet formation, and pointed out that infrequent forms possessed only one accessory aperture; these forms may belong to *Globigerinita pera* (Todd).

In the Lindi area, *Globigerinita pera* ranges from at least the upper part of the *Truncorotaloides rohri* Zone, Middle Eocene, throughout the Upper Eocene, to the top of the *Globigerina oligocaenica* Zone, Oligocene. It has not been observed in any Aquitanian sediments, either in the Caribbean, East Africa, Sicily, Europe, or in the Fina-Sisu formation of Saipan, which is also believed to be of Aquitanian age.

Globigerinita unicava (Bolli, Loeblich & Tappan) *emended*

Globigerinita unicava unicava (Bolli, Loeblich & Tappan) (Plate XIV M, N)

Catapsydrax unicavus Bolli, Loeblich & Tappan, 1957, *Bull. U.S. Nat. Mus.* **215**, 37, pl. 7, figs. 9 a–c.

C. unicavus Bolli, Loeblich & Tappan, Blow, 1959, *Bull. Amer. Paleont.* **39**, no. 178, p. 204, pl. 15, figs. 94 a–c.

DESCRIPTION

The fairly small test consists of about three whorls of fairly rapidly enlarging, moderately inflated, partially embracing chambers. The early whorls consist of five chambers, but this reduces to four (or four and a half) in the last whorl; they are arranged in a low trochospire. In equatorial profile, the test is subcircular to subquadrate, and the equatorial periphery is broadly lobulate. In axial profile the test is suboval; the dorsal side is slightly more convex than the ventral side. In axial profile, the chambers are subglobular to slightly ovoid, the ventral and dorsal sides being almost equally inflated. The axial periphery is broadly rounded and there is no distinct dorso-peripheral shoulder. In dorsal view the sutures are distinctly depressed, and they become broadly depressed by the beginning of the last whorl at least. The dorsal intercameral sutures are slightly curved to subradial throughout ontogeny. The spiral suture is lobulate. The ventral intercameral sutures are subradial and deeply depressed. In dorsal view, the chambers are initially approximately as broad as long, but during ontogeny their length increases relative to their breadth, the chambers becoming reniform to semicircular in outline. The ventral umbilicus is broad and deep but is almost entirely covered by a very weakly inflated bulla. Shallow sutures mark the attachment of the bulla to the last-formed primary chamber and the first two primary chambers of the last whorl. A single narrow accessory aperture extends along the margin of the bulla for the full breadth of the penultimate chamber only. This accessory aperture is usually furnished with a very weak lip or thin rim along the whole free margin of the bulla. The bulla is of subquadrate outline, and it covers the umbilicus; it protrudes little if at all beyond the limits of the umbilicus, whether seen in axial or ventral view. The concealed primary aperture is an intraumbilical low arch. The wall of the primary chambers is fairly thin compared to *Globigerinita pera*, and it is uniformly and moderately coarsely perforate. The surface of the test is punctate and may be weakly hispid. The wall of the bulla is much thinner, distinctly more finely perforate and much smoother than that of the primary chambers.

REMARKS AND STRATIGRAPHICAL RANGE

Our figured hypotype comes from the type locality of the *Globigerina ampliapertura* Zone (see Bolli, 1957b, p. 100), lower Cipero formation, southern Trinidad; the subspecies is known to range from this Zone up into the *Globigerinita stainforthi* Zone of the lower Cipero formation, southern Trinidad, and it has also been observed in the Aquitanian of south-east Sicily (Ragusa limestone formation) and Tanganyika. It has not been found to occur in either the Eocene or the Oligocene of the Lindi area, and it has not been seen in the Upper Eocene of south Trinidad.

Globigerinita unicava primitiva Blow & Banner subsp.nov.
(Plate XIV J–L)

Catapsydrax unicavus Bolli, Loeblich & Tappan, Bolli, 1957, *Bull. U.S. Nat. Mus.* **215**, 166, pl. 37, figs. 7a–b.

DIAGNOSIS

The fairly small test consists of about three whorls of fairly rapidly enlarging, weakly inflated, partially embracing, depressed chambers. The early whorls consist of five chambers, but this is reduced to four in the last whorl. They are coiled in a low trochospire. In equatorial profile, the test is subcircular to subquadrate, and the equatorial periphery is broadly and weakly lobulate. In axial profile the test is tumid and broadly oval, but the ventral side is much more strongly vaulted than the weakly convex dorsal side. In axial profile the chambers are ovoid, their dorsal surfaces being weakly convex and their ventral sides strongly inflated. The axial periphery is broadly rounded, no distinct dorso-peripheral shoulder being present. The dorsal sutures are initially obscure, but later become increasingly distinctly and broadly depressed. The dorsal intercameral sutures are initially curved, meeting the weakly lobulate spiral suture at broad angles, but they become subradial between the last two or three chambers only. The ventral intercameral sutures are weakly curved to subradial. In dorsal view the chambers are initially reniform; during ontogeny they become increasingly depressed and increasingly longer than broad. The ventral umbilicus is broad and deep, but it is covered by a broad and inflated bulla. Deeply depressed sutures mark the attachment of the bulla to the last-formed chamber, the earliest chamber, and the posterior part of the second chamber of the last whorl. A single broad accessory aperture extends along the margin of the bulla for the full breadth of the penultimate chamber, and almost the whole breadth of the antepenultimate chamber. This accessory aperture is of almost uniform height throughout its length, usually being highest above the surface of the penultimate chamber; it is unrestricted by any lip or rim-like thickening of the bulla. The bulla covers the umbilicus, except where it is truncated by the accessory aperture; this truncation gives the bulla a subtriangular outline. The concealed primary aperture is an intra-umbilical arch. The wall of the primary chambers is fairly thin compared to *Globigerinita pera*; it is uniformly and moderately coarsely perforate. The surface of the test is punctate and weakly hispid, the hispidity being strongest in the vicinity of the accessory aperture; no clear reticulate (cancellate) pattern is present. The wall of the bulla is thinner, distinctly more finely perforate and much smoother than that of the primary chambers. Maximum diameter of holotype: 0·34 mm; from sample FCRM 1645, *Globigerapsis semi-involuta* Zone, Upper Eocene, Lindi area.

REMARKS

Globigerinita unicava primitiva differs from *G. unicava unicava* in possessing more strongly vaulted ventral surfaces to the primary chambers, a more inflated bulla (which protrudes above the umbilicus) and in possessing more depressed later chambers. *G. unicava primitiva* differs essentially from *G. pera* in possessing a more restricted bulla of subtriangular outline, which incompletely covers the umbilicus, and which possesses a longer accessory aperture; in addition, the wall of *G. unicava primitiva* is thinner, more finely perforate, and lacks a cancellate surface.

The specimen of *Globigerinita unicava* illustrated by Bolli (1957c, pl. 37, figs. 7a–b) from the *Truncorotaloides rohri* Zone, Navet formation, southern Trinidad, appears to show all the characters of *Globigerinita unicava primitiva*.

STRATIGRAPHICAL RANGE

In the Lindi area, this subspecies ranges from the *Truncorotaloides rohri* Zone, Middle Eocene, throughout the Upper Eocene, to the top of the *Globigerina oligocaenica* Zone, Oligocene. It does not seem to occur in the Aquitanian of either Trinidad (lower Cipero formation) or East Africa.

Genus: *Globigerinoides* Cushman, 1927

Type species: *Globigerina rubra* d'Orbigny, 1839 (lectotype selected, Banner & Blow, 1960*a*)

Globigerinoides quadrilobatus (d'Orbigny)

Globigerinoides quadrilobatus primordius Blow & Banner subsp.nov.
(Plate IXDd–Ff; Fig. 14 (iii–viii))

DIAGNOSIS

The test consists of about nine chambers arranged in a low, fairly loose trochospire of two to three whorls, each whorl consisting of about four chambers. The chambers are fairly rapidly and almost uniformly enlarging, and are distinctly inflated. The degree of embrace between the chambers decreases during ontogeny. The equatorial profile is ovoid and the equatorial periphery is lobulate. The axial profile is suboval, and the axial periphery is broadly and smoothly rounded. The chambers are subglobular and the intercameral sutures become increasingly deeply depressed during ontogeny, becoming almost incised between the last few chambers. The dorsal intercameral sutures, like the ventral ones, are subradial throughout the greater part of the test. The spiral suture is lobulate. The umbilicus is open, but is narrow and shallow; as the apertural face is of equal convexity to the rest of the chamber surface, the umbilical margins are not sharply delimited. Only two apertures are present on the test exterior. The primary aperture is a low intraumbilical arch, lacking a distinct lip or rim. The single dorsal supplementary sutural aperture is a small hole or very low arch situated in the wall of the last chamber at the junction of the last intercameral suture and the spiral suture; it lacks a lip or rim. The relict aperture of the penultimate chamber may sometimes be observed within the ventral umbili-

cus. The wall is fairly thick and is uniformly perforate; the surface is markedly cancellate and punctate, and this is strongest on the dorsal surfaces of the earlier chambers. Maximum diameter of holotype: 0·41 mm; from the type locality of the *Globorotalia kugleri* Zone, Cipero formation, southern Trinidad (see Bolli, 1957*b*, p. 100).

REMARKS

As this form possesses a dorsal sutural supplementary aperture, together with a distinct ventral (umbilical) primary aperture, it is referable to the genus *Globigerinoides* Cushman, 1927, as emended by Bolli *et al.* (1957) (see also Banner & Blow, 1959, 1960*a*). It is distinguished from *G. quadrilobatus quadrilobatus* (d'Orbigny) (see Banner & Blow, 1960*a*, pp. 17–19, pl. 4, for description of lectotype) by its possession of only one single dorsal sutural supplementary aperture in the adult test. We consider that this form is worthy of formal taxonomic recognition because it has a very restricted stratigraphical occurrence, and because it clearly demonstrates the transition between the genera *Globigerina* and *Globigerinoides*. It is distinguished from advanced forms of its immediate ancestor, *Globigerina praebulloides occlusa*, solely on its possession of a dorsal supplementary aperture; it also differs from typical (i.e. Oligocene) forms of *G. praebulloides occlusa* by being a little less tightly coiled and more coarsely cancellate, with broader punctae on the surface of the test, into which open pores still closely comparable to those of *G. praebulloides occlusa* (see also §VII, p. 136, and Fig. 14).

STRATIGRAPHICAL RANGE

This form is confined to the *Globorotalia kugleri* Zone in southern Trinidad, and it also occurs in stratigraphically sharply limited horizons in the Ragusa area of south-east Sicily and in East Africa, where it occurs in the Aquitanian, probably near the middle part. It does not occur, so far as is known, in the lower Aquitanian of Escornebéou or Moulin de l'Église, Aquitaine, nor has it ever been observed in any Oligocene samples from Germany, France or East Africa.

Genus: *Globoquadrina* Finlay, 1947

Type species: *Globorotalia dehiscens* Chapman, Parr & Collins, 1934

Globoquadrina dehiscens (Chapman, Parr & Collins)

Globoquadrina dehiscens praedehiscens Blow & Banner subsp.nov.
(Plate XV Q–s)

Globoquadrina rohri (Bolli), Blow, 1959, *Bull. Amer. Paleont.* **39**, no. 178, p. 185, pl. 11, figs. 57a–c (*not Globigerina rohri* Bolli, 1957).

DIAGNOSIS

The moderately large test consists of about three whorls of rapidly enlarging, moderately inflated, partially embracing chambers, coiled in a low trochospire. Four chambers are typically present in each of the early whorls, characteristically reducing to three and a half or three in the last whorl. The equatorial profile of the test is subcircular and the equatorial periphery is very weakly and broadly lobulate. In axial profile the test is globosely sub-conical. The dorsal surface is only very slightly convex, whereas the ventral side is highly vaulted. The axial periphery is smoothly and broadly rounded; the dorsal surface of each chamber, especially in late ontogeny, is both narrow and slightly convex, and the broadly rounded dorso-peripheral shoulder constitutes the greatest dia-meter of the test. Dorsally the chambers are initially semicircular to reniform, but they rapidly become much longer than broad and distinctly depressed. In axial view the adult chambers clearly show this depression, and are narrowly ovoid. Only three chambers are typically visible ventrally in the adult. The dorsal sutures are initially obscure, but they later become distinctly and narrowly de-pressed. The dorsal intercameral sutures are initially curved, meeting the moderately lobulate spiral suture at broad angles, but they later become even more weakly curved or subradial. The ventral intercameral sutures are fairly deeply but narrowly depressed. In the adult the umbilicus is small but deep and open; it is approximately triangular in shape, being sharply delimited by the steeply sloping umbilical margins of the chambers of the last whorl. The apertural face of the last chamber is narrow, distinctly flattened, faces the umbilicus,

and is quite sharply delimited from the rest of the chamber surface. The final (primary) aperture is a very low arch extending to the extreme limits of the umbilicus; it is bordered by a flap-like lip which becomes narrowly triangular at its mid-point, and which projects into the umbilicus as an umbilical 'tooth'. Early ontogenetic stages possess low, arched primary apertures which are umbilical-extraumbilical in extent, bordered by thickened rim-like lips (cf. *Globigerina oligocaenica*, Plate X G, L–N). The wall of the test is thick and moderately coarsely perforate, and it possesses a rough, 'granular' surface. The test is distinctly hispid in the vicinity of the umbilicus. Maximum diameter of holotype: 0·49 mm; from the type locality of the *Globorotalia kugleri* Zone, Cipero formation, southern Trinidad (see Bolli, 1957*b*, p. 100).

REMARKS

This subspecies differs from *Globoquadrina dehiscens dehiscens* (Chapman, Parr and Collins) (*syn. Globorotalia quadraria* Cushman & Ellisor, 1939) in possessing an intraumbilical aperture at an earlier ontogenetic stage. The ontogenetic change from an umbilical-extraumbilical to an intraum-bilical apertural position, occurring later in the ontogeny of *G. dehiscens dehiscens*, indicates that the evolution from *Globigerina tripartita tripartita*, through *Globoquadrina dehiscens praedehiscens* to *G. dehiscens dehiscens*, is proterogenetic.

Globoquadrina dehiscens praedehiscens also differs from *G. dehiscens dehiscens* in possessing more rapidly enlarging chambers, typically with only three visible ventrally in the last whorl, a more triangular umbilicus bordered, in regularly deve-loped specimens, by a lower and narrower apertural face, which directly faces the umbilicus. The shape of the test is more rounded and less quadrate in *G. dehiscens praedehiscens*.

STRATIGRAPHICAL RANGE

In southern Trinidad, this subspecies first occurs in about the middle of the *Globigerina ouachitaensis ciperoensis* Zone, and ranges up to about the lower part of the *Globigerinita stainforthi* Zone. It has been observed in equivalent horizons in eastern

Falcón, Venezuela, and in about the middle part of the Aquitanian of south-east Sicily, Tanganyika and New Zealand. It is likely that Bolli's stratigraphically higher records of *Globigerina rohri* and stratigraphically lower records of *Globoquadrina dehiscens* both refer to this subspecies (see p. 97).

Subfamily: GLOBOROTALIINAE Cushman, 1927, emended Banner & Blow, 1960

Genus: *Globorotalia* Cushman, 1927

Type species: *Pulvinulina menardii* (d'Orbigny) var. *tumida* Brady, 1877 (lectotype selected, Banner & Blow, 1960*a*)

Subgenus: *Turborotalia* Cushman & Bermúdez, 1949

Type species: *Globorotalia centralis* Cushman & Bermúdez, 1937

Globorotalia (*Turborotalia*) *centralis* Cushman & Bermúdez
(Plates XII K–M, XVII B, G; Fig. 12*c*, *d*)

Globorotalia centralis Cushman & Bermúdez, 1937, *Contr. Cushman Lab. Foram. Res.* **13**, 26, pl. 2, figs. 62–65.

G. (*Turborotalia*) *centralis* Cushman & Bermúdez, Cushman & Bermúdez, 1949, *Contr. Cushman Lab. Foram. Res.* **25**, 42.

Acarinina centralis (Cushman & Bermúdez), Subbotina, 1953, *Trudi, VNIGRI*, N.S., **76**, 237–239, pl. 25, figs. 7a–9c (?*not* figs. 10–11).

G. centralis Cushman & Bermúdez, Bolli, Loeblich & Tappan, 1957, *Bull. U.S. Nat. Mus.* **215**, 41, pl. 10, figs. 4a–c.

G. centralis Cushman & Bermúdez, Bolli, 1957, *Bull. U.S. Nat. Mus.* **215**, 169, pl. 39, figs. 1–4.

DESCRIPTION

The fairly large test consists of about three whorls of slowly enlarging, moderately inflated, partially embracing chambers coiled in a low trochospire. The adult test typically possesses four to five chambers in the last whorl, the number often reducing slightly in ontogeny. The chambers are only very slightly convex dorsally, but they are relatively very convex ventrally, being rounded-triangular in axial view. The equatorial profile is subcircular and the equatorial periphery is weakly lobulate. In axial profile the test shows marked dorsal flattening and a strongly convex ventral side; the axial periphery is broadly and smoothly rounded, but a distinct dorso-peripheral shoulder is present. The ventral intercameral sutures are distinctly but not strongly depressed, and are subradial. Dorsally, the sutures are initially indistinct, but they become increasingly depressed during ontogeny. The dorsal intercameral sutures are curved, meeting the spiral suture at acute angles and the periphery tangentially. The spiral suture is weakly lobulate. In dorsal aspect, the chambers are typically semicircular in outline, but they become increasingly depressed and longer than broad during ontogeny; the anterior part of the chamber is often narrower than the posterior part. The umbilicus is very small but deep, and it is often partially or wholly covered by the umbilical margin of the last chamber. The aperture is interiomarginal, umbilical–extraumbilical, a fairly high arch extending almost to the periphery; a very weak lip or rim may be present. Typically, the relict aperture of the penultimate chamber may be observed only through the last-formed aperture, and it may become merely a septal aperture. The apertural face is narrow, somewhat flattened, and often hispid. The wall of the test is uniformly and finely perforate (relative to the size of the test). The surface of the test is smooth, except for weakly hispid areas around the umbilicus and aperture.

REMARKS

This species should be compared to *Globorotalia* (*T.*) *increbescens* (see p. 118, Plate XIII T–V) and to *Globigerina pseudoampliapertura* (see p. 95, Plate XII A–C).

STRATIGRAPHICAL RANGE

This species ranges from beds at least as old as Lutetian to high within the *Cribrohantkenina danvillensis* Zone, Upper Eocene. It does not appear to range either in the uppermost part of the *C. danvillensis* Zone or in the overlying *Globigerina turritilina turritilina* Zone, Upper Eocene, of the Lindi area. In the Caribbean, *Globorotalia centralis* occurs in the Middle and Upper Eocene. Subbotina's records (1953) of this species as high as the *Bolivina* Zone (in the Caucasus) probably refer to *Globigerina pseudoampliapertura* sp.nov. (*q.v.*).

Globorotalia (*Turborotalia*) *cerro-azulensis* (Cole)
(Plate XII D–F; Figs. 12*d*, *e*)

Globigerina cerro-azulensis Cole, 1928, *Bull. Amer. Paleont.* **14**, no. 53, p. 217, pl. 32, figs. 11–13.

Globorotalia cocoaensis Cushman, 1928, *Contr. Cushman Lab. Foram. Res.* **4**, pt. 3, p. 75, pl. 10, figs. 3a–c.

G. cocoaensis Cushman, Cushman, 1935, *U.S. Geol. Surv., Prof. Paper*, **181**, 50, pl. 21, figs. 1a–3c.

G. cocoaensis Cushman, Bolli, 1957, *Bull. U.S. Nat. Mus.* **215**, 169, pl. 39, figs. 5a–7b.

REMARKS

The evolution of this species from *Globorotalia* (*Turborotalia*) *centralis* is discussed elsewhere (p. 135).

STRATIGRAPHICAL RANGE

In both southern Trinidad and the Lindi area, this species ranges from the base of the *Globigerapsis semi-involuta* Zone, Upper Eocene. In East Africa, it ranges from this zone to the top of the *Cribrohantkenina danvillensis* Zone, Upper Eocene, but not into the overlying *Globigerina turritilina turritilina* Zone of the uppermost Eocene. In Trinidad, Bolli (1957c) used this species to define his 'Globorotalia cocoaensis Zone' for the 'San Fernando formation', but, as previously discussed (p. 81), we consider that the 'San Fernando formation' as used by Bolli (1957c) (and consequently the 'G. cocoaensis Zone' also) contains a largely reworked series of beds (and derived faunas) in its upper part.

Globorotalia (*Turborotalia*) *increbescens* (Bandy)
(Plates XIII T–V, XVII D, K; Fig. 9 (xiii–xv))

Globigerina increbescens Bandy, 1949, *Bull. Amer. Paleont.* **32**, no. 131, p. 120, pl. 23, figs. 3a–c.

DESCRIPTION

The fairly small test consists of about three whorls of slowly enlarging, inflated, moderately embracing chambers, with about four and a half chambers in each whorl of a low trochospire. The chambers are slightly convex dorsally and are rather more strongly convex ventrally. The equatorial profile of the test is subcircular and the equatorial periphery is weakly lobulate. The axial profile is suboval and the axial periphery is broadly and smoothly rounded. The dorsal surface of the test is distinctly convex. The sutures are distinctly but not strongly

depressed. The ventral intercameral sutures are nearly radial, whilst the dorsal intercameral sutures are initially curved, becoming subradial during ontogeny. The dorsal surface is completely evolute, and the spiral suture is uniformly lobulate. In dorsal aspect, the earlier chambers are reniform to semicircular in outline; in later ontogenetic stages they remain distinctly longer than broad, but approach a quadrangular outline. The ventral umbilicus is very small and shallow, and it is often almost completely closed. The umbilical–extra-umbilical primary aperture extends almost to the periphery; it is a high arch, highest at its anterior end, and it is bordered by a very narrow, rim-like lip. The apertural face is weakly inflated and somewhat flattened. The wall is thick relative to the size of the test, and it is uniformly and fairly coarsely perforate. The surface of the test has a slight 'granular' appearance, and it is most strongly cancellate and mamelonate around the umbilical margins.

REMARKS

This species is distinguished from *Globorotalia* (*T.*) *centralis* by its more coarsely perforate and thicker wall (relative to over-all test size) which possesses a typically 'granular' appearance; *G.* (*T.*) *centralis* has a characteristically 'polished' look, especially dorsally. The chambers of *G.* (*T.*) *increbescens*, when seen in dorsal aspect, are more semicircular than the relatively longer chambers of *G.* (*T.*) *centralis*. The dorsal intercameral sutures of *G.* (*T.*) *increbescens* are only slightly curved, or subradial, meeting the spiral suture at right angles, whilst those of *G.* (*T.*) *centralis* are strongly curved throughout ontogeny, meeting the spiral suture acutely and the periphery tangentially; the dorsal appearance of *G.* (*T.*) *centralis* is often of the pattern which reaches an extreme form in *G.* (*T.*) *scitula* (see Banner & Blow, 1960a), whereas *G.* (*T.*) *increbescens* has a more globigerinid pattern. In axial profile, the dorso-peripheral shoulder is strongly developed in *G.* (*T.*) *centralis*, with an abrupt change in convexity from the dorsal to the ventral surfaces; this contrasts with the more globose *G.* (*T.*) *increbescens*, which has chambers that are almost hemispherical in axial profile. In ventral view, the chambers of *G.* (*T.*) *centralis* are distinctly broader posteriorly than anteriorly, whilst in *G. (T.) increbescens* the chambers are of symmetrical shape (as in *G. (T.) opima*). Studies of assemblages of populations of *G. (T.) centralis* and *G. (T.) increbescens* show that the latter is comparable in size to *G. (T.) opima*, and does not reach the relatively large size normal for fully developed *G.(T.) centralis*.

We are grateful to Dr Orville L. Bandy for metatypic specimens of *G. (T.) increbescens* from Little Stave Creek, Alabama (Jacksonian, Upper Eocene) (see Fig. 9), with which we have compared our specimens from the Upper Eocene and Oligocene of East Africa.

STRATIGRAPHICAL RANGE

Specimens referable to this species range from the topmost part of the *Globigerapsis semi-involuta* Zone, Upper Eocene, to the top of the *Globigerina oligocaenica* Zone, Oligocene, in the Lindi area. It has not been observed in the *G. ampliapertura* Zone of the Caribbean or in the Aquitanian of southwestern France (Escornebéou and Moulin de l'Église).

Globorotalia (Turborotalia) opima Bolli

Globorotalia (Turborotalia) opima opima Bolli

Globorotalia opima opima Bolli, 1957, *Bull. U.S. Nat. Mus.* **215**, 117, pl. 28, figs. 1a–2.

REMARKS

We have examined topotype material of this subspecies from southern Trinidad, lower Cipero formation, where it is restricted to its Zone (see Bolli, 1957b, pp. 99, 100) and is considered by us to be of lower Aquitanian age. We have failed to obtain this subspecies from Europe, the Mediterranean or East Africa.

Globorotalia (Turborotalia) opima nana Bolli
(Plate XIIIq–s)

Globorotalia opima nana Bolli, 1957, *Bull. U.S. Nat. Mus.* **215**, 118, pl. 28, figs. 3a–c.

Globigerina sp. Batjes, 1958, *Mém. Inst. Sci. nat. Belg.* no. 143, pp. 161–162, pl. 9, figs. 7a–c.

DESCRIPTION

The fairly small test consists of about three whorls of slowly enlarging, inflated, moderately embracing chambers, with about four and a half chambers in each whorl of a very low trochospire. The chambers are slightly convex dorsally, but fairly strongly convex ventrally. The equatorial profile of the test is subcircular, and the equatorial periphery is weakly lobulate. The axial profile is suboval to subovoid and the axial periphery is broadly and smoothly rounded. The dorsal surface of the test is flattened. The sutures are distinctly but not strongly depressed. The ventral intercameral sutures are nearly radial, whilst the dorsal intercameral sutures are initially curved, becoming subradial only in the later parts of the last whorl. The spiral suture is uniformly lobulate where the dorsal surface is completely evolute, but many specimens show a slight but definite overlap of the chambers of the last whorl on to the dorsal surfaces of earlier chambers, so causing the test to become slightly involute dorsally. In these cases, the later parts of the spiral suture are much less lobulate. In dorsal aspect, the chambers are initially reniform to semicircular in outline; later, they remain distinctly longer than broad, but approach a quadrangular outline. The ventral umbilicus is very small, almost completely closed and shallow. The umbilical-extraumbilical aperture extends almost to the periphery; it is a low (almost slit-like) arch, bordered by a distinct, broad lip. The apertural face is weakly flattened, but remains convex. The wall is thick and coarsely perforate relative to the size of the test. The surface has a distinct 'granular' appearance, and it is most strongly cancellate and mamelonate around the umbilical margins.

REMARKS

We have compared our specimens from the Oligocene of East Africa with topotypic specimens from southern Trinidad, and the Oligocene assemblages show a higher proportion of dorsally fully evolute forms; otherwise, they are identical. This subspecies is closely similar to *Globorotalia (T.) increbescens* (Bandy); the major distinctions are the tighter coiling, even more restricted umbilicus, much lower aperture with a much stronger lip, and much flatter dorsal surface of *G. (T.) opima nana*. These distinctions may be considered to be of less than specific value; however, it is our intention to retain the taxonomic *status quo* whenever possible, and to use the existing taxonomy to define stratigraphically useful, although often arbitrary, stages in rapidly evolving lineages. It should be noted (Lineage A, p. 130) that the phylogenetic end-forms of the *G. opima–Globigerina ampliapertura* lineages are very distinct both morphologically and stratigraphically.

STRATIGRAPHICAL RANGE

Globorotalia (*T.*) *opima nana* was first described by Bolli (1957*b*) from the *G. opima opima* Zone, Cipero formation, Trinidad. It occurs in the Lindi area from beds at least as old as Middle Eocene (Lutetian), and ranges to the top of the Oligocene. Elsewhere in East Africa it occurs in the lower and middle parts of the Aquitanian, but not in the higher Aquitanian and Burdigalian. In south-eastern Sicily (Ragusa area) it occurs in the Aquitanian, whilst rare specimens have been found by us in the lower Aquitanian of Escornebéou. It also occurs in the Oligocene of Elmsheim and Offenbach (Mainz Basin), and it has been recorded by Batjes (as *Globigerina* sp.) from the Oligocene of Belgium.

Globorotalia (*Turborotalia*) *permicra* Blow & Banner sp.nov.
(Plate XII N–P)

DIAGNOSIS

The minute test consists of about two whorls of slowly enlarging, slightly embracing, inflated chambers, arranged in a low trochospire with five to six chambers in each whorl, the number per whorl decreasing during ontogeny. The chambers are subovoid, and enlarge more rapidly in the later whorls. In equatorial profile the test is subovate and the equatorial periphery is lobulate. In axial profile the test is subovate, the dorsal side being flattened and the ventral side being relatively strongly convex; the axial periphery is broadly rounded. In axial view, the chambers are ovoid. The sutures are distinctly depressed. The dorsal intercameral sutures are curved, meeting the weakly lobulate spiral suture and the periphery of the test at broad angles. The ventral sutures are subradial or very slightly curved. In dorsal view the chambers

are initially semicircular to reniform, but they broaden relatively more rapidly in the later ontogenetic stages. The ventral umbilicus is narrow but open and deep; the umbilical margins of the adjacent chambers are steeply convex. The final aperture extends from the umbilicus to the periphery; it is distinctly broadest at its anterior end, where it forms a high arch. The aperture is furnished with a thin and narrow lip throughout its length, often broadest at the mid-point. The wall of the test is thin and usually translucent; it is uniformly and fairly densely perforate. The surface of the test is very weakly hispid in the vicinity of the umbilicus, and on the earlier chambers (of the last whorl) which face the aperture; elsewhere the surface is smooth. Maximum diameter of holotype: 0·155 mm; from sample FCRM 1922, Oligocene, Lindi area.

REMARKS

This species differs from *Globorotalia* (*T.*) *minutissima* Bolli by its very different appearance in axial view, the dorso-ventral convexity being almost equal in Bolli's species but being distinctly unequal in *G.* (*T.*) *permicra*, which also possesses more rapidly enlarging chambers and a much higher aperture.

STRATIGRAPHICAL RANGE

In the Lindi area, this species ranges from the *Globigerina turritilina turritilina* Zone, uppermost Eocene, to the *G. oligocaenica* Zone, Oligocene. It has been observed in the Rupelian of Elmsheim, Mainz Basin, in the *G. ampliapertura* Zone, Cipero formation (lower Aquitanian), southern Trinidad, and in the lower Aquitanian of Escornebéou, south-western France.

Globorotalia (*Turborotalia*) *postcretacea* (Myatliuk)
(Plate XII G–J)

Globigerina postcretacea Myatliuk, 1950, *Mikrofauna SSSR*, Sb. IV, p. 280, pl. 4, figs. 3 a–b.

G. postcretacea Myatliuk, Subbotina, 1953, *Trudi, VNIGRI*, N.S., **76**, 60–61, pl. 2, figs. 16a–20c.

DESCRIPTION

The minute test consists of about two whorls of slowly enlarging, inflated, slightly embracing chambers arranged in a very low trochospire with five to six chambers in each whorl, the number of chambers per whorl increasing during ontogeny. The chambers are subglobular and enlarge more slowly in the later whorls. In equatorial profile the test is subcircular and the equatorial periphery is moderately strongly lobulate. In axial profile the test is somewhat compressed, the dorsal and ventral sides both being flattened; the dorsal surface varies from being weakly convex to slightly concave, depending upon the height of the trochospire; in some cases the spire is so low that the dorsal surfaces of the later chambers form a level higher than that of the proloculum and first few chambers. The axial periphery is broadly and smoothly rounded. The sutures are distinctly depressed. The dorsal intercameral sutures are initially curved, meeting the lobulate spiral suture almost at right angles, but before the end of the first whorl they become subradial. The ventral sutures are almost radial or very slightly curved. In both dorsal and ventral views the chambers are approximately as broad as long. The ventral umbilicus is narrow but open and deep; the umbilical margins of the adjacent chambers are smoothly convex. The final aperture extends from the umbilicus to the periphery; it is distinctly broadest at its anterior end, where it forms a high arch. The aperture is furnished throughout its length with a distinct, uniform, thin and narrow lip. The wall of the test is thin and usually translucent (sometimes transparent in the later chambers); it contains uniformly but fairly sparsely distributed small pores. The surface of the test is weakly hispid in the vicinity of the umbilicus and the aperture; elsewhere the surface is smooth and polished.

REMARKS

As noted by Myatliuk (1950), this species bears a remarkable superficial resemblance to many Lower Cretaceous species now referred to *Praeglobotruncana* (*Hedbergella*) (see Banner & Blow, 1959) (e.g. *P.* (*H.*) *infracretacea* (Glaessner) and *P.* (*H.*) *planispira* (Tappan)). However, the porticus, the fundamentally distinctive character of the Hantkeninidae, to which family *Hedbergella* belongs, is lacking in *Globorotalia* (*T.*) *postcretacea*.

We have compared this species with metatypes of *Globorotalia* (*T.*) *minutissima* Bolli, and the latter

species is distinguished by its more rapidly enlarging chambers, its more closed umbilicus, its narrow, slit-like aperture, and more densely perforate wall.

STRATIGRAPHICAL RANGE

Myatliuk first described this species from beds thought to be of Oligocene age in the Ukraine, and it has subsequently been described by Subbotina (1953) from the lower part of the Oligocene of the Caucasus, where she believes that it ranges up from her *Bolivina* Zone (which may be uppermost Eocene).

In the Lindi area, this species has been observed only in the Oligocene. It has also been observed by us in a sample from the Rupelian of Elmsheim, Germany. It does not seem to occur in the *Globigerina ampliapertura* Zone of the Cipero formation, Trinidad, or in the lower Aquitanian of southwestern France (Escornebéou, Moulin de l'Église).

Genus: *Truncorotaloides* Brönnimann &
Bermúdez, 1953

Type species: *Truncorotaloides rohri* Brönnimann &
Bermúdez, 1953

Truncorotaloides rohri Brönnimann & Bermúdez

Truncorotaloides rohri Brönnimann & Bermúdez, 1953, *J. Paleont.* 27, no. 6, pp. 818–819, pl. 87, figs. 7–9.

T. rohri Brönnimann & Bermúdez, Bolli, Loeblich & Tappan, 1957, *Bull. U.S. Nat. Mus.* 215, 42, pl. 10, figs. 5a–c (holotype redrawn).

T. rohri Brönnimann & Bermúdez, Bolli, 1957, *Bull. U.S. Nat. Mus.* 215, 170, pl. 39, figs. 8–12c.

REMARKS

We have compared our specimens from East Africa with material from the Navet formation of southern Trinidad, and no significant differences occur. At present, we do not differentiate between the 'varieties' *piparoensis*, *mayoensis* and *guaracaensis* proposed by Brönnimann & Bermúdez (1953).

STRATIGRAPHICAL RANGE

This species is typically associated with Lutetian larger foraminifera in East Africa. However, the upper part of its range has not yet been fully evaluated with regard to the ranges of larger index foraminifera, and for the time being it is convenient to draw the Auversian/Lutetian boundary at the top of the *Truncorotaloides rohri* Zone.

Genus: *Turborotalita* Blow & Banner gen.nov.

Type species: *Turborotalita humilis* (Brady) = *Truncatulina humilis* Brady, 1884, *Rep. Voy. 'Challenger' Zool.* **9**, pt. 22, pp. 665–666, pl. 94, figs. 7 a–c; lectotype selected and described by Banner & Blow, 1960 *a*, *Contr. Cushman Found. Foram. Res.* **11**, pt. 1, p. 36, pl. 8, figs. 1 a–c.

DIAGNOSIS

Turborotalita comprises those members of the subfamily Globorotaliinae, family Globigerinidae (see Banner & Blow, 1959, pp. 6, 7, 16), which possess a primary aperture (of umbilical–extraumbilical position) covered by a bulla (see Banner & Blow, 1959, p. 26); the bulla possesses accessory apertures at its suture with the ventral sides of the primary chambers of the last whorl. The bulla may take the *apparent* form of a modified final chamber (often with reduced pore-size), which spreads ventrally partially or wholly to conceal the ventral umbilicus. The accessory apertures may or may not be situated at the ends of tunnel-like extensions from the main body of the bulla, and they may or may not be restricted by lips. No supplementary sutural apertures are present. No peripheral true carinae are known.

REMARKS

This genus is distinguished from *Globigerinita* Brönnimann (*em.*) by the umbilical–extraumbilical extent of the primary aperture. It is distinguished from *Globorotalia (Turborotalia)* by possessing a bulla and accessory apertures.

Globigerina cristata Heron-Allen & Earland, 1929 (lectotype selected and figured by Banner & Blow, 1960 *a*), and *Globigerinita parkerae* Loeblich & Tappan, 1957, are referable to this genus. We thank Dr F. L. Parker for supplying us with recent specimens of *Turborotalita parkerae*.

STRATIGRAPHICAL RANGE

As distinct from *Globigerinita*, which ranges from Eocene to Recent, *Turborotalita* is a Neogene genus; its species are most abundant in recent seas, but its representatives may have occurred since Upper Miocene times.

Subfamily: GLOBOROTALOIDINAE, Banner & Blow, 1959

We retain and use this subfamily because, unlike the 'Catapsydracinae', the members of which may now be referred to the subfamilies of their parent genera, the species contained within the Globorotaloidinae are characterized by possessing a primary aperture which migrates during ontogeny from an umbilical–extraumbilical position to an intra-umbilical position. The primary tests of *Catapsydrax*, *Tinophodella* and *Globigerinita* are simply those of *Globigerina*, while the primary test of *Turborotalita* is that of a *Turborotalia*, but the primary test of *Globorotaloides* is unique. Not only does the stratigraphical range of *Globorotaloides* preclude a direct relationship with *Globoquadrina*, but, unlike *Globoquadrina*, within any population of, for example, *Globorotaloides suteri*, adult specimens of comparable size may be found whose primary aperture beneath the bulla is either intra-umbilical or partly extraumbilical (Fig. 11, v, ix). Therefore, the form of the primary test in the Globorotaloidinae is intermediate between those of the Globigerininae and the Globorotaliinae, and this, considered together with the characteristic bulla, indicated that the forms included here cannot be placed satisfactorily in any other subfamily.

Genus: *Globorotaloides* Bolli, 1957

Type species: *Globorotaloides variabilis* Bolli, 1957

Globorotaloides suteri Bolli
(Plate XIII N–P; Fig. 11 (v, ix))

Globorotaloides suteri Bolli, 1957, *Bull. U.S. Nat. Mus.* **215**, 116, 166; pl. 27, figs. 9–13b, cf. 14a–c; pl. 37, figs. 10a–12.

Globigerina globularis Roemer, Batjes, 1958 (part; ?*not* Roemer, 1838), *Mém. Inst. Sci. nat. Belg.* no. 143, pp. 161–162, pl. 11, figs. 3a–c, 5a–c (*not* figs. 4a–c).

Globigerina sp. Batjes, 1958, *ibid.* pl. 11, figs. 8a–c.

REMARKS

Our specimens of the species have been carefully compared to topotypes from the type locality of the *Globigerina ampliapertura* Zone, Cipero formation, southern Trinidad. Our specimens show the progressive ontogenetic restriction of the umbilical–extraumbilical aperture of early ontogeny (with five or six chambers per whorl) to the intraumbilical aperture of later ontogeny (with four chambers per whorl), where a simple bulla is typically acquired. The acquisition of a bulla may occur at the end of the earlier (neanic?) ontogenetic stage, where the primary aperture is still umbilical–extraumbilical; these forms, where the bulla covers a globorotaliid aperture, are often comparable in size to other forms where a globigerinid aperture is developed at or before acquisition of a bulla. These two forms co-exist throughout the whole stratigraphical range of the species, and they may represent one of the rare cases of sexual dimorphism to be found in the Globigerinidae. Independently of this, the bulla seems to be acquired at varying growth stages (i.e. in tests of different sizes, independently of the onto-genetic stage attained); this may be due to ecological factors or to genetically controlled variation in the metabolic rates of individuals.

The wall structure and surface of *Globorotaloides suteri* are strikingly different from those of any known species of *Globigerinita*. The surface is very coarsely punctate, the punctae often coalescing to produce a cancellate ('reticulate') pattern reminiscent of *Globigerinoides quadrilobatus*; as in *G. quadrilobatus*, the pores of *Globorotaloides suteri* appear to be of much smaller diameter than the pit-like superficial punctae into which they individually open. The test was probably very strongly hispid during life.

STRATIGRAPHICAL RANGE

This species ranges from the upper Lutetian to about the middle part of the Aquitanian in both East Africa and southern Trinidad. It is common in the Oligocene of Tanganyika, and has been recorded by Batjes (1958, as *Globigerina globularis* Roemer) from the Belgian Oligocene. It has also been observed by us in the Aquitanian of south-eastern Sicily, where it occurs associated with *Miogypsina s.s.* in marls from the Ragusa lime-stone.

Subfamily: ORBULININAE (Schultze, 1854)

Genus: *Globigerapsis* Bolli, Loeblich & Tappan, 1957

Type species: *Globigerapsis kugleri* Bolli, Loeblich & Tappan, 1957

REMARKS

The genus *Globigerapsis* possesses a primary chamber development differing from *Globigerina*, *Globigerinita* or *Globigerinoides*; the early whorls of regularly enlarging, trochospirally coiled, *Globigerina*-like chambers are followed in the last whorl by chambers which abruptly start to enlarge very rapidly, become more ventrally extensive (more tightly coiled), and increasingly embrace the umbilicus of the earlier whorls. The final chamber of the adult, whilst retaining its position in the modified growth-spiral, completely covers the umbilicus and the previous primary aperture, and develops sutural 'supplementary' apertures. The final chamber may completely envelop the ventral surface of the earlier test. The bulla of *Globigerinita* is characterized by the fact that it does not form a part of the normal growth-spire, and even when it does *appear* to do so (e.g. *Globigerinita howei* and *Turborotalita* spp.) it is clearly distinct from the regularly developing primary test. In *Globigerapsis* the embracing last chamber is naturally preceded by a series of chambers which also have gradually increased their mutual embrace, concomitantly with an increase in the tightness of the growth-spire. These characters enable *Globigerapsis* to be recognized even when the fully developed adult characters are not present (cf. Fig. 11, vi–viii, with the growth stages of a *Globigerinita*; *Globigerinatheka* and *Globigerapsis* have similar primary test developments).

The adult *Globigerapsis* was originally distinguished from *Globigerinoides* (by Bolli *et al.* 1957, pp. 33–34—see also Banner & Blow, 1959, p. 5) by its lack of a single, distinct, principal aperture, the multiple sutural apertures being of approximately equal size and shape. These apertures appear to be homologous to the sutural apertures of *Porticula-sphaera*, but this genus differs in possessing such apertures in the spiral suture between earlier chambers, as well as in the suture of the last. Because the last chamber of *Globigerapsis* appears to belong to the normal growth spire, and because the multiple final apertures appear functionally to replace the single primary aperture of earlier ontogeny (without supplementing it or being accessory to it), *Globigerapsis*, like *Porticulasphaera* and *Candeina*, appears to be referable to the subfamily Orbulininae. The regular change in the growth-spire is only clearly recognizable in the morphologically more primitive members of this sub-family, the change of coiling occurring more abruptly in the morphologically more advanced forms. For example, *Biorbulina* shows a relatively abrupt change in coiling at the formation of the penultimate chamber, whilst *Orbulina* shows this abrupt change in the positioning of the final chamber; as would be expected, the most advanced forms possess the greatest number of sutural apertures, and the reduction in size of these apertures may be accompanied by the acquisition of areal apertures also (see Blow, 1956; Banner & Blow, 1959, p. 4).

Globigerapsis tropicalis Blow & Banner sp.nov.
(Plate XV D–F)

Globigerinoides conglobatus (H. B. Brady), Glaessner, 1937 (*not Globigerina conglobata* Brady, 1879), *Studies in Micropalaeontology, Moscow Univ.* **1**, fasc. 1, p. 29, pl. 1, fig. 3.

G. conglobatus (H. B. Brady), Subbotina, 1953 (*not* Brady, 1879), *Trudi, VNIGRI,* **76**, N.S., 91, pl. 14, figs. 2a–5c.

Globigerapsis index (Finlay), Bolli, 1957 (*not Globigerinoides index* Finlay, 1939), *Bull. U.S. Nat. Mus.* **215**, 165, pl. 36, figs. 14a–18b.

Globigerinoides conglobatus (H. B. Brady), Shutskaya, 1958 (*not* Brady, 1879), *Vopr. Mikropal.* no. 2, p. 88, pl. 2, figs. 1–5.

DIAGNOSIS

The moderately large test consists of three whorls of chambers arranged trochospirally. The first two whorls of chambers are arranged in a very low trochospire, with about five chambers in each whorl; the chambers are moderately inflated and partially embracing, the amount of embrace apparently decreasing slowly and regularly when seen in dorsal aspect. The succeeding whorl consists of three to four chambers, which are abruptly and rapidly enlarging, are coiled in an increasingly high trochospire, and which increasingly embrace the earlier test ventrally. The last-formed adult chamber envelops all or most of the earlier ventral surface of the test. The penultimate chamber possesses a single, intraumbilical, arched aperture, which may be bordered by a thickened rim; the

true umbilicus of the earlier chambers may be seen within the penultimate primary aperture. The adult test possesses three or four arched semicircular apertures placed in the basal suture of the last chamber at its junctions with the intercameral sutures of the earlier chambers. These adult apertures are of approximately equal size, and each possesses a thickened rim. The adult test is globose, the final chamber constituting approximately one-third to one-half of the total test. The initial sutures are weakly depressed, the dorsal intercameral sutures being curved, and the spiral suture weakly lobulate; during ontogeny, the sutures become increasingly depressed, the intercameral sutures become subradial, and the spiral suture and periphery become moderately lobulate. The wall is thin, compared to *Globigerapsis index* (Finlay); it is finely and uniformly perforate, but it sometimes becomes weakly cancellate, especially dorsally. Maximum diameter of holotype: 0·37 mm; from sample FCRM 1645, Upper Eocene, Lindi area.

REMARKS

Globigerapsis tropicalis differs from *G. index* (Finlay) in lacking the very thick walls, deeply incised sutures, and heavily granular surface characteristic of the latter species. Hornibrook (1958, p. 34, pl. 1, figs. 11–14) re-illustrated and briefly discussed Finlay's type specimens of '*Globigerinoides*' *index*. F. C. Dilley has kindly shown us topotypic specimens from the 'Bortonian' stage of the Middle Eocene of South Island, New Zealand; these specimens confirm the presence of thick walls and 'deeply cleft sutures' in this species, as stated by both Finlay (1939) and Hornibrook (1958). *Globigerapsis tropicalis* not only lacks these characters, but is morphologically indistinguishable from specimens seen by us in the Hospital Hill Marl (Navet formation) of southern Trinidad; such specimens were illustrated by Bolli (1957c), and are believed to be synonymous. Specimens which are correctly referable to *G. index* (Finlay) occur abundantly in coastal Tanganyika, at horizons which may safely be correlated with the lower Middle Eocene. *G. index* appears to be absent in the Upper Eocene of East Africa and Trinidad, and it is probable that other records from the Upper

Eocene (e.g. from Cuba, Syria, Israel and south-eastern Australia—Hornibrook, 1958, p. 34) should refer to *G. tropicalis*.

Comparison of Finlay's holotype of *G. index* (re-illustrated by Hornibrook, 1958) with Bolli's (1957c, figs. 15–18) equally immature specimens, shows that the distinctions between *G. index* and *G. tropicalis* are apparent even before the adult stage of ontogeny is reached (Plate VII c, D).

Globigerapsis tropicalis differs from *G. semi-involuta* (Keijzer) in possessing less embracing and better separated chambers throughout all stages of ontogeny, a less tightly coiled trochospire, slightly greater hispidity in the juvenile stage, smaller multiple apertures in the adult stage, less distinctive apertural lips in both the juvenile and adult stages, and a much less inflated and embracing final chamber.

'*Globigerinoides*' *subconglobatus* Shutskaya (given as 'Chalilov MS' by Shutskaya, 1958, pp. 83–88, pl. 1, figs. 1–11), a form of *Globigerapsis* from the Middle and Upper Eocene of the Caucasus, is claimed by Shutskaya to be ancestral to the form referred to by us as *G. tropicalis*. We have seen specimens similar to *G. index subconglobatus* (Shutskaya) in the Middle Eocene of East Africa, these forms differing from both *G. index index* and *G. tropicalis* in possessing less rapidly enlarging and inflated chambers, smaller adult apertures and a more compact and subglobular test.

It is highly likely that *Globigerapsis tropicalis* is ancestral to *G. semi-involuta*.

STRATIGRAPHICAL RANGE

In the area of coastal Tanganyika, *Globigerapsis tropicalis* ranges from the Middle Eocene to the top of the *G. semi-involuta* Zone (lower part of the Upper Eocene). Subbotina (1953) and Glaessner (1937) recorded a similar range for this form in the Caucasus, and the former author used it to define her '*Globigerinoides conglobatus* Zone' of the lower Upper Eocene. Bolli (1957c) records a similar range in southern Trinidad. *Globigerapsis index* (Finlay) appears to possess a distinctly different stratigraphical distribution; all records which can be verified suggest that it does not occur above the Middle Eocene, and additional evidence from East Africa suggests that it may occur as low as the uppermost part of the Lower Eocene.

Globigerapsis semi-involuta (Keijzer)
(Plate XV J–L)

Globigerinoides semi-involutus Keijzer, 1945, *Univ. Utrecht, Geogr. Geol. Med. Phys.-Geol. Reeks*, ser. 2, no. 6, p. 206, pl. 4, figs. 58 a–e.
Globigerapsis semi-involuta (Keijzer), Bolli, 1957, *Bull. U.S. Nat. Mus.* **215**, 165, pl. 36, figs. 19–20.

REMARKS

Globigerapsis semi-involuta should be distinguished from the similar, but morphologically more primitive form, *G. tropicalis*, from which it is probably descended. The smoother walls, larger multiple adult apertures (which are furnished with strong lips) and the much larger, more embracing and inflated final chamber, serve to separate this form from *G. tropicalis*.

STRATIGRAPHICAL RANGE

This species is often abundant in, and is strictly limited to, its Zone within the lower part of the Upper Eocene, both in southern Trinidad (Hospital Hill Marl, Navet formation) and in the Lindi area, East Africa.

Genus: *Globigerinatheka* Brönnimann, 1952

Type species: *Globigerinatheka barri* Brönnimann, 1952

Globigerinatheka lindiensis Blow & Banner sp.nov.
(Plate XV o, P; Fig. 11 (vi–viii))

DIAGNOSIS

The moderately large test consists of about three whorls of fairly rapidly enlarging, inflated, partially embracing chambers, arranged four in each whorl in a low trochospire. The equatorial profile of the test (excluding the bullae) is subcircular to subquadrate, and the equatorial periphery is broadly and weakly lobulate. In axial profile, the test is subcircular and the periphery is broadly rounded. At the end of the second whorl the chambers rapidly increase their rate of enlargement, becoming increasingly inflated and embracing ventrally, so that the last-formed primary chamber envelops the whole of the ventral surface of the earlier test. In dorsal view the chambers are initially semicircular to reniform in outline, but they become increasingly depressed and longer than broad during the last whorl. The dorsal sutures are initially indistinct, but, by the second whorl, they become distinctly but broadly and weakly depressed; the dorsal intercameral sutures are subradial or slightly curved, meeting the weakly lobulate spiral suture almost at right angles. The initially subglobular primary chambers become, in axial view, increasingly ovoid in the last whorl; they are weakly inflated dorsally but very strongly inflated ventrally, and there is no distinct dorso-peripheral shoulder. In the adult, about three highly arched apertures are present in the suture between the last chamber and the ventral surface of the earlier test. These multiple apertures are each covered by small, strongly inflated, subglobular bullae, which are attached along part or whole of their breadth to the surface of the last primary chamber. Each bulla may possess a single accessory aperture, which may be almost as highly and broadly arched as the primary aperture beneath it; alternatively, the bulla may arch over from the last primary chamber, to unite with the ventral surface of the adjacent earlier primary chamber, to restrict or divide the accessory aperture into two or more smaller accessory openings. The accessory apertures are typically furnished with narrow lips or rim-like thickenings. The wall of the primary chambers is fairly thick, and it is uniformly and finely perforate; its surface may be densely and very finely hispid. The walls of the bullae are similar to the wall of the primary chambers, but are distinctly thinner; the surface of the bullae is comparable to that of the primary chambers. Maximum diameter of holotype: 0·37 mm; from sample FCRM 1645, *Globigerapsis semi-involuta* Zone, Upper Eocene, Lindi area.

REMARKS

This species is closely related to, and is very probably directly descended from, *Globigerapsis semi-involuta* (Keijzer); the structure, form and disposition of the primary chambers is closely similar in the two species, but even the juveniles may be distinguished (see Fig. 11, vi–vii) by the presence of bullae in *Globigerinatheka lindiensis*. Also, throughout later ontogeny, the chambers of *G.*

lindiensis are more depressed, embracing, and more tightly coiled than in *Globigerapsis semi-involuta*.

Globigerinatheka lindiensis is distinguished from *G. barri* Brönnimann by its possession of primary chambers which are more inflated dorsally and more strongly embracing ventrally, and by the weaker depression of the sutures. The wall of *G. lindiensis* is much more finely perforate (not coarsely punctate or cancellate) and more finely hispid than in *G. barri*. We are grateful to Dr J. P. Beckmann for supplying us with near-topotypes of *G. barri*.

Globigerinatheka lindiensis appears to demonstrate the polyphyletic nature of the genus *Globigerinatheka*, since *G. lindiensis* is very probably descended from *Globigerapsis semi-involuta* while *Globigerinatheka barri* probably originated from another species of *Globigerapsis*.

STRATIGRAPHICAL RANGE

In southern Trinidad and the Lindi area, *Globigerinatheka lindiensis* appears to be limited to the *Globigerapsis semi-involuta* Zone, Upper Eocene.

Family: HANTKENINIDAE Cushman, 1927, emended Banner & Blow, 1959

Subfamily: HANTKENININAE Cushman, 1927, emended Banner & Blow, 1959

Genus: *Hantkenina* Cushman, 1925

Type species: *Hantkenina alabamensis* Cushman, 1925

Hantkenina alabamensis Cushman
(Plate XVI c, D, J, K)

Hantkenina alabamensis Cushman, 1925, *Proc. U.S. Nat. Mus.* **66**, no. 2567, art. 30, p. 3, pl. 1, figs. 1–6; pl. 2, fig. 5; text-fig. 1.

H. alabamensis Cushman, Cushman, 1935, *U.S. Geol. Surv., Prof. Paper*, **181**, pp. 49–50, pl. 13, figs. 1–5.

H. alabamensis Cushman subsp. *compressa* Parr, 1947, *Proc. Roy. Soc. Vict.*, N.S., **58**, pts. 1–2, p. 46, text-figs. 4–7a (?text-figs. 1–3).

H. (Hantkenina) alabamensis Cushman, Brönnimann, 1950, *J. Paleont.* **20**, no. 4, p. 414, pl. 56, figs. 10, 14–16.

H. alabamensis Cushman; Bolli, Loeblich & Tappan, 1957, *Bull. U.S. Nat. Mus.* **215**, 26–28, pl. 2, fig. 8.

REMARKS

This species is of great importance both stratigraphically and taxonomically, as it is known to be of world-wide distribution in the Middle and

126

Upper Eocene and it is the type species of the genus *Hantkenina*, and thus of higher taxonomic categories. Its gross morphology has been described and illustrated on many occasions. The details of the apertural system have, however, not been fully discussed, even though Brönnimann (1950) has reviewed them.

Throughout later ontogeny, at least, the aperture of *Hantkenina alabamensis* is an interiomarginal opening extending the full length of the basal suture; the aperture is very highly and narrowly arched in the apertural (terminal) face, being narrowly elongate from the area above the periphery of the preceding whorl to approximately half the height of the apertural face. This apertural extension is much higher than broad, and its lateral margins are almost straight, subparallel, or slowly converging towards the narrowly arched, highest point of the aperture. The aperture is furnished throughout its length with a strongly developed imperforate porticus, which is narrowest over the highest (middle) part of the aperture, but which greatly broadens towards the lower margins of the aperture (forming the 'lateral lip-like projections' of Brönnimann, 1950) and continues into the umbilici. The porticus projects nearly at right angles from the apertural face, but at its broadest part (near the base of the terminal face) the two lateral flanges of the porticus curve inwards towards each other and often partially restrict the aperture, thus giving it an external (superficial) trilobed or tripartite appearance (cf. Cushman, 1925, fig. 1, where the porticus is not distinguished from the primary chamber wall, and the aperture is drawn as possessing a primarily trilobed shape). This basal restriction of the primary aperture appears to be characteristic of advanced species of the genus *Hantkenina*; the stratigraphically earlier forms, with a relatively broad and unrestricted primary aperture, may be referred to the subgenus *Aragonella* Thalmann, 1942. The restriction of the primary aperture reaches its greatest development in *Cribrohantkenina* (*q.v.*), where the primary aperture itself has significantly changed its form (see p. 145, Fig. 19).

The early whorls of *Hantkenina alabamensis* are usually distinctly hispid.

The porticus of *Hantkenina* may be very finely hispid, but it is distinctly imperforate (see Plate XVIJ, K); the dense perforations of the primary chamber wall end abruptly against the limits of the porticus. The porticus does not appear to be a simple reflexed or thickened continuation of the primary chamber (as is the lip in *Globigerina*), but appears to be an additional structure of fundamental taxonomic importance (see Banner & Blow, 1959), and may be analogous to the apertural flap-like structures (tenons, *auct.*) of the Discorbidae. The porticus typically possesses its own marginal rim-like thickening, which is particularly prominent in the most advanced forms (e.g. *Cribrohantkenina*).

Hantkenina primitiva Cushman & Jarvis
(Plate XVIA, B; Fig. 9 (ix))

Hantkenina alabamensis Cushman var. *primitiva* Cushman & Jarvis, 1929, *Contr. Cushman Lab. Foram. Res.* **5**, pt. 1, no. 72, p. 16, pl. 3, figs. 2, 3.

H. (*Hantkeninella*) *primitiva* Cushman & Jarvis, Brönnimann, 1950, *J. Paleont.* **20**, no. 4, pp. 416–417, pl. 56, figs. 4, 26, 27.

H. alabamensis Cushman var. *primitiva* Cushman & Jarvis, Bolli, Loeblich & Tappan, 1957, *Bull. U.S. Nat. Mus.* **215**, 26–27, pl. 2, figs. 7a–b.

REMARKS

This species differs from *Hantkenina alabamensis* in possessing less inflated, more slowly enlarging chambers, coiled in a more evolute planispire and, consequently, in possessing less depressed sutures and broader umbilici. The early whorls of *H. primitiva* possess more numerous globular chambers, and their hispidity persists later in ontogeny. The tubulospines seem to appear at a later stage in the ontogeny of *H. primitiva*, and this species may be considered to be morphologically more primitive (although it occurs stratigraphically later!) than *H. alabamensis*. *H. primitiva* was described by Brönnimann (1950), when it was referred by him to his new and monotypic subgenus *Hantkeninella*; *Hantkeninella* was erected on the bases of a 'spine-less early stage' and the (theoretically) more primitive position of the tubulospines in the later chambers of this species. While we agree with Brönnimann that the tubulospines usually migrate to the anterior of each chamber during the ontogeny of *Hantkenina s.l.* spp., we also agree with Bolli *et al.* (1957, p. 27) that this migration does not clearly occur in phylogeny, and that young individuals of *Hantkenina alabamensis* also may lack tubulospines on their hispid early chambers. The form of the primary aperture in *H. primitiva*, even after its restriction by the porticus, does not appear to differ essentially from that of *H. alabamensis.* Consequently, we see no justification in theory or practice for retaining the subgenus *Hantkeninella.*

STRATIGRAPHICAL RANGE

In the Lindi area, *Hantkenina primitiva* ranges from the base of the *Globigerapsis semi-involuta* Zone to the top of the *Cribrohantkenina danvillensis* Zone, Upper Eocene. In southern Trinidad, it occurs in the uppermost Navet formation (Hospital Hill marl) and also in the 'San Fernando formation'. It is also known in the Jacksonian of the American Gulf States and in the Upper Eocene of Russia (Shokhina, 1937, 'The genus *Hantkenina* and its stratigraphical distribution in the Northern Caucasus', *Moscow Univ. Pal. Lab., Probl. Pal.*, **2**/3, 425–441).

Genus: *Cribrohantkenina* Thalmann, 1942

Type species: *Hantkenina (Cribrohantkenina) bermudezi* Thalmann, 1942

Cribrohantkenina danvillensis (Howe & Wallace)
(Plate XVI G, H; Fig. 19 (i–vii))

Hantkenina danvillensis Howe & Wallace, 1934, *J. Paleont.* **8**, 37, pl. 5, figs. 14, 17.

H. (Sporohantkenina) brevispina Cushman, Bermúdez, 1937 (*not H. brevispina* Cushman, 1925), *Mem. Soc. Cubaña Hist. Nat., Havana*, **11**, 151–152, pl.19, figs.7–10.

H. (Cribrohantkenina) bermudezi Thalmann, 1942, *Amer. J. Sci.* **240**, no. 11, pp. 812, 815, pl. 1, figs. 5–6 (after Bermúdez, 1937, figs. 7 and 9).

Cribrohantkenina bermudezi Thalmann; Bolli, Loeblich & Tappan, 1957, *Bull. U.S. Nat. Mus.* **215**, 28–29, pl. 2, figs. 9a–11b.

REMARKS

The evolution of this species from *Hantkenina alabamensis* Cushman is discussed in §VII (see p. 146, Fig. 19). Throughout the later ontogeny, at least, the primary aperture of *Cribrohantkenina danvillensis* is an interiomarginal opening, extending the full length of the basal suture, which be-

comes highly and broadly arched in the apertural (terminal) face of the last chamber. The breadth of the primary aperture is maintained, no matter how complex and complete its subsequent restriction by the imperforate porticus becomes. In the morphologically simplest form of *C. danvillensis* (Fig. 19, i), the broad primary aperture is furnished with a narrow porticus around its margin; the imperforate porticus is formed in the same plane as the apertural face of the primary chamber wall (cf. *Hantkenina*), and it is distinguished by an abrupt change in structure, the dense perforations of the primary wall ending abruptly against the porticus. The free edge of the porticus possesses a narrow reflexed rim ('lip' of authors) of its own. In progressively more advanced forms (Fig. 19, ii–v), the porticus gradually extends to enclose partially the primary aperture; in so doing, its free margin becomes corrugated and scalloped, and the crenulations so produced fuse to enclose isolated accessory apertures within the area of the imperforate porticus, each accessory aperture possessing a rim like that of the free margin. The encroachment of the porticus almost always occurs laterally, at first restricting the primary aperture to an opening similar in size and shape to that of the primary aperture of *Hantkenina* (Fig. 19, iv, v). Ultimately, the free margins of the lateral flanges of the porticus fuse medially, isolating further subcircular accessory apertures within the area of the porticus (Fig. 19, vi, vii), and restricting the opening of the primary aperture to a very low, interiomarginal, equatorial arch, which often becomes slit-like. This equatorial apertural arch may become still further restricted and divided (Fig. 19, vi) if the porticus fuses with a tubulospine of the preceding whorl. This morphological grade attains greatest complexity in the youngest populations, where the area of the primary aperture (apart from the basal equatorial portion) is completely enclosed by the fused porticus, which is itself fully occupied and crowded with subcircular accessory apertures and their rims (Plate XVI G, H).

We do not believe that true supplementary apertures (i.e. apertures situated within the primary perforate chamber wall, supplementing the primary aperture, as in *Globigerinoides*) occur in *Cribro-*

hantkenina. Single round holes, which occur in the medial plane of the chambers, in an anterio-peripheral position, mark the sites of protruding tubulospines which have been broken off. It is significant that these holes lack any imperforate margin or rim, and are perforate right to their borders (see, for example, Bolli *et al.* 1957, pl. 2, figs. 11, a–b).

The taxa *Hantkenina inflata* Howe, 1928, and *H. mccordi* Howe & Wallace, 1932, may prove to be prior synonyms of *H. danvillensis* Howe & Wallace, 1934; as they are inadequately figured and described (the holotype of *mccordi* seems to be broken), we prefer to assign the forms discussed here to the earliest well-documented specific name.

STRATIGRAPHICAL RANGE

In the Lindi area, primitive forms of *Cribrohantkenina danvillensis* first occur in about the middle part of the *Globigerapsis semi-involuta* Zone, and the species ranges, with increasing apertural complexity, to the top of the *C. danvillensis* Zone. It does not appear to occur in the uppermost Eocene *Globigerina turritilina turritilina* Zone. In southern Trinidad, *C. danvillensis* first occurs in the uppermost Navet formation (Hospital Hill marl), and it is also known in the 'San Fernando formation' and in derived blocks of the 'San Fernando conglomerate'. It is known from the Upper Eocene Jackson formation of the American Gulf States, and in equivalent beds throughout the Caribbean.

Subfamily: PLANOMALININAE Bolli, Loeblich & Tappan, 1957, emended Banner & Blow, 1959

Genus: *Pseudohastigerina* Banner & Blow, 1959

Type species: *Nonion micrus* Cole, 1927

Pseudohastigerina micra (Cole)
(Plate XVIE, F; Fig. 9 (x))

Nonion micrus Cole, 1927, *Bull. Amer. Paleont.* **14**, no. 51, p. 22, pl. 5, fig. 12.

?Globigerinella micra (Cole), Glaessner, 1937, *Pal. Lab., Moscow Univ., Studies in Micropaleontology*, **1**, fasc. 1, p. 30, text-fig. 2 (*?not* pl. 1, figs. 4, a–b).

?Nonion iota Finlay, 1940, *Trans. Roy. Soc. N.Z.* **69**, pt. 1, p. 456, pl. 65, figs. 108–110.

Globigerinella micra (Cole), Subbotina, 1953, *Trudi, VNIGRI*, N.S., **76**, 88, pl. 13, figs. 16, 17.

Hastigerina micra (Cole), Bolli, 1957, *Bull. U.S. Nat. Mus.* **215**, 161, pl. 35, figs. 1a–2b.

?'Globigerinella' iota (Finlay), Hornibrook, 1958, *Micropaleontology*, **4**, no. 1, p. 34, pl. 1, figs. 22–24.

Pseudohastigerina micra (Cole), Banner & Blow, 1959, *Palaeontology*, **2**, pt. 1, pp. 19–20, pl. 3, figs. 6, a–b; text-fig. 4, g–i.

DESCRIPTION

The fairly small test is planispirally coiled throughout, consisting of two to three whorls of slowly enlarging, slightly compressed, weakly inflated, partially embracing chambers, with six to seven chambers in each whorl. In equatorial profile the test is subcircular and the periphery is weakly lobulate. In axial profile the test is suboval, compressed and biumbilicate; the periphery is rounded, becoming increasingly narrowly rounded during ontogeny as the chambers become increasingly compressed. In equatorial view the chambers are subtriangular; the intercameral sutures are initially subradial, but they become increasingly curved at their distal ends during ontogeny (as the chambers become compressed) although they still meet the periphery at broad angles. The umbilici are shallow, and reveal the earlier whorls. The aperture is an interiomarginal, equatorial, symmetrical, low arch, extending along the full length of the basal suture of the last chamber. Small relict portions of the primary apertures of the penultimate and antepenultimate chambers are usually visible, with their portici. The porticus of the last chamber is narrowest at its mid-point (above the periphery of the preceding whorl) and broadens laterally to form narrow flaps (homologous to the 'lateral flanges' of *Hantkenina*) directed approximately towards the centre of each umbilicus. The wall of the test is fairly thin, densely and finely perforate; its surface is smooth, except for the earliest chambers of the last whorl (facing the aperture) which are weakly and finely hispid.

REMARKS

The specimens illustrated here come from the Oligocene; they appear to be indistinguishable from forms previously illustrated by us from the Upper Eocene of Tanganyika (1959, text-fig. 4, g–i). However, the Oligocene populations of *Pseudohastigerina micra* often contain individuals which display (separately or severally) reduced hispidity on the early chambers, smaller umbilici,

129

enlarged portici, fewer relict apertures and fewer chambers per whorl. These forms appear to fall within the range of variation of *P. micra*, but seem to display a trend comparable, in part at least, to that which we believe to have occurred during the evolution of *Hantkenina* from *Pseudohastigerina*.

Also present in the Oligocene populations (but not in those assemblages known to us from the Eocene) are rarer specimens which differ from *P. micra* by possessing a more finely perforate test, a higher primary aperture, and much broader umbilici in which no relict apertures can be seen. These are referred to here as *Pseudohastigerina* aff. *micra* (Fig. 9, xi–xii), and they compare well with the specimen illustrated by Batjes (1958, pl. 11, figs. 6, a–b) from the Oligocene (Chattian) of Pietzpuhl, Germany. They may prove to be closely related to '*Globigerinella*' *praemicra* Subbotina, 1960, which has been recorded from the Oligocene of Russia.

STRATIGRAPHICAL RANGE

In the Lindi area, this species is common in both the Upper Eocene and the Oligocene. It has been observed by us in Rupelian samples from Elmsheim and Offenbach, Mainz Basin, Germany. It occurs commonly in the Middle and Upper Eocene of the whole Caribbean area and in the American Gulf States. It is not known to occur in the *Globigerina ampliapertura* Zone (lowest Aquitanian) of the Cipero formation, Trinidad, or in the Aquitanian of south-western France. However, rare occurrences in the lower Vicksburg (e.g. Red Bluff Clay) are believed to be due to reworking of Eocene material, as is certainly the origin of the specimens of *Hantkenina* with which such examples of *Pseudohastigerina micra* are typically associated.

VII. The Evolution of some Upper Eocene to Lower Miocene Globigerinaceae

The Globigerinaceae are a rapidly evolving group in which stratigraphically useful and biologically significant lineages may be accurately traced. The clarification of these lineages can be of great help in solving biostratigraphic and taxonomic problems.

Within the Miocene, the evolutionary lineage of *Globorotalia fohsi* and *G. lobata* (Bolli, 1950, 1957*b*;

Banner & Blow, 1959) is of particular value in the stratigraphical subdivision of the Burdigalian. The evolution of *Orbulina* (Blow, 1956) is of great help in recognizing the topmost part of the Aquitanian. The evolution of *Globigerinoides quadrilobatus* from its ancestral *Globigerina praebulloides occlusa* (see p. 136) takes place in the middle part of the Aquitanian and forms another datum of the greatest value. The coincidence of correlations based upon such evolutionary lineages within different subfamilies of the Globigerinaceae confirms the separate isochroneity of the stages of each individual lineage. Such isochroneity may be recognized over widely separated regions. The distribution in space and time of a progressively and uniformly evolving community is controlled only by the limits of faunal migration and genetic interchange. The planktonic habit of the Globigerinaceae ensures rapid faunal (and genetic) migration; the observed density of living populations, which are free to migrate, indicates rapid reproduction in this superfamily (as also observed in other foraminiferal groups) and easy genetic exchange. All these factors reduce impediments to faunal migration and genetic interchange to a minimum.

The detailed sampling made in the Lindi area, Tanganyika, together with the meticulous work done in the Caribbean area (Bolli, 1957*a–c*; Blow, 1959) now enables some further lineages to be clarified and applied to biostratigraphical studies. Banner & Blow (1959, 1960*a*, *b*) have discussed the iterative nature of many of the biocharacters which have been used in the Globigerinaceae for supraspecific taxonomy. Blow (1959) has discussed some lineages deduced from a study of the Miocene faunas of eastern Falcón, Venezuela; these lineages are related here to others described below. The full morphological details of the forms comprising the following lineages has been given in the section dealing with the Systematic Palaeontology (§VI, p. 81).

LINEAGE A (see Fig. 12*a*)

[The evolution of *Globorotalia* (*Turborotalia*) *increbescens* (Bandy) from *G.* (*T.*) *opima nana* Bolli, and *Globigerina ampliapertura* Bolli from *Globorotalia* (*Turborotalia*) *increbescens* (Fig. 12*b*).]

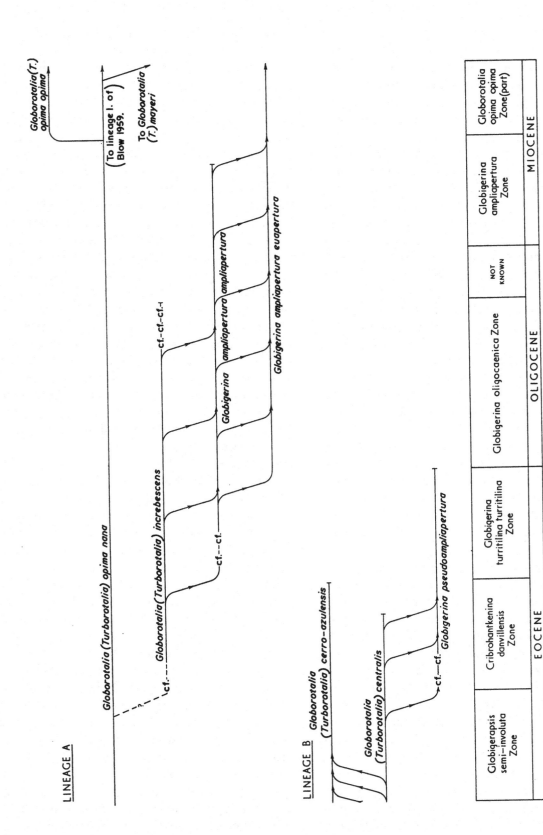

Fig. 12a. The evolution of the *Globorotalia opima* → *Globigerina ampliapertura* and the *Globorotalia centralis* → *Globigerina pseudoampliapertura* lineages (lineages A and B)

131

Globorotalia (*Turborotalia*) *opima nana* Bolli is a small form with four to five chambers in the last whorl; the chambers are subspherical in shape and are rather embracing. This form has a low, slit-like, umbilical–extraumbilical aperture furnished with a distinct lip. The test shows a tendency to become more tightly coiled, involute and larger with the passage of time. '*Globigerina*' *increbescens* Bandy also has four to five chambers in the last whorl and is, in general, very similar to *Globorotalia* (*Turborotalia*) *opima nana*. *G. increbescens* is distinguished by being more loosely coiled, with slightly more inflated and better separated chambers, and in possessing a more rounded open aperture furnished only with a narrow rim. The test of *G. increbescens* is typically slightly larger than that of *G. opima nana*. A number of transitional forms between *G. increbescens* and *G. opima nana* occur in the lower part of the *Cribrohantkenina danvillensis* Zone, and *Globorotalia increbescens* has not been seen by us below this level which is the type horizon for Bandy's species. Records of occurrences of this species below this level are probably the closely similar and ancestral form *G. opima nana*. Both *G. increbescens* and *G. opima nana* have very similar, coarsely punctate 'granular' walls with a very similar surface texture (see Plate XVII, and compare *G. centralis* below). As the test became more globose and the aperture more rounded and open (with a reduced lip), *G. opima nana* evolved into *G. increbescens*.

Bolli (1957b) described *Globigerina ampliapertura* from the zone of this name in the basal part of the Cipero formation. We have topotype material of this zone and a detailed analysis of this topotype material shows that *G. ampliapertura ampliapertura* has a similar wall texture and structure to *Globorotalia increbescens* (see Plate XVII). In the upper part of the *Globigerina turritilina turritilina* Zone and in the *G. oligocaenica* Zone many transitional forms are common showing all stages and gradation (see Fig. 12b) between the two forms. The test of '*increbescens*' becomes more loosely coiled and, concomitantly, the aperture becomes umbilical in position. The coiling also becomes looser with time even within the forms ascribable to *G. ampliapertura ampliapertura*. Neither *G. ampliapertura*

ampliapertura nor *Globorotalia increbescens* possesses a lip but only a slightly thickened rim. The evolution does not seem to take place instantaneously, but, during an appreciable interval of time, there seems to be an evolving plexus involving the intermediates between two extreme morphological forms, one ascribable to the genus *Globorotalia* and the other to the genus *Globigerina*. Therefore, for any one population in time, there is a curve of distribution whose median characterizes the evolutionary level of the population. From a biological point of view there seems little doubt that during this period of transition both '*increbescens*' and '*ampliapertura ampliapertura*' are one species. However, in the *Cribrohantkenina danvillensis* Zone, only forms referable to *Globorotalia* (*Turborotalia*) *increbescens* are present and this form has become extinct by the beginning of the Aquitanian when only *Globigerina ampliapertura ampliapertura* persists. This affords a reason for retaining the two extremes as separate taxa and demonstrates the necessity of examination of the full development of a species in space and time before its taxonomy can be fully evaluated. The transition demonstrated here between the genera *Globorotalia* and *Globigerina* illustrates that it is unwise to separate them at family level (cf. Bolli *et al.* 1957, and Banner & Blow, 1959) and also illustrates the value of the subgenus *Turborotalia*, which should not be considered as a useless synonym of the carinate *Globorotalia s.s.* The Linnéan system of nomenclature was originally designed for the taxonomy of forms existing on one particular time-plane; it has obvious disadvantages when applied to evolving species and its rigid application should not be allowed to mask biological affinities. The various taxa employed in palaeontology may not always represent true biological entities but represent useful groupings to which the 'Rules of Zoological Nomenclature' may be uniformly applied to avoid taxonomic confusion. It should not be surprising or dismaying to find evolutionary lineages which cross arbitrary taxonomic groups which, however, still remain invaluable.

Globigerina ampliapertura euapertura (Jenkins) develops from *G. ampliapertura ampliapertura* by the final chamber tending to embrace more of the

132

earlier test so that the whole test tends to become more globular. The aperture of the final chamber in *G. ampliapertura euapertura* becomes a low arch as compared with the almost circular opening of *G. ampliapertura ampliapertura*. *G. ampliapertura euapertura* persists to a higher stratigraphical level than *G. ampliapertura ampliapertura*.

Previously, *Globorotalia (T.) increbescens* has often been confused with *G. (T.) centralis*. Bandy, who has kindly sent us metatypic specimens of *G. (T.) increbescens*, has pointed out in a private letter (dated 11 May 1959) that *increbescens* is a more globose species than *G. centralis*. Further distinctions from *G. (T.) centralis* are the characteristically

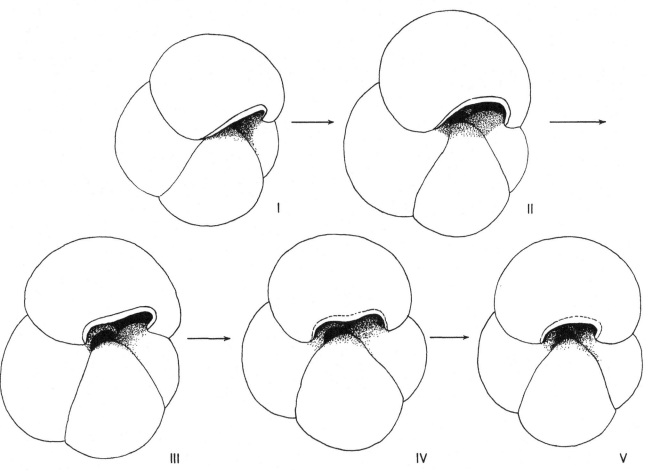

Fig. 12b. Specimens illustrating the morphological transition from *Globorotalia (Turborotalia) increbescens* to *Globigerina ampliapertura*; all *ca.* × 100, from sample FCRM 1650, *G. turritilina* Zone, Upper Eocene, Lindi area, Tanganyika. Specimens deposited in the British Museum (Natural History), registered nos. P.44431–P.44435.

Lineage A leads on via *Globorotalia opima nana* to the evolution of *G. (T.) mayeri* etc., in the lower Aquitanian (see Blow, 1959, p. 96, text-fig. 1, where the 'Oligocene' is now to be considered as still within the Aquitanian).

LINEAGE B (Fig. 12a)

[The evolution of *Globigerina pseudoampliapertura* sp. nov. from *Globorotalia (Turborotalia) centralis* Cushman and Bermúdez, and *Globorotalia (Turborotalia) cerroazulensis* (Cole) from *G. (T.) centralis* (Figs. 12c, 12d, 12e).]

smaller size, *relatively* much more coarsely punctate and 'granular' walls (see Plate XVII), and the more highly arched and open aperture of *G. (T.) increbescens*.

Typical *Globorotalia (Turborotalia) centralis* does not seem to range to the top of the *Cribrohantkenina danvillensis* Zone. In about the middle part of this zone transitional forms from *Globorotalia (T.) centralis* to *Globigerina pseudoampliapertura* appear; these forms superficially resemble *Globo-*

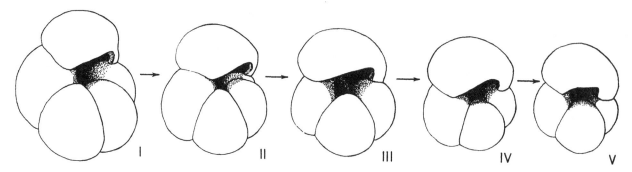

Fig. 12c. Specimens illustrating the morphological transition from *Globorotalia* (*Turborotalia*) *centralis* to *Globigerina pseudoampli-apertura*; all *ca.* × 60, from sample FCRM 1923, *Cribrohantkenina danvillensis* Zone, Upper Eocene, Lindi area, Tanganyika. Specimens deposited in the British Museum (Natural History), registered nos. P.44436–P.44440.

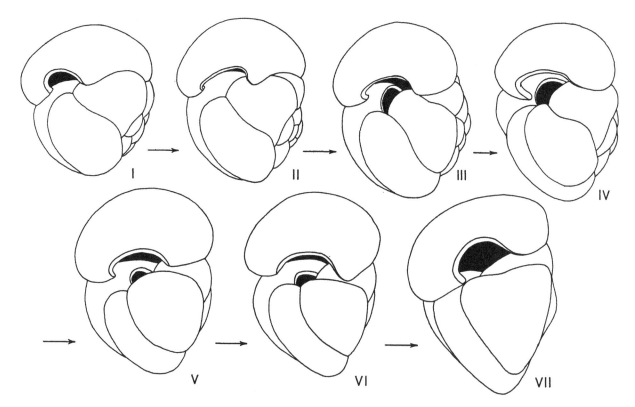

Fig. 12d. Specimens illustrating the morphological transition from *Globorotalia* (*Turborotalia*) *centralis* to *G.* (*T.*). *cerro-azulensis*; all *ca.* × 80, from sample FCRM 1645, *Globigerapsis semi-involuta* Zone, Upper Eocene, Lindi area, Tanganyika. Specimens deposited in the British Museum (Natural History), registered nos. P.44441–P.44447.

rotalia (*T.*) *centralis* but their tests have a more convex and smoothly rounded dorso-peripheral shoulder and this is accompanied by an intra-umbilical migration of the aperture. Later, during this evolution, the aperture widens becoming much higher compared to its length; the coiling becomes looser and the test almost globose with a smoothly rounded and convex dorsal side. There is very little, if any, general and over-all increase of test size in this lineage. In the uppermost part of the *Cribro-hantkenina danvillensis* Zone, *Globigerina pseudo-ampliapertura* becomes established as a distinct species and replaces *Globorotalia* (*T.*) *centralis* in the fossil assemblage. *Globigerina pseudoampli-*

134

apertura persists throughout the *G. turritilina turritilina* Zone and this species appears to become extinct at the close of Eocene times.

It is interesting to note that the trends exemplified by the *Globorotalia increbescens* → *Globigerina ampliapertura* and the *Globorotalia centralis* →

morphological changes from a *Globorotalia* (*Turborotalia*) to a *Globigerina* is yet a further example of an iteration of biocharacters within the Globigerinaceae (see Banner & Blow, 1959, 1960 *a*).

Also noted in lineage B, Figs. 12 *d, e*, is the development of *Globorotalia* (*Turborotalia*) *cerro-azulensis*

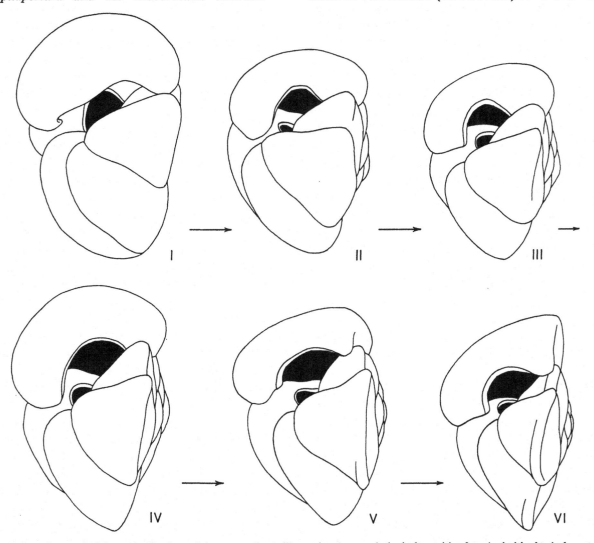

Fig. 12*e*. Specimens of *Globorotalia* (*Turborotalia*) *cerro-azulensis*, illustrating the morphological transition from 'primitive' to 'advanced' forms of the species; all *ca.* × 80, from sample FCRM 1923, *Cribrohantkenina danvillensis* Zone, Upper Eocene, Lindi area, Tanganyika. Specimens deposited in the British Museum (Natural History), registered nos. P. 44448–P. 44453.

Globigerina pseudoampliapertura lineages (lineages A and B) seem to be paralleled by a third and much later evolution in the late Pliocene and/or Pleistocene; thus *Globorotalia* (*Turborotalia*) *inflata* (d'Orbigny) (a gross pseudomorph of *G.* (*T.*) *centralis*) gives rise to a *Globigerina* species via *Globigerina oscitans* Todd. This repetition of

(Cole) from *G.* (*T.*) *centralis*. This evolution appears to occur in the uppermost part of the *Truncorotaloides rohri* Zone and the lowermost part of the *Globigerapsis semi-involuta* Zone by the development of more circumferentially elongate chambers as seen in dorsal aspect, also by the umbilicus becoming smaller and sometimes almost closed.

135

There seems to be a change in the wall structure and texture as the evolutionary line to *cerro-azulensis* is followed, the wall becoming less coarsely punctate, more shiny and smooth (i.e. less hispid). Indeed, at the base of the *G. semi-involuta* Zone it is only the difference in the wall character that allows an unambiguous separation of the two forms to be made. In the later evolution of *Globorotalia (T.) cerro-azulensis* there is a general over-all increase of test size and two additional subsidiary lines seem to be followed within this species. In one of these (Fig. 12*d*), the test remains distinctly vaulted on the ventral side with a rounded peripheral margin. In the second of the subsidiary lines (Fig. 12*e*) the test tends to become dorso-ventrally compressed with an acute margin. The acute margin sometimes appears as if it might be truly carinate, but thin sections (Plate XVI M) show that the acute margin is fully perforate, not imperforate as in *Globorotalia (s.s.)* (Banner & Blow, 1959). It is considered that the peripheral margin is secondarily thickened and that the 'keel' is no more than a pseudocarina (Banner & Blow, 1959).

LINEAGE C (Fig. 13)

[The evolution of the *Globigerina praebulloides* stock, and the origin of *Globigerinoides quadrilobatus* (d'Orbigny) (Fig. 13).]

The primitive form of this lineage seems to be the Middle Eocene to Aquitanian form *Globigerina praebulloides occlusa* subsp.nov. This form is described in detail on p. 93 of this work. *G. praebulloides occlusa* possesses a small but distinct umbilicus, together with a low intraumbilical aperture which has only a faint lip or rim. The chambers are moderately embracing and slightly inflated. The embrace of the last chamber especially gives a 'compact' appearance to the test.

Bolli (1957*b*) recorded '*Globigerina* cf. *trilocularis* d'Orbigny' from the Eocene and Oligocene (*vel* Aquitanian) sediments of southern Trinidad. We believe that his Eocene specimens (Bolli, 1957*c*, p. 163, pl. 36, figs. 3a–b) together with one of his Aquitanian specimens (Bolli, 1957*b*, p. 110, pl. 22, figs. 9a–c, *not* figs. 8a–c) are referable to *G. praebulloides occlusa*. Bolli (1957*b*) notes that *Globigerinoides trilobus* (Reuss) develops from '*Globi-*

gerina cf. *trilocularis* d'Orbigny', and Blow (1959) has also recorded this evolution which takes place in the middle part of the *Globorotalia kugleri* Zone both in southern Trinidad and eastern Falcón, Venezuela. However, Banner & Blow (1960*b*) pointed out that *Globigerinoides quadrilobatus* (d'Orbigny) has taxonomic priority over, and is morphologically ancestral to, the form known as '*Globigerinoides trilobus trilobus* (Reuss)'.

During the evolution of *Globigerina praebulloides occlusa*, there appears to be a gradual development of punctae, as the exterior ends of the pores broaden and develop into small pit-like depressions. This results in a gradual development of a cancellate wall surface with the acquisition of a weakly favose appearance due to the confluence of the margins of adjacent punctae. There appears to be no significant increase in the diameters of the pores themselves, which open into the punctae. At the base of the *Globorotalia kugleri* Zone, these advanced forms of *Globigerina praebulloides occlusa* develop a single dorsal supplementary aperture in the spiral suture of the ultimate chamber; these forms are now referable to the genus *Globigerinoides* Cushman, 1927, and are the immediate precursors of the whole *G. quadrilobatus* stock (see Blow, 1959, p. 105, lineage 5, text-fig. 5; Banner & Blow, 1960*b*, text-fig. 1). These forms with a single dorsal supplementary aperture are here referred to as *G. quadrilobatus primordius* subsp.nov. and comprise a short-lived stratigraphical index for the middle Aquitanian *Globorotalia kugleri* Zone, and this enables the zone to be recognized even in the absence of its name-fossil. The acquisition of additional dorsal supplementary apertures in the earlier chambers of the test, together with a further coarsening of the punctations of the test surface (in conjunction with the retention of the compact test shape and rather low, short primary aperture similar to that of *Globigerina praebulloides occlusa*) gives rise to *Globigerinoides quadrilobatus quadrilobatus* by the upper part of the *Globorotalia kugleri* Zone.

Globigerinoides ruber appears a little later than the *G. quadrilobatus* stock. It appears that it may develop independently from *Globigerina praebulloides occlusa* via *G. woodi* (Jenkins, in the

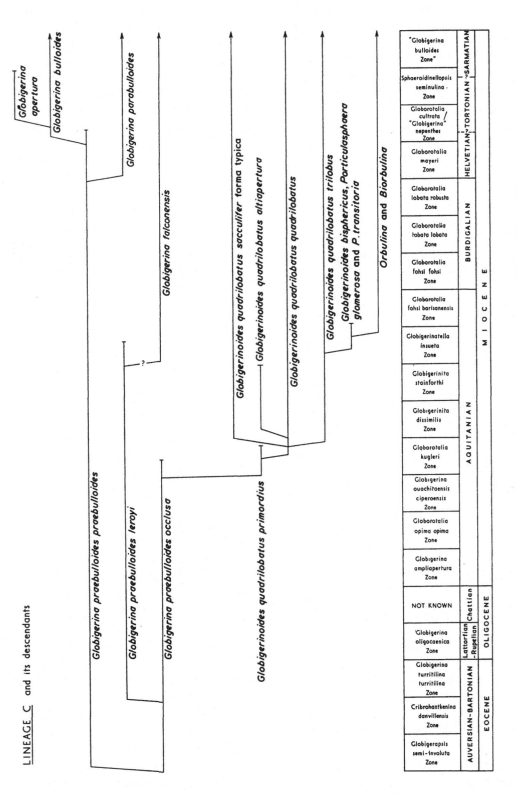

Fig. 13. The evolution of the *Globigerina praebulloides → Globigerinoides quadrilobatus* lineage (lineage C).

137

press). In this evolution the major trends seem to be those of reduction of the number of chambers in the last whorl, from four to three, and of the primary aperture becoming symmetrical with respect to the intercameral suture between the penultimate and antepenultimate chambers.

Globigerina praebulloides leroyi subsp.nov. develops from *G. praebulloides occlusa* in the uppermost Eocene (base of the *G. turritilina turritilina*

attaining a looser mode of coiling, a wider umbilicus, better separated and less embracing chambers together with a more open and higher arched aperture which lacks any distinct lip or rim. *Globigerina praebulloides praebulloides* gives rise to *G. parabulloides* Blow and *G. bulloides* d'Orbigny (which, in turn, is ancestral to the short-lived *G. apertura* Cushman) in the Helvetian and becomes extinct at this time.

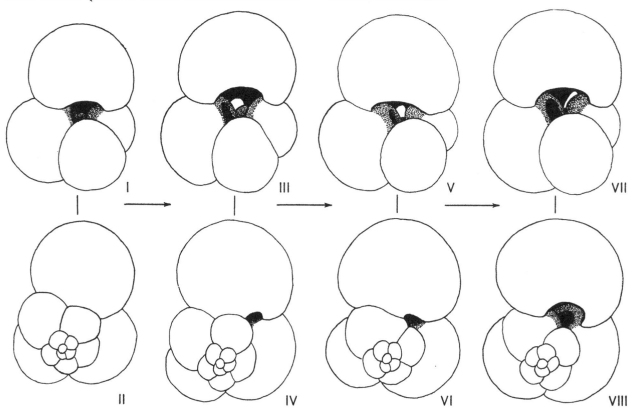

Fig. 14. Specimens illustrating the evolutionary transition from *Globigerina praebulloides occlusa* (i, ii) to *Globigerinoides quadrilobatus primordius* (iii–viii). (i), (iii), (v) and (vii) are ventral views, and (ii), (iv), (vi) and (viii) are dorsal views of the same specimens. The increased tightness of coiling, which typically occurs concomitantly with the acquisition of a dorsal supplementary aperture of advanced type, may be seen. All specimens × 125, from the type locality and horizon of the *Globorotalia kugleri* Zone, Cipero formation (Aquitanian) of Trinidad, W.I.; deposited in the British Museum (Natural History), registered nos. P.44454–P.44457.

Zone) by developing more inflated and embracing chambers, an almost closed umbilicus, and a more laterally restricted aperture which develops a distinct lip. *G. praebulloides leroyi* persists until, at least, the upper Aquitanian (*Globigerinatella insueta* Zone) where it may give rise to *Globigerina falconensis* Blow.

Globigerina praebulloides praebulloides Blow (*emend.*) develops from *G. praebulloides occlusa* at the base of the *Globigerapsis semi-involuta* Zone by

As a whole the forms included in lineage c retain the simple unlipped primary aperture characteristic of the progenitor *Globigerina praebulloides occlusa*, even if they develop into *Globigerinoides, Porticulasphaera* or *Orbulina*. The exceptions (*Globigerina praebulloides leroyi, G. parabulloides*) are specialized and either short-lived or, where descendants are present, the lips remain a constant feature. This may suggest that the ancestor of the family Globigerinidae lacked an apertural lip and that the characteristically strongly lipped Palaeocene and Lower Eocene *Globigerina* species are in themselves too specialized to be ancestral to the Neogene stocks.

Again, as in lineage A, lineage C illustrates what appears to be a fundamental phenomenon, i.e. the increase of somatic volume with the passage of time.

LINEAGE D (Fig. 15)

[The independent evolution of *Globigerina angustiumbilicata* Bolli and the *Globigerina ouachitaensis* Howe and Wallace → *Globigerina angulisuturalis* Bolli stock from the ancestral *Globigerina officinalis* Subbotina (Fig. 16).]

The progenitor of this lineage is the Middle Eocene to Aquitanian form *Globigerina officinalis* Subbotina. This characteristically small species possesses a fairly tightly coiled trochoid test with four chambers in the last whorl and an almost closed and shallow umbilicus; the very low aperture, although usually intraumbilical in position, sometimes shows a tendency to become somewhat extraumbilical. The aperture possesses a very faint thickened rim but lacks a distinct lip.

The only distinctions between *Globigerina officinalis* and *G. angustiumbilicata* Bolli are the presence of five chambers in the last whorl of the latter form and its slightly better developed apertural lip. Transitional forms between *G. officinalis* and *G. angustiumbilicata* are common in the uppermost part of the *Cribrohantkenina danvillensis* Zone and in the lowermost part of the *Globigerina turritilina turritilina* Zone; at these levels it is very difficult to separate the two forms. However, by the upper part of the *G. turritilina turritilina* Zone, *G. angustiumbilicata* becomes a distinct species and persists as such until at least the uppermost Miocene. During the range of *G. angustiumbilicata* its over-all test size and apertural lip enlarges with the passage of time.

Globigerina ouachitaensis Howe & Wallace is another form similar to *G. officinalis*, but it possesses an open, distinctly quadrate umbilicus and a more highly arched aperture. The test is distinctly more loosely coiled in *G. ouachitaensis* (*s.s.*) and the chambers are inflated and less embracing; the more open umbilicus is related to this laxity of coiling, and the primary aperture similarly becomes clearly intraumbilical. Bolli (1957c, pl. 36, figs. 7a–c)

figured a *hypotype* of *Globigerina parva* Bolli from the *Globigerapsis semi-involuta* Zone (southern Trinidad); we think that this form is not conspecific with Bolli's *holotype* of *Globigerina parva* (= *G. officinalis* Subbotina, 1953, see p. 88), but is referable to *G. ouachitaensis ouachitaensis* which develops from *G. officinalis*, with every gradational stage present, in about the middle part of this zone.

The typically four-chambered *Globigerina ouachitaensis* (*s.s.*) persists until the lower or middle part of the *Globorotalia opima opima* Zone, but the five-chambered *Globigerina ouachitaensis ciperoensis* (Bolli) appears with every gradation in the lower part of the Oligocene section of East Africa. Concomitant with the adoption of a five-chambered last whorl the coiling becomes looser and the increasing laxity of coiling continues within the general evolution of *G. ouachitaensis ciperoensis* itself. Thus the populations of *G. ouachitaensis ciperoensis* from the *G. oligocaenica* Zone contain very many specimens which are more tightly coiled than those observed in populations from the *G. ampliapertura ampliapertura* Zone; similarly populations from this latter zone contain more tightly coiled specimens than is seen in populations of *G. ouachitaensis ciperoensis* from the overlying *Globorotalia opima opima* and *Globigerina ouachitaensis ciperoensis* Zones. The increasing laxity of coiling is accompanied by a greater separation of the chambers and an apparent decrease in the height of the spire.

The trend towards more distinct separation of the chambers is further expressed in the development of *Globigerina angulisuturalis* Bolli which appears at the base of the *Globorotalia opima opima* Zone. In *Globigerina angulisuturalis* the intercameral sutures become sharply incised and, in their extreme development, become 'U'-shaped. *G. angulisuturalis* may also be distinguished from *G. ouachitaensis ciperoensis* by its distinctly hispid chamber walls (except in the sutures), its smaller aperture and more restricted umbilicus.

We regard the evolutionary change between *G. ouachitaensis ouachitaensis* and *G. ouachitaensis ciperoensis* as being due to a changed growth-rate which is genetically inheritable and based on a fixed

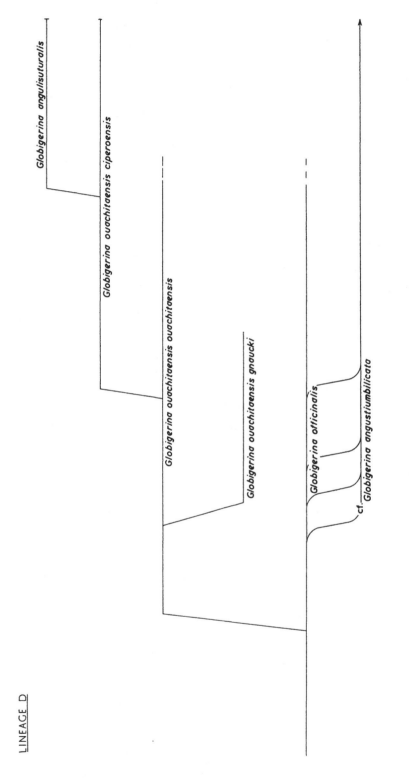

Fig. 15. The evolution of the *Globigerina officinalis* → *Globigerina ouachitaensis* lineage (lineage D).

140

genetic pattern. However, the distinction is in one biocharacter only, and this is susceptible, during its appearance, to full morphological grade; we therefore consider the two forms to be distinct at subspecific level only. In contrast, *G. angulisuturalis* differs from *G. ouachitaensis* (*s.l.*) in at least three apparently unrelated biocharacters; we believe this to be due to mutation and thus to a change in the genetic pattern which indicates the appearance of a distinct species.

LINEAGE E (Fig. 17)

[The evolution of the *Globigerina yeguaensis* → *Globigerina tripartita* stock and the origin of *Globoquadrina dehiscens s.l.* (Fig. 18).]

Although *Globigerina yeguaensis yeguaensis* does not appear to be morphologically primitive, it ranges from about the upper part of the Lower Eocene to high in the Oligocene and appears to be ancestral to the species discussed below. *G. yegua-*

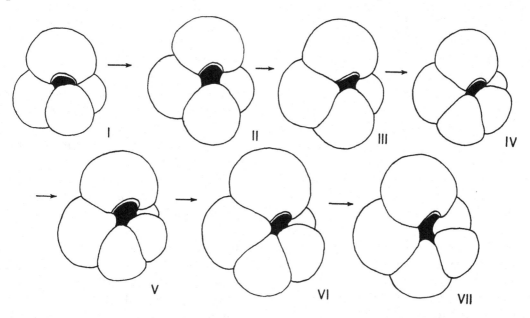

Fig. 16. Specimens illustrating the morphological transition from *Globigerina officinalis* to *G. angustiumbilicata*; all *ca.* × 175, from sample FCRM 1964, *G. oligocaenica* Zone, Oligocene (Rupelian/Lattorfian), Lindı area, Tanganyika. Specimens deposited in the British Museum (Natural History), registered nos. P.44458–P.44464.

Globigerina ouachitaensis gnaucki subsp.nov. is another off-shoot from *G. ouachitaensis ouachitaensis*, this time appearing at the base of the *G. turritilina turritilina* Zone and persisting to the top of the *G. oligocaenica* Zone. *G. ouachitaensis gnaucki* develops from *G. ouachitaensis ouachitaensis* by an increase in the height of the trochospire, by the development of a distinct apertural lip and by the coiling becoming looser. It is probably due to this last factor that the chambers become better separated; four and a half chambers now comprise the last whorl, and the umbilicus becomes wider and loses its quadrate outline. The chambers appear to be more hispid than in the ancestral *G. ouachitaensis ouachitaensis*.

ensis yeguaensis may have its origin in a form close to the one described by Bolli (1957*a*) as *G. triangularis* White from the Palaeocene and Lower Eocene.

In samples from the *Truncorotaloides rohri* Zone, *Globigerapsis semi-involuta* Zone and the lower *Cribrohantkenina danvillensis* Zone, every intermediate stage is present between *Globigerina yeguaensis yeguaensis* and *G. tripartita tripartita*. At these levels, in spite of the fact that the two ends of the gradation are very different, the gradation is complete (see Fig. 18). At higher levels this is not the case, and by the upper part of the *Cribrohantkenina danvillensis* Zone, *Globigerina yeguaensis yeguaensis* and *G. tripartita tripartita* are without morphological intergrades. Although *G. yeguaensis yeguaensis* does not survive the Oligocene, *G. tripartita tripartita* persists until about the top of the *G. ouachitaensis ciperoensis* Zone, lower Aquitanian.

Globigerina yeguaensis yeguaensis evolves into *G. tripartita tripartita* by the increased appression of increasingly embracing chambers; the chambers become less distinctly inflated and enlarge more rapidly during ontogeny so that the last-formed chamber occupies an increasingly greater proportion of the total volume of the test. The form illustrated by Bolli (1957*c*, pl. 35, figs. 15a–b) as *G. yeguaensis* represents one of the earlier stages in this evolution; his illustration shows the more slit-like and laterally elongate aperture which develops during this evolution. In the later stages there is a gradual reduction of the number of chambers in the last whorl from four to three, and finally the test assumes the tripartite appearance typical of *G. tripartita tripartita*.

During the evolution the 'umbilical tooth' becomes less distinct and in the subspecies *Globigerina tripartita tapuriensis* subsp.nov. it becomes rudimentary. Concomitant with the reduction of the flap-like lip ('umbilical tooth') to a rim-like thickening, the aperture becomes slightly higher and the umbilicus remains open.

Globigerina oligocaenica sp.nov. develops from *G. tripartita tapuriensis* by the still more rapid enlargement of the chambers during ontogeny; the chambers embrace more strongly on the ventral side and become less inflated dorsally, producing a smaller umbilicus and a flatter dorsal surface. The chambers become more depressed, reniform, and less inflated and they characteristically develop marked lobes lateral to the umbilicus, so that the aperture becomes more restricted laterally and appears to be situated in a re-entrant in the terminal face. *G. oligocaenica* grades from *G. tripartita tapuriensis* at the base of the Oligocene and rapidly becomes a distinct species which does not survive the Oligocene.

In about the middle part of the *Globigerina ouachitaensis ciperoensis* Zone, forms developing from *G. tripartita tripartita* show morphological changes which are the opposite of those displayed in the *G. oligocaenica* evolution discussed above.

This time the dorsal surface of the test becomes smoother by decreased depression of the sutures, although the forms retain their dorsal convexity. Ventrally, although the umbilicus becomes smaller and deeper, the aperture becomes less restricted and slightly extraumbilical in position. The umbilical tooth (which has persisted since its origin in *G. yeguaensis yeguaensis*) becomes accentuated, long and often sharply pointed. The extraumbilical migration of a fundamentally intraumbilical aperture indicates that a major morphological change is taking place which must rate at generic level. The forms are now to be referred to the genus *Globoquadrina* Finlay, 1947 (see Banner & Blow, 1959, p. 5). We believe that these forms are ancestral to *G. dehiscens dehiscens* which appears in the middle part of the Aquitanian (*Globigerinita dissimilis* Zone, upper part) and in consequence they are referred to here as *Globoquadrina dehiscens praedehiscens* subsp.nov.

Globigerina venezuelana Hedberg appears, in the type figure, to have an aperture furnished with a distinct 'umbilical tooth'. Topotype specimens from the Carapita formation have been seen and these, together with specimens from the same stratigraphic level elsewhere in Venezuela, as well as in Trinidad, also show the presence of the 'umbilical tooth'. This umbilical tooth may be rudimentary or strongly developed within a population. It may even be absent on the last chamber of a particular specimen although at least traces of it can be observed on the penultimate and earlier chambers. These forms with pointed 'umbilical teeth' appear to have a close relationship with *G. tripartita tripartita* and appear to differ essentially from the latter form only by the presence of four chambers in the last whorl and a generally wider umbilicus. Forms strictly referable to *G. venezuelana* appear to develop from *G. tripartita tripartita* in the upper part of the *Globorotalia opima opima* Zone.

In the Middle and Upper Eocene forms which are superficially similar to *Globigerina venezuelana* Hedberg are common, and they appear to develop directly and repeatedly from *G. yeguaensis yeguaensis*. We refer to these forms as *G. yeguaensis pseudovenezuelana*, and the specimen illustrated by Bolli (1957*c*, pl. 35, figs. 14a–c) from the Middle Eocene

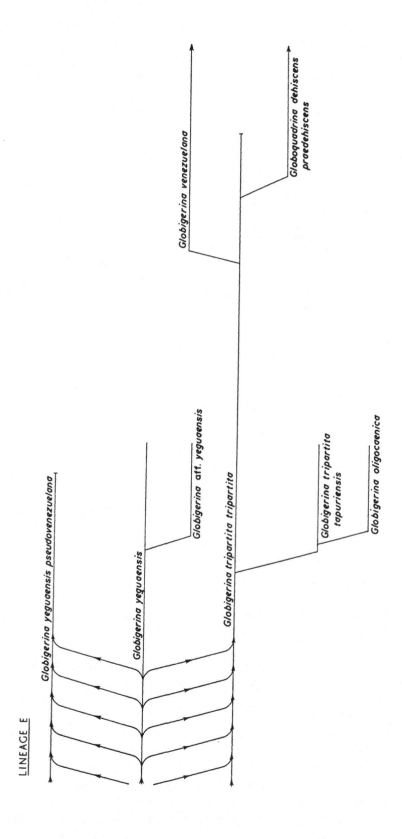

LINEAGE E

Fig. 17. The evolution of the *Globigerina yeguaensis* → *Globigerina tripartita* lineage (lineage E).

143

Porticulasphaera mexicana Zone seems to represent one of the morphologically intermediate stages in this evolution. The broad 'umbilical tooth' of *Globigerina yeguaensis yeguaensis* is maintained in its subspecies *pseudovenezuelana* and it may even become broader still (a distinction from *G. venezuelana*). The test becomes more tightly coiled and

between the two forms are discussed at length in the appropriate part of §VI (see p. 100).

Globigerina aff. *yeguaensis* is a rare form which has been observed so far only in the Oligocene. It seems to be a phylogerontic end-form of *G. yeguaensis yeguaensis*, which has developed loose coiling, a wide umbilicus and a slower rate of

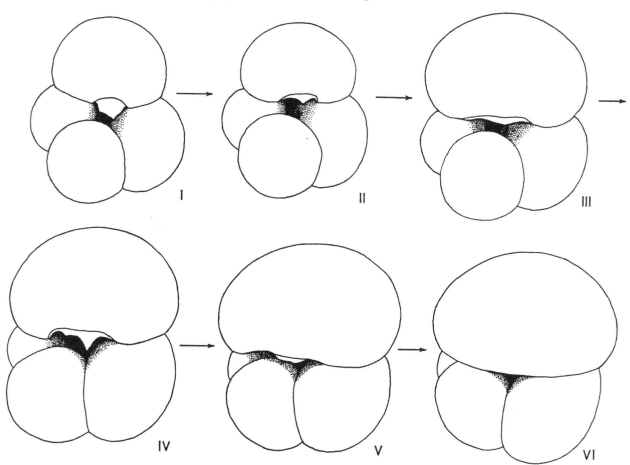

Fig. 18. Specimens illustrating the morphological transition from *Globigerina yeguaensis yeguaensis* to *G. tripartita tripartita*; all *ca.* × 125, from sample FCRM 1923, *Cribrohantkenina danvillensis* Zone, Upper Eocene, Lindi area, Tanganyika. Specimens deposited in the British Museum (Natural History), registered nos. P.44465–P.44470.

the umbilicus becomes relatively smaller and sometimes almost closed; the chambers are depressed, less inflated and more slowly enlarging than in *G. yeguaensis yeguaensis*. Concomitantly, the sutures of *pseudovenezuelana* become less strongly depressed and the periphery of the test less lobulate. This form very closely pseudomorphs *G. venezuelana* Hedberg, but it is separated from Hedberg's species by a considerable stratigraphic interval and a distinct evolutionary history. The distinctions

chamber enlargement, without modification of the shape of the chambers or 'umbilical teeth'.

LINEAGE F

[The origin of *Globigerinita unicava* (Bolli, Loeblich & Tappan)]

An evolutionary line leading from *Globigerinita pera* (Todd) through *G. unicava primitiva* nov. subsp. to *G. unicava unicava* (Bolli, Loeblich & Tappan) may exist. However, it is difficult to

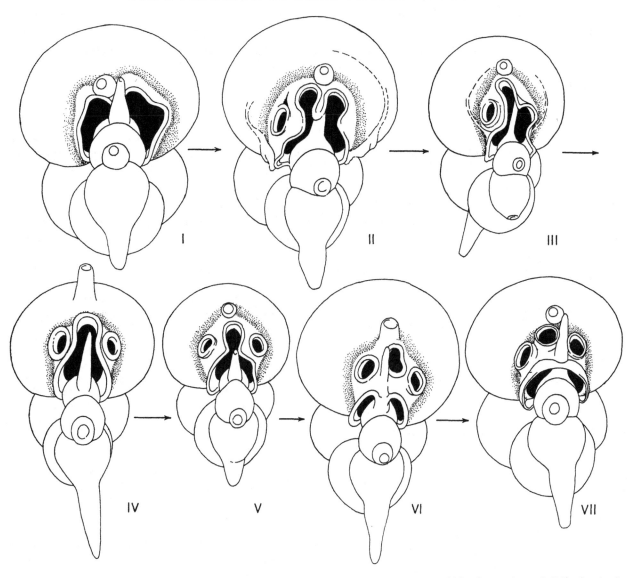

Fig. 19. Specimens illustrating the development of complex cribrate accessory apertures within the porticus of *Cribrohantkenina danvillensis*. All specimens *ca*. × 85, all from sample FCRM 1923, *C. danvillensis* Zone, Upper Eocene, Lindi area, Tanganyika. The stippled area of the last chamber represents the limit of the perforate primary chamber wall, adjacent to the apertural porticus; the final chambers of specimens (ii) and (iii) have been removed, illustrating the permanency of the portici over apertures which no longer open directly to the exterior. Specimens deposited in the British Museum (Natural History), registered nos. P.44471–P.44477.

confirm this with the material available at present, and it is possible that the species referred to *Globigerinita* may have had independent origins from separate species of *Globigerina* (see p. 107).

It has been suggested by Hofker (1959) that a bulla may be a reproductive feature, perhaps associated with sexual dimorphism (or trimorphism) of a *Globigerina*. However, the examples quoted by Hofker of such hypothetical dimorphism (e.g. *Globigerinita dissimilis* and *Globigerina venezuelana*) are not convincing, because of the stratigraphical differences in range of the supposed dimorphic pairs. It may be noted that it is possible that Hofker is confusing true bullae with aborted final chambers. A bulla is constant in shape and position for its species, and this alone would indicate that it is genetically controlled. A recognizably constant and regularly inherited morphological character must be considered in classification whether or not it be associated solely with reproduction (compare

145

placental and implacental mammals, brood pouches in the Crustacea and ovicells in the Bryozoa).

Therefore we consider that the bullate forms of the Globigerinidae are worthy of generic distinction; this is discussed at length in §VI (p. 102).

LINEAGE G (Fig. 19)
[The possible origin of *Hantkenina*, and the evolution of the genus *Cribrohantkenina*]

As discussed by Banner & Blow (1959, p. 19) *Hantkenina* is a morphologically more advanced form than *Pseudohastigerina* and may well have evolved from it. The occurrence at relatively high stratigraphical levels of a morphologically comparatively primitive species, i.e. *Hantkenina primitiva*, suggests that either the species has regressed or that it has arisen independently from *Pseudohastigerina*. We suggest that the latter view is more likely since many small specimens of *Hantkenina primitiva* from low in the *Globigerapsis semiinvoluta* Zone show distinctly evolute coiling, very weakly developed tubulospines present only on the later chambers and a relatively low aperture together with a chamber shape and hispidity which is closely comparable to *Pseudohastigerina micra* (Fig. 9, ix–xii).

In the upper part of the *Globigerapsis semiinvoluta* Zone, *Cribrohantkenina* first appears. The broadening of the primary aperture (as noted by Barnard, 1954, fig. 1) is accompanied by a concomitant broadening and strengthening of the porticus (see Banner & Blow, 1959) so that the principal opening to the exterior is still comparable in size and shape to that of *Hantkenina alabamensis*, which appears to be the ancestral form. As imperforate shell material is added to the margins of the porticus, still further enclosing the primary aperture, the porticus gains structural strength by corrugating and scalloping its margin. The processes so formed coalesce distally and laterally to form a series of isolated openings within the area of the porticus. Ultimately, in the most advanced specimens (common in the *Cribrohantkenina danvillensis* Zone) the primary aperture becomes almost completely enclosed by a cribrate sheet of imperforate shell material, which consists of the two fused lateral flanges (portici), and is reduced to a very low symmetrical interiomarginal arch. The cribrate openings within the area of the porticus are thus accessory apertures within a structure modifying the external shape of the single primary aperture. Barnard's (1954) observations were essentially correct in as far as such a morphological series was noted. However, he did not appear to appreciate the significance of the fundamental distinction between the nature of the primary aperture and that of the accessory apertures, or between that of the perforate true shell wall and the imperforate apertural accessory structures (i.e. portici); perhaps in consequence of this Barnard failed to note the important stratigraphical value of the genus *Cribrohantkenina*.

POSTSCRIPT. Whilst in page proof we received a copy of Dr Borsetti's paper (1959, 'Tre nuovi foraminiferi planctonici dell'Oligocene Piacentino', *Ann. Mus. Geol. Bologna, Giorn. Geol.* (2), **27**, 205–212, pl. 1). The kind loan of topotypes and paratypes indicates that, throughout our work above, *Globigerina turritilina* should be referred to as *G. gortanii* (Borsetti), and *G. oligocaenica* should be referred to as *G. sellii* (Borsetti). The occurrence of these two species, however, in the Lower Oligocene of North Italy, is in accord with and confirms our stratigraphical conclusions.

References

The references given here are largely additional to those included in the Bibliography to Part 1.

Akers, W. H. (1955). Some planktonic foraminifera of the American Gulf Coast and suggested correlations with the Caribbean Tertiary. *J. Paleont.* **29**, 647–664.

Banner, F. T. & Blow, W. H. (1959). The classification and stratigraphical distribution of the Globigerinaceae. *Palaeontology*, **2**, pt. 1, 1–27, pls. 1–3 (*cum bibl.*).

Banner, F. T. & Blow, W. H. (1960a). Some primary types of species belonging to the superfamily Globigerinaceae. *Contr. Cushman Found. Foram. Res.* **11**, 1–41, pls. 1–8.

Banner, F. T. & Blow, W. H. (1960b). The taxonomy, morphology and affinities of the genera included in the subfamily Hastigerininae. *Micropaleontology*, **6**, 19–31, text-figs.

Bailey, E. B. & Weir, J. (1933). Submarine faulting in Kimmeridgian times: East Sutherland. *Trans. Roy. Soc. Edinb.* **57**, 429.

Barnard, T. (1954). *Hantkenina alabamensis* Cushman and some related forms. *Geol. Mag.* **91**, 384–390, text-figs.

Batjes, D. A. J. (1958). Foraminifera of the Oligocene of Belgium. *Mém. Inst. Sci. nat. Belg.*, no. 143.

Beckmann, J. P. (1953). Die Foraminiferen der Oceanic Formation (Eocaen–Oligocaen) von Barbados, Kl. Antillen. *Ecl. geol. Helv.* **46**, 301–412, 15 pls., 26 text-figs.

REFERENCES

Beckmann, J. P. (1957). Correlation of pelagic and reefal faunas from the Eocene and Paleocene of Cuba. *Ecl. geol. Helv.* **51**, 415–422, text-figs.

Blow, W. H. (1956). Origin and evolution of the foraminiferal genus *Orbulina* d'Orbigny. *Micropaleontology*, **2**, 57–70, text-figs.

Blow, W. H. (1957). Trans-Atlantic correlation of Miocene sediments. *Micropaleontology*, **3**, 77–79.

Blow, W. H. (1959). Age, correlation and biostratigraphy of the Upper Tocuyo (San Lorenzo) and Pozón formations, eastern Falcón, Venezuela. *Bull. Amer. Paleont.* **39**, no. 178 (*cum bibl.*).

Bolli, H. M. (1950). The direction of coiling in the evolution of some Globorotaliidae. *Contr. Cushman Found. Foram. Res.* **1**, 82–89, pl. 15.

Bolli, H. M. (1951). Notes on the direction of coiling of rotalid foraminifera. *Contr. Cushman Found. Foram. Res.* **2**, 139–143.

Bolli, H. M. (1957*a*). The genera *Globigerina* and *Globorotalia* in the Paleocene—Lower Eocene Lizard Springs formation of Trinidad, B.W.I. *Bull. U.S. Nat. Mus.* **215**, 61–81, pls. 15–20.

Bolli, H. M. (1957*b*). Planktonic foraminifera from the Oligocene–Miocene Cipero and Lengua formations of Trinidad, B.W.I. *Bull. U.S. Nat. Mus.* **215**, 97–123, pls. 22–29.

Bolli, H. M. (1957*c*). Planktonic foraminifera from the Eocene Navet and San Fernando formations of Trinidad, B.W.I. *Bull. U.S. Nat. Mus.* **215**, 155–172, pls. 35–39.

Bolli, H. M., Loeblich, A. R. & Tappan, H. (1957). Planktonic foraminiferal families Hantkeninidae, Orbulinidae, Globorotaliidae and Globotruncanidae. *Bull. U.S. Nat. Mus.* **215**, 3–50, pls. 1–11.

Brönnimann, P. (1950). The genus *Hantkenina* Cushman in Trinidad and Barbados, B.W.I. *J. Paleont.* **24**, 397–420.

Brönnimann, P. (1951). *Globigerinita naparimaensis*, n.gen., n.sp., from the Miocene of Trinidad, B.W.I. *Contr. Cushman Found. Foram. Res.* **2**, 16–18, text-figs.

Bykova, N. K., *et al.* (1958). Novie rodi i vidii foraminifer. *Mikrofauna SSSR, Trudi, VNIGRI*, Sb. 9, Leningrad, 5–81, pls. 1–12.

Cox, L. R. (1927). Mollusca. In *Report on the Palaeontology of the Zanzibar Protectorate*. London: H.M.S.O.

Cushman, J. A. & Stainforth, R. M. (1945). The foraminifera of the Cipero Marl formation of Trinidad, B.W.I. *Spec. Publ. Cushman Lab. Foram. Res.* **14**, 1–75, pls. 1–16.

Davies, A. M. (1927). Foraminifera. In *Report on the Palaeontology of the Zanzibar Protectorate*. London: H.M.S.O.

Dixey, F. (1956). Some aspects of the geomorphology of Central and Southern Africa. Alex. du Toit Memorial Lecture No. 4, *Geol. Soc. S. Africa*, annex to vol. **4**.

Drooger, C. W. & Batjes, D. A. J. (1959). Planktonic foraminifera in the Oligocene and Miocene of the North Sea Basin. *Proc. K. Akad. Wet. Amst.* B, **62**, no. 3, pp. 172–186.

Fornasini, C. (1898). Le sabbie gialle Bolognese e le ricerche di J. B. Beccari. *Rend. R. Accad. Sci. Ist. Bologna*, n.s., **2** (1897–98), fasc. 1, 3–8, pl. 1.

Glaessner, M. F. (1937). Planktonforaminiferen aus der Kreide und dem Eozän und ihre stratigraphische Bedeutung. *Moscow Univ., Pal. Lab., Studies Micropal.*, Moscow, **1**, fasc. 1, 27–46, pls. 1, 2, text-figs.

Grimsdale, T. F. (1951). Correlation, age determination and the Tertiary pelagic foraminifera. *Proc. 3rd Wld Petroleum Congr.*, sect. 1, Leiden.

Hofker, J. (1956*a*). Tertiary foraminifera of coastal Ecuador: Pt. 2, additional notes on the Eocene species. *J. Paleont.* **30**, 891–958, text-figs.

Hofker, J. (1956*b*). Foraminifera Dentata: foraminifera of Santa Cruz and Thatch Island, Virginia Archipelago, West Indies. *Spolia Zool. Musei Hauniensis*, **15**, Skrifter, 1–237, pls. 1–35.

Hornibrook, N. de B. (1958). New Zealand Upper Cretaceous and Tertiary foraminiferal zones and some overseas correlations. *Micropaleontology*, **4**, 25–38, text-figs.

Howe, H. V. & Wallace, W. E. (1932). Foraminifera of the Jackson Eocene at Danville Landing on the Ouachita, Catahoula Parish, Louisiana. *Geol. Bull., New Orleans*, no. 2, 18–79, pls. 1–15.

Koch, R. (1926). Mitteltertiäre Foraminiferen aus Bulongan, Öst-Borneo. *Ecl. geol. Helv.* **19**, 722–751, text-figs.

Kugler, H. G. (1953). Jurassic to Recent sedimentary environments in Trinidad. *Bull. Ass. suisse Géol. Ing. Petrole*, **20**, no. 59, 27–60, text-figs.

Kugler, H. G. (1954). The Miocene/Oligocene boundary in the Caribbean region. *Geol. Mag.* **91**, no. 5, 410–414.

Le Calvez, Y. (1949). Revision des Foraminifères Lutetiens du Bassin de Paris. II. Rotaliidae et familles affines. *Mém. Carte géol. Fr.* (Paris, Imp. Nat.), pp. 1–54, pls. 1–6.

Loeblich, A. R. & Tappan, H. (1957*a*). The new planktonic foraminiferal genus *Tinophodella*, and an emendation of *Globigerinita* Brönnimann. *J. Wash. Acad. Sci.* **47**, 112–116, 1 plate.

Loeblich, A. R. & Tappan, H. (1957*b*). Planktonic foraminifera of Paleocene and early Eocene age from the Gulf and Atlantic Coastal Plains. *Bull. U.S. Nat. Mus.* **215**, 173–198, pls. 40–64.

Myatliuk, E. V. (1953). *Trudi, VNIGRI*; *Mikrofauna SSSR*, n.s., Sb. 51 (*non vidi*, quoted from Subbotina, 1953, *q.v.*).

Pokorný, V. (1955). *Cassigerinella boudecensis* n.gen., n.sp. (Foraminifera, Protozoa) from the Oligocene of the Zdanice Flysch. *Věstnik, Ŭstřed. Ŭst. Geol.* **30**, c. 3, 136–140.

Pokorný, V. (1958). *Grundzüge der zoologischen Mikropaläontologie*, Bd. 1, VEB Deutscher Verlag der Wissenschaften, Berlin.

Rey, M. (1954). Description de quelques espèces nouvelles de Foraminifères dans le Nummulitique Nord-Marocain. *Bull. Soc. géol. Fr.* (6), tome **4**, fasc. 4–6, 209.

Roemer, F. H. (1838). Die Cephalopoden des Nord-Deutschen Tertiären Meersandes. *Neues Jb. Min. Geog. Geol. Petrefakten*, Hft. 3, 381–394, pl. 3.

Shokhina, V. A. (1937). The genus *Hantkenina* and its stratigraphical distribution in the Northern Caucasus. *Moscow Univ., Pal. Lab., Probl. Pal.* 2/3, 425–441, pls. 1–2.

Shutskaya, E. K. (1958). Izmenchivosti nekotorikh nizhne-paleogenovikh planktonnikh foraminifer severnogo Kavkaza. *Voprosi Mikropaleont., Akad. Nauk SSSR, Otdel. Geol.-Geog. Nauk, Geol. Inst.*, no. 2, 84–90, pls. 1–3.

Stainforth, R. M. (1948). Description, correlation and paleo-ecology of the Tertiary Cipero Marl formation, Trinidad, B.W.I. *Bull. Amer. Ass. Petrol. Geol.* **32**, 1292–1330.

Stainforth, R. M. (1960). Current status of Trans-Atlantic Oligo-Miocene correlation by means of planktonic foraminifera. *Rev. Micropal.* **2**, 219–230, text-figures (*cum bibl.*).

Stockley, G. M. (1927). Echinoidea. In *Report on the Palaeontology of the Zanzibar Protectorate*. London: H.M.S.O.

Subbotina, N. N. (1953). Iskopaemi foraminiferi S.S.S.R.; Globigerinidae, Hantkeninidae, i Globorotaliidae. *Trudi, Vses. Neft. Nauchno-Issled. Geol.-Razved. Inst. (VNIGRI)*, N.S., no. 76, 1–239, 40 pls.

Subbotina, N. N. (1960). Mikrofauna Oligotsenovikh i Miotsenovikh otlozhenii R. Vorotishchye (PredKarpatye). *Mikrofauna SSSR, Trudi VNIGRI*, Sb. **11**, vol. **153**, 157–264, pls. 1–10.

Subbotina, N. N., Glushko, V. V. & Pishvanova, L. S. (1955). O vosraste nizhnei vorotishchenskoi suiti predKarpatskogo kraevogo progiba. *Dokl. Akad. Nauk SSSR*, **104**, no. 4, 605–607.

Subbotina, N. N., Pishvanova, L. S. & Ivanova, L. V. (1960). Stratigrafiya Oligotsenovikh i Miotsenovikh otlozhenii PredKarpatya po foraminiferam. *Mikrofauna SSSR, Trudi VNIGRI*, Sb. **11**, vol. **153**, 5–156, pls. 1–14.

Suter, H. H. (1951–52). The general and economic geology of Trinidad, B.W.I. *Colon. Geol. Min. Res.* **2**, nos. 3–4; **3**, no. 1. London: H.M.S.O.

Todd, R. (1957). Smaller Foraminifera. In *Geology of Saipan, Mariana Islands. U.S. Geol. Survey, Prof. Paper*, **280**-H.

EXPLANATION OF PLATES

PLATE VIII

Diagnostic larger foraminifera from the type localities of the *Cribrohantkenina danvillensis* (J–L) and *Globigerina turritilina* Zones (E–H), Upper Eocene, and from the cotype locality of the *G. oligocaenica* Zone (A–D), Oligocene; specimens identified by W. J. Clarke.

A, B. *Nummulites fichteli* Michelotti, 1841; Lattorfian–Rupelian; sample FCRM 1576; × 12·5; BM(NH), nos. P. 44495 and P. 44496.

C, D. *Lepidocyclina (Eulepidina) dilatata* (Michelotti), 1861; Rupelian–Aquitanian; sample FCRM 1576; × 12·5; BM(NH), no. P. 44497.

E. *Chapmanina gassinensis* (A. Silvestri), 1905; Upper Eocene; sample FCRM 1650; × 30; BM(NH), no. P. 44498.

F, G. *Nummulites hormoensis* Nuttall & Brighton, 1931; Upper Eocene; sample FCRM 1650; × 12·5; BM(NH), nos. P. 44499 and P. 44500.

H. *Discocyclina* sp.; Eocene; sample FCRM 1650; × 30; BM(NH), no. P. 44501.

J–L. *Pellatispira madaraszi* (Hantken), 1875; Upper Eocene; sample FCRM 1932; fig. J, × 15; figs. K, L, × 20; BM(NH), nos. P. 44502, P. 44503 and P. 44504.

PLATE IX

A–C. *Globigerina officinalis* Subbotina, hypotype, × 100; sample FCRM 1964, *G. oligocaenica* Zone, Oligocene, Lindi area; BM(NH), no. P. 44505.

D. *G. ouachitaensis ouachitaensis* Howe & Wallace, hypotype, × 100; FCRM 1627, *G. oligocaenica* Zone, Oligocene, Lindi area; a small specimen with a reduced final chamber, comparable to the holotype of the species; BM(NH), no. P. 44506.

E–G. *G. ouachitaensis ciperoensis* (Bolli), hypotype, × 100; FCRM 1964, *G. oligocaenica* Zone, Oligocene, Lindi area; BM(NH), no. P. 44507.

H–K. *G. ouachitaensis ouachitaensis* Howe & Wallace, hypotype; × 100; FCRM 1627, *G. oligocaenica* Zone, Oligocene, Lindi area; BM(NH), no. P. 44508.

L–N. *G. ouachitaensis gnaucki* Blow & Banner, new subspecies, holotype, × 100; FCRM 1965, *G. oligocaenica* Zone, Oligocene, Lindi area; BM(NH), no. P. 44509.

O–Q. *G. praebulloides praebulloides* Blow, ideotype, × 100; FCRM 1964, *G. oligocaenica* Zone, Oligocene, Lindi area; BM(NH), no. P. 44510.

R–T. *G. praebulloides leroyi* Blow & Banner, new subspecies, holotype, × 100; FCRM 1965, *G. oligocaenica* Zone, Oligocene, Lindi area; BM(NH), no. P. 44511.

U–W. *G. praebulloides occlusa* Blow & Banner, new subspecies, holotype, × 100; FCRM 1922, *G. oligocaenica* Zone, Oligocene, Lindi area; BM(NH), no. P. 44512.

X–Z. *G. angustiumbilicata* Bolli, hypotype, × 100; FCRM 1964, *G. oligocaenica* Zone, Oligocene, Lindi area; BM(NH), no. P. 44513.

Aa–Cc. *G. angulisuturalis* Bolli, hypotype, × 100; sample WACR 1326, Lower Ragusa limestone formation, Aquitanian, Monte Alia, south-east Sicily; BM(NH), no. P. 44514.

Dd–Ff. *Globigerinoides quadrilobatus primordius* Blow & Banner, new subspecies, holotype, × 100; from the same locality and horizon as Bolli sample Bo 274, type locality of the *Globorotalia kugleri* Zone, San Fernando Bypass Road, co-ordinates N: 225700; E: 361900 links, Trinidad, W.I., Aquitanian, Cipero formation; BM(NH), no. P. 44515.

Gg. *Globigerina quadripartita* Koch, holotype, × 75; this is the specimen illustrated by Koch, 1926, as fig. 20, obtained by him from the 'Middle Tertiary', 'unterer Teil der Globigerinenmergel', sample S. At. Dunok A. 491, south-east Bulongan, East Borneo; the ventral side is destroyed, and an attempt is made here to reconstruct the outline of the last two chambers, of which traces only remain; deposited in the Naturhistorisches Museum, Basel, Switzerland.

PLATE X

A–C. *Globigerina tripartita* Koch, holotype, × 75; this is the specimen illustrated by Koch, 1926, as fig. 21, obtained by him from the 'Middle Tertiary', 'unterer Teil der Globigerinenmergel', Sadjau-Njak, south-east Bulongan, East Borneo; the final chamber of this specimen is deformed; slide numbered '59', deposited in the Naturhistorisches Museum, Basel, Switzerland.

D–F. *G. tripartita tripartita* Koch, hypotype, ×75; sample FCRM 1923, *Cribrohantkenina danvillensis* Zone, Upper Eocene, Lindi area; this specimen is not deformed, and shows the umbilical and apertural characters believed to be typical of Koch's species; BM(NH), no. P.44516.

G. *G. oligocaenica* Banner & Blow, new species, paratype, ×75; FCRM 1964, *G. oligocaenica* Zone, Oligocene, Lindi area; showing the high aperture which may occur in this species; BM(NH), no. P.44517.

H–K. *G. tripartita tapuriensis* Blow & Banner, new subspecies, holotype, ×75; FCRM 1964, *G. oligocaenica* Zone, Oligocene, Lindi area; BM(NH), no. P.44518.

L–N. *G. oligocaenica* Blow & Banner, new species, holotype, ×75; FCRM 1964, *G. oligocaenica* Zone, Oligocene, Lindi area; BM(NH), no. P.44519.

PLATE XI

A–C. *Globigerina ampliapertura ampliapertura* Bolli, hypotype, ×100; sample FCRM 1965, *G. oligocaenica* Zone, Oligocene, Lindi area; BM(NH), no. P.44520.

D. *G. ampliapertura ampliapertura*, hypotype, ×100; oblique ventral view; FCRM 1965, *G. oligocaenica* Zone, Oligocene, Lindi area; BM(NH), no. P.44521.

E–G. *G. ampliapertura euapertura* (Jenkins), hypotype, ×100; sample FCRM 1965, *G. oligocaenica* Zone, Oligocene, Lindi area; BM(NH), no. P.44522.

H. *G. linaperta linaperta* Finlay, hypotype, ×100; FCRM 1923, *Cribrohantkenina danvillensis* Zone, Upper Eocene, Lindi area; BM(NH), no. P.44523.

J–L. *G. yeguaensis pseudovenezuelana* Blow & Banner, new subspecies, holotype, ×75; FCRM 1923, *Cribrohantkenina danvillensis* Zone, Upper Eocene, Lindi area; BM(NH), no. P.44524.

M. *G. linaperta pseudoeocaena* (Subbotina), hypotype, ×100; FCRM 1923, *Cribrohantkenina danvillensis* Zone, Upper Eocene, Lindi area; BM(NH), no. P.44525.

N, O. *G. yeguaensis pseudovenezuelana* Blow & Banner, new subspecies, paratype, ×75; FCRM 1965, *G. oligocaenica* Zone, Oligocene, Lindi area; BM(NH), no. P.44526.

P, Q. *Globigerina* sp. aff. *yeguaensis* Weinzierl & Applin, hypotype, ×75; FCRM 1922, *G. oligocaenica* Zone, Oligocene, Lindi area; BM(NH), no. P.44527.

R–T. *G. senilis* Bandy, hypotype, ×100; FCRM 1922, *G. oligocaenica* Zone, Oligocene, Lindi area; regularly developed specimen, with final aperture typical of the species; BM(NH), no. P.44528.

U. *G. senilis* Bandy, hypotype, ×100; FCRM 1922, *G. oligocaenica* Zone, Oligocene, Lindi area; umbilical (ventral) view of a specimen with a reduced final chamber, comparable to that of the holotype of the species; BM(NH), no. P.44529.

PLATE XII

A–C. *Globigerina pseudoampliapertura* Blow & Banner, new species, holotype, ×75; sample FCRM 1923, *Cribrohantkenina danvillensis* Zone, Upper Eocene, Lindi area; BM(NH), no. P.44530.

D–F. *Globorotalia (Turborotalia) cerro-azulensis* (Cole), hypotype ×75; FCRM 1923, *Cribrohantkenina danvillensis* Zone, Upper Eocene, Lindi area; BM(NH), no. P.44531.

G–J. *G. (Turborotalia) postcretacea* (Myatliuk), hypotype, ×300; FRCM 1964, *Globigerina oligocaenica* Zone, Oligocene, Lindi area; BM(NH), no. P.44532.

K–M. *G. (Turborotalia) centralis* Cushman & Bermúdez, hypotype, ×75; FCRM 1644, *Globigerapsis semi-involuta* Zone, Upper Eocene, Lindi area; BM(NH), no. P.44533.

N–P. *G. (Turborotalia) permicra* Blow & Banner, new species, holotype, ×300; FCRM 1922, *Globigerina oligocaenica* Zone, Oligocene, Lindi area; BM(NH), no. P.44534.

PLATE XIII

A–C. *Globigerina turritilina praeturritilina* Blow & Banner, new species, new subspecies, holotype, ×75; FCRM 1645, *Globigerapsis semi-involuta* Zone, Upper Eocene, Lindi area; BM(NH), no. P.44535.

D. *G. turritilina turritilina* Blow & Banner, new species, paratype, ×75; FCRM 1922, *G. oligocaenica* Zone, Oligocene, Lindi area; this specimen possesses an abnormally placed final chamber; BM(NH), no. P.44536.

E–G. *G. turritilina turritilina* Blow & Banner, new species, holotype, ×75; FCRM 1964, *G. oligocaenica* Zone, Oligocene, Lindi area; this specimen has lost a final abortive chamber and displays a normal, regularly formed umbilicus and aperture; BM(NH), no. P.44537.

H, J. *G. yeguaensis yeguaensis* Weinzierl & Applin, hypotype, ×75; FCRM 1922, *G. oligocaenica* Zone, Oligocene, Lindi area; this specimen, although having lost its later chambers, is closely comparable to the holotype; BM(NH), no. P.44538.

K–M. *G. yeguaensis yeguaensis* Weinzierl & Applin, hypotype, ×100; typical specimen, showing the 'umbilical teeth' projecting into the umbilicus; FRCM 1965, *G. oligocaenica* Zone, Oligocene, Lindi area; BM(NH), no. P.44539.

N–P. *Globorotaloides suteri* Bolli, hypotype, ×100; FCRM 1922, *Globigerina oligocaenica* Zone, Oligocene, Lindi area; BM(NH), no. P.44540.

Q–S. *Globorotalia (Turborotalia) opima nana* Bolli, hypotype, ×100; FCRM 1965, *Globigerina oligocaenica* Zone, Oligocene, Lindi area; BM(NH), no. P.44541.

T–V. *Globorotalia (Turborotalia) increbescens* (Bandy), hypotype, ×100; FCRM 1647, *Globigerina turritilina turritilina* Zone, Upper Eocene, Lindi area; BM(NH), no. P.44542.

PLATE XIV

A–C. *Globigerinita dissimilis ciperoensis* Blow & Banner, new subspecies, holotype, ×75; from the same locality and horizon as the type *Globigerina ampliapertura* Zone (of Bolli), Cipero formation, Aquitanian, co-ordinates N:237850; E:357560 links, Trinidad, W.I.; BM(NH), no. P.44543.

D. *Globigerinita dissimilis dissimilis* (Cushman & Bermúdez), hypotype, ×100; sample FCRM 1645, *Globigerapsis semi-involuta* Zone, Upper Eocene, Lindi area; BM(NH), no. P.44544.

149

E. *Globigerinita pera* (Todd), hypotype, × 100; FCRM 1922, *Globigerina oligocaenica* Zone, Oligocene, Lindi area; BM(NH), no. P.44545.

F–H. *Globigerinita pera* (Todd), hypotype, × 100; FCRM 1922, *Globigerina oligocaenica* Zone, Oligocene, Lindi area; BM(NH), no. P.44546.

J–L. *Globigerinita unicava primitiva* Blow & Banner, new subspecies, holotype, × 100; FCRM 1645, *Globigerapsis semi-involuta* Zone, Upper Eocene, Lindi area; BM(NH), no. P. 44547.

M, N. *Globigerinita unicava unicava* (Bolli, Loeblich & Tappan), hypotype, × 100; from the same horizon and locality as the type *Globigerina ampliapertura* Zone (of Bolli), Cipero formation, Aquitanian, co-ordinates N:237850; E:357560 links, Trinidad, W.I.; BM(NH), no. P.44548.

O. *Globigerinita martini martini* Blow & Banner, new species, holotype, × 200; FCRM 1932, *Cribrohantkenina danvillensis* Zone, Upper Eocene, Lindi area; BM(NH), no. P.44549.

P–R. *Globigerinita howei* Blow & Banner, new species, holotype, × 100; FCRM 1645, *Globigerapsis semi-involuta* Zone, Upper Eocene, Lindi area; BM(NH), no. P.44550.

S–U. *Globigerinita globiformis* Blow & Banner, new species, holotype, × 100; FCRM 1645, *Globigerapsis semi-involuta* Zone, Upper Eocene, Lindi area; BM(NH), no. P.44551.

V–X. *Globigerinita martini scandretti* Blow & Banner, new species, new subspecies, holotype, × 200; FCRM 1922, *Globigerina oligocaenica* Zone, Oligocene, Lindi area; BM(NH), no. P.44552.

PLATE XV

A–C. *Globigerinita africana* Blow & Banner, new species, holotype, × 100; sample FCRM 1645, *Globigerapsis semi-involuta* Zone, Upper Eocene, Lindi area; BM(NH), no. P.44553.

D, E. *Globigerapsis tropicalis* Blow & Banner, new species, holotype, dorsal and peripheral views, × 100; FCRM 1645, *G. semi-involuta* Zone, Upper Eocene, Lindi area; BM(NH), no. P.44554.

F. *G. tropicalis* Blow & Banner, new species, paratype, ventral view of an immature specimen, an ontogenetic stage earlier than that of multiple primary apertures; × 100; FCRM 1645, *G. semi-involuta* Zone, Upper Eocene, Lindi area; BM(NH), no. P.44555.

G, H. *G. index* (Finlay), hypotypes, × 100; G, ventral view of an immature individual; H, dorsal view of an adult individual; sample RS 24, Middle Eocene, excavation 200 yards west of the prison at Kilwa Masoko, Kilwa district, Tanganyika; BM(NH), nos. P.44556 and P.44557.

J–L. *G. semi-involuta* (Keijzer), hypotypes, × 100; J, K, dorsal and peripheral views of an adult specimen; L, ventral view of an immature specimen; FCRM 1645, *Globigerapsis semi-involuta* Zone, Upper Eocene, Lindi area; BM(NH), nos. P.44558 and P.44559.

M, N. *Cassigerinella chipolensis* (Cushman & Ponton), hypotype, × 300; ventral (M) and peripheral (N) views; FCRM 1964, *Globigerina oligocaenica* Zone, Oligocene, Lindi area; BM(NH), no. P.44560.

O, P. *Globigerinatheka lindiensis* Blow & Banner, new species, holotype, × 100; O, dorsal view; P, peripheral view; FCRM 1645, *Globigerapsis semi-involuta* Zone, Upper Eocene, Lindi area; BM(NH), no. P.44561.

Q–S. *Globoquadrina dehiscens praedehiscens* Blow & Banner, new subspecies, holotype, × 75; from the same locality and horizon as Bolli sample Bo 274, type for the *Globorotalia kugleri* Zone, Aquitanian, Cipero Formation, San Fernando Bypass Road, co-ordinates N:225700; E:361900 links, Trinidad, W.I.; BM(NH), no. P.44562.

PLATE XVI

A, B. *Hantkenina primitiva* Cushman & Jarvis, hypotype, × 75; one side of the porticus has been removed (on the right as seen in B), disclosing the narrowly elongate primary aperture typical of *Hantkenina (Hantkenina)*; FCRM 1932, *Cribrohantkenina danvillensis* Zone, Upper Eocene, Lindi area; BM(NH), no. P.44563.

C, D. *Hantkenina alabamensis* Cushman, hypotype, × 75; FCRM 1932, *Cribrohantkenina danvillensis* Zone, Upper Eocene, Lindi area; BM(NH), no. P.44564.

E, F. *Pseudohastigerina micra* (Cole), hypotype, × 150; FCRM 1964, *Globigerina oligocaenica* Zone, Oligocene, Lindi area; BM(NH), no. P.44565.

G, H. *Cribrohantkenina danvillensis* (Howe & Wallace), hypotype, × 75; FCRM 1932, *C. danvillensis* Zone, Upper Eocene, Lindi area; BM(NH), no. P.44566.

J, K. The wall of *Hantkenina alabamensis* Cushman; J, × 120, the area of the aperture, showing the abrupt termination of the pores against the imperforate porticus, which itself possesses a lip-like rim; K, part of the same photograph, × 500, showing the shape, size and distribution of the pores in the primary wall, and part of the imperforate porticus; specimen from FCRM 1932, *Cribrohantkenina danvillensis* Zone, Lindi area.

L. *Globorotalia (Turborotalia) centralis* Cushman & Bermúdez, hypotype, × 150; axial section, showing the uniformly perforate nature of the wall and the extra-umbilical extent of the early apertures, with their enlargement due to probable resorption; FCRM 1923, *Cribrohantkenina danvillensis* Zone, Upper Eocene, Lindi area; BM(NH), no. P.44567.

M. *Globorotalia (Turborotalia) cerro-azulensis* (Cole), hypotype, × 150; axial section, showing the uniformly perforate nature of the wall and the extraumbilical extent of the early apertures; FCRM 1923, *Cribrohantkenina danvillensis* Zone, Upper Eocene, Lindi area; BM(NH), no. P.44568.

PLATE XVII

The wall structures of the species involved in the evolutionary lineages *Globigerina pseudoampliapertura–Globorotalia (Turborotalia) centralis* and *Globigerina ampliapertura–Globorotalia (Turborotalia) increbescens*.

EXPLANATION OF PLATES

A–D, ×500; E–J, ×1000.

A, E. *Globigerina pseudoampliapertura* Blow & Banner, new species; FCRM 1923, Upper Eocene, Lindi area.

B, G. *Globorotalia (T.) centralis* Cushman & Bermúdez; FCRM 1645, Upper Eocene, Lindi area.

These photographs illustrate the thick, smooth-surfaced walls, with parallel-sided pores, characteristic of this lineage, diagrammatically shown in section by F.

C, H. *Globigerina ampliapertura* Bolli, from the type locality and horizon of the *G. ampliapertura* Zone, Cipero Formation, Trinidad, W.I. (Aquitanian).

D, K. *Globorotalia (T.) increbescens* (Bandy); FCRM 1965, Oligocene, Lindi area.

These photographs illustrate the thinner, 'granular'-surfaced walls, with pores which are proximally narrow but distally rapidly widening, characteristic of this lineage. The bundles of radial crystals, in the areas between the pores, typically form small mamelones, diagrammatically shown in section by J, giving the mamelonated, 'granular' appearance to the surface.

(*Note:* Illustrations on Plates IX–XVI are shaded camera-lucida drawings by F. T. Banner.)

151

PLATE VIII

PLATE IX

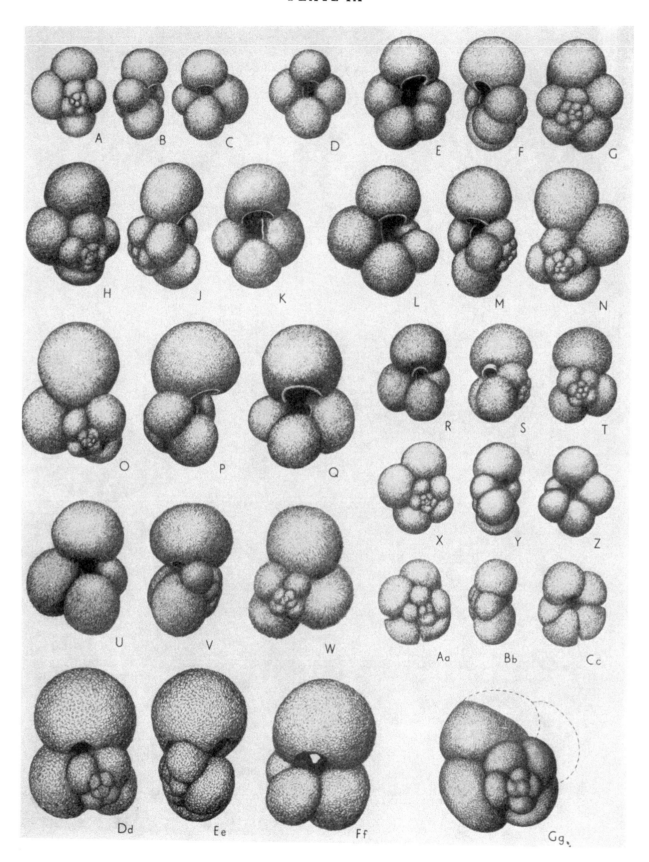

PLATE X

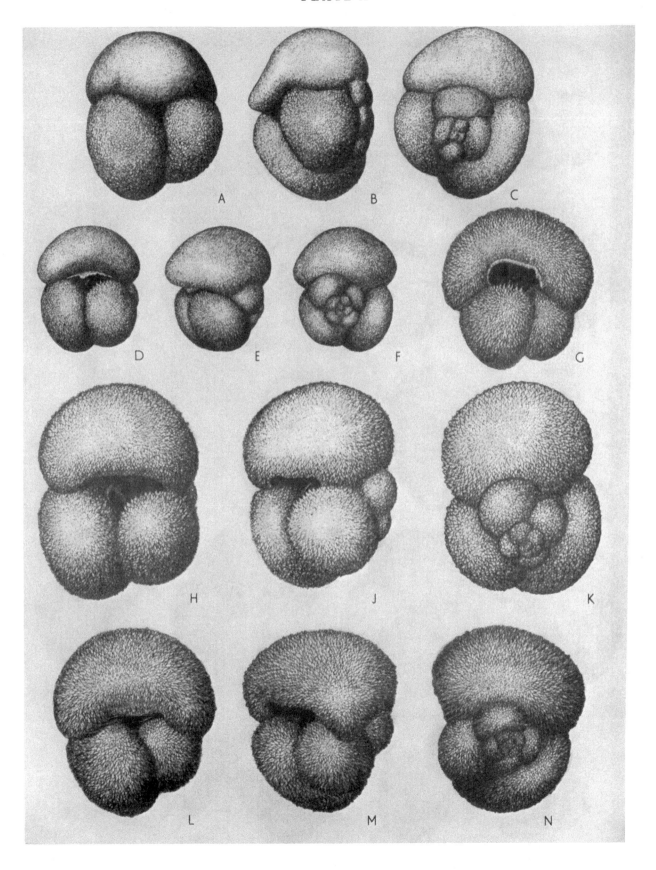

PLATE XI

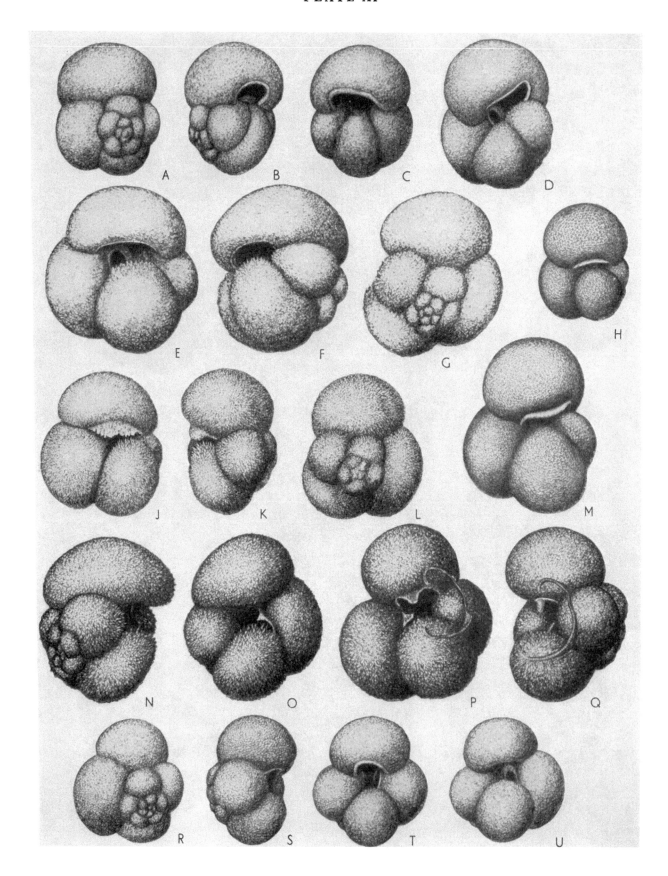

PLATE XII

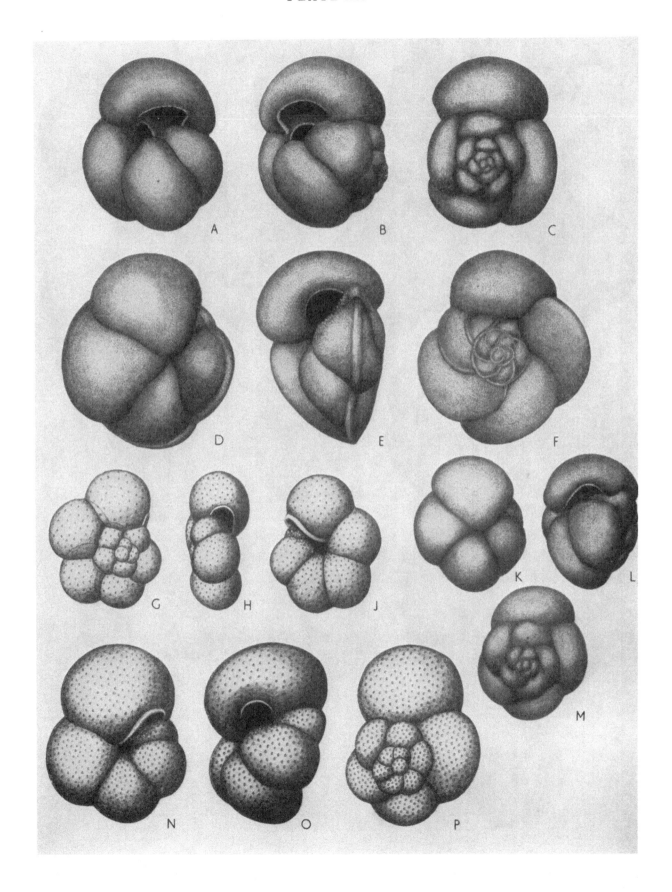

PLATE XIII

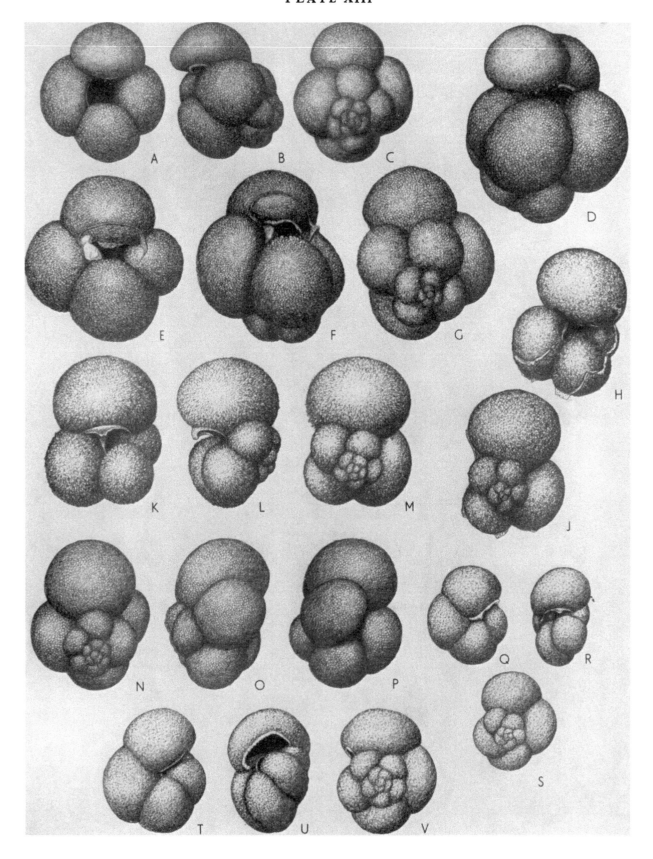

PLATE XIV

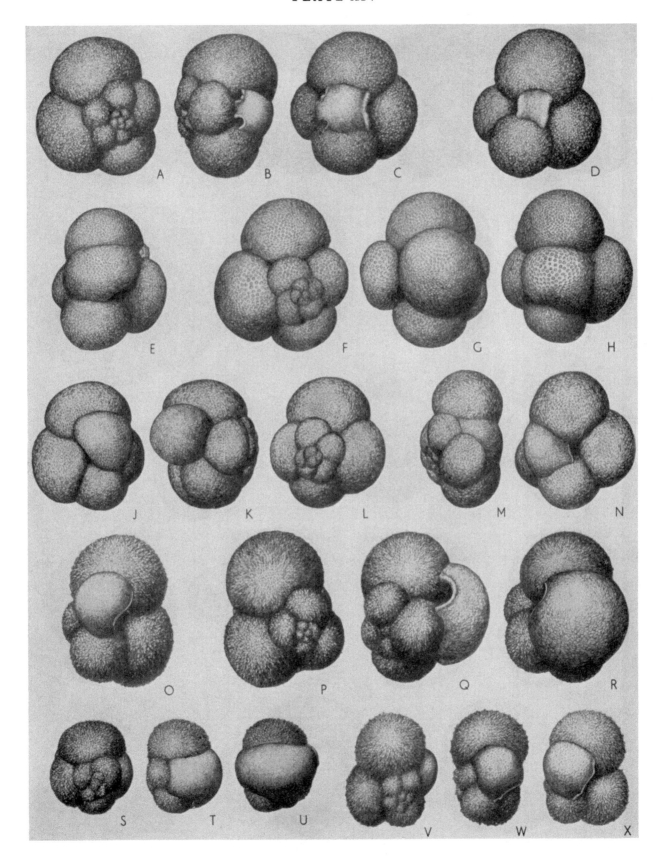

PLATE XV

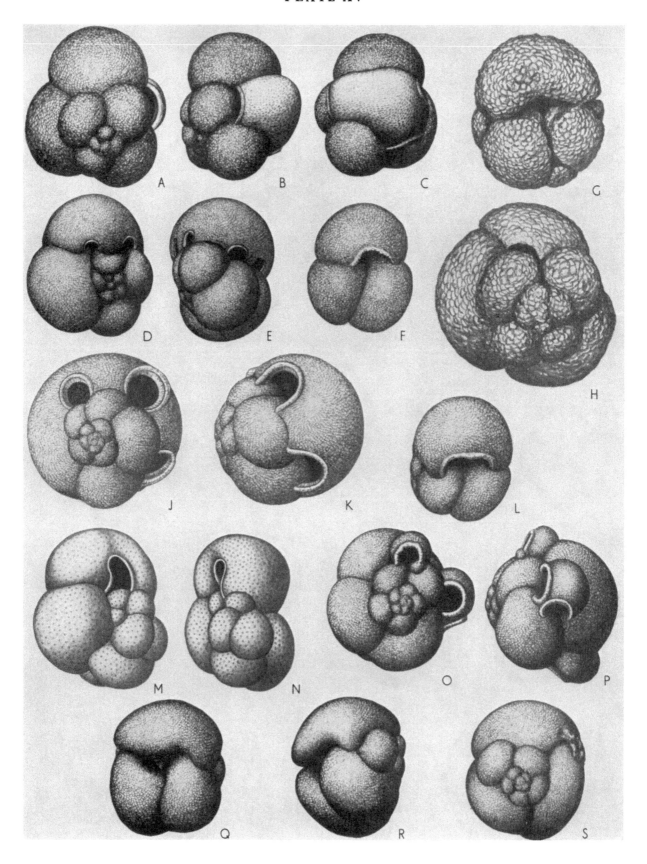

PLATE XVI

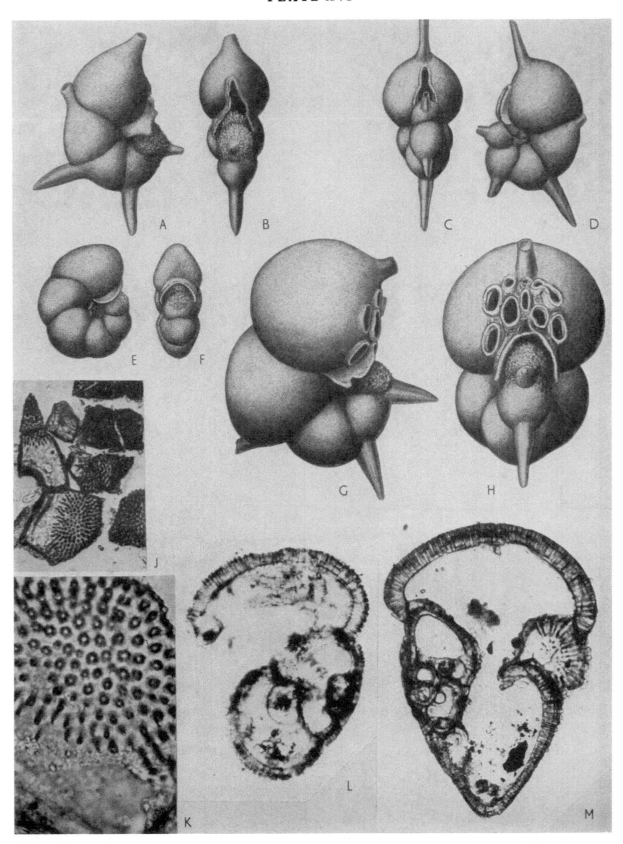

PLATE XVII

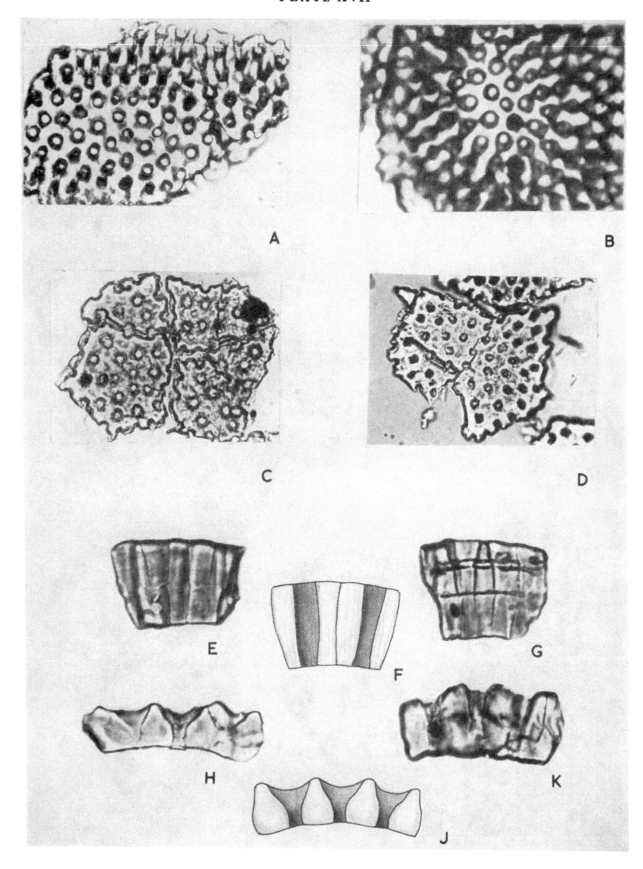

A

B

C

D

E

F

G

H

K

J

INDEXES

Bold numerals indicate pages of fuller treatment, italic numerals indicate pages on which Figures or Plate explanations appear and Figure-numbers unaccompanied by page-references indicate the pull-outs.

GEOGRAPHICAL INDEX

STRATIGRAPHICAL INDEX

157

STRATIGRAPHICAL INDEX

PALAEONTOLOGICAL INDEX

159